Geometry and Martingales in Banach Spaces

Geometry and Martingales in Banach Spaces

Wojbor A. Woyczyński
Case Western Reserve University

CRC Press is an imprint of the
Taylor & Francis Group, an **informa** business

CRC Press
Taylor & Francis Group
6000 Broken Sound Parkway NW, Suite 300
Boca Raton, FL 33487-2742

First issued in paperback 2020

ISBN-13: 978-1-138-61637-0 (hbk)
ISBN-13: 978-0-367-65704-8 (pbk)

Library of Congress Cataloging-in-Publication Data

Names: Woyczyânski, W. A. (Wojbor Andrzej), 1943- author.
Title: Geometry and martingales in Banach spaces / Wojbor A. Woyczynski (Case Western Reserve University).
Description: Boca Raton, Florida : CRC Press, [2018] | Includes bibliographical references and index.
Identifiers: LCCN 2018026561| ISBN 9781138616370 (hardback : alk. paper) | ISBN 9780429462153 (ebook)
Subjects: LCSH: Martingales (Mathematics) | Geometric analysis. | Banach spaces.
Classification: LCC QA274.5 .W69 2018 | DDC 519.2/36--dc23
LC record available at https://lccn.loc.gov/2018026561

Visit the Taylor & Francis Web site at
http://www.taylorandfrancis.com

and the CRC Press Web site at
http://www.crcpress.com

Contents

Introduction

In this volume we are providing a compact exposition of the results explaining the interrelations existing between the metric geometry of Banach spaces and probability theory of random vectors with values in those Banach spaces. In particular martingales and random series of independent random vectors are studied.

Chapter 1 introduces the basic geometric and probabilistic concepts in Banach spaces. Chapter 2 concentrates on the geometric concept of dentability and provides an exposition of the results originally due to M.A. Rieffel, H.B. Maynard, S.D. Chatterjii, and others. The concept of dentability turns out to be very natural in the context of martingales even though it was originally introduced in a study of the Radon-Nikodym theorem in Banach spaces. The chapter ends with an exposition of the theory of sub-martingales with values in Banach lattices and the related issues of the lattice bounded operators.

Chapter 3 deals with the two classical concepts of metric geometry in Banach spaces, namely, the uniform smoothness and the uniform convexity. Here, the works of G. Pisicr and P. Assuad (also, see the important 560-page long recent monograph, *Martingales in Banach Spaces*, by G. Pisier) showed that some of the results obtained earlier by the author for sums of independent random vectors in such Banach spaces carry over to the more general situation of martingales, and even provide a complete characterization of those geometric properties in the language of martingales. Finite tree property and super-reflexivity, the notions introduced by R.C. James, turn out to be the properties that are most intimately related to the martingale theory as shown by results of S. Kwapień, and G. Pisier, which are discussed in this chapter.

In Chapters 4 through 9 we concentrate on martingales with independent increments, that is sums of independent random vectors with values in Banach spaces, and the relationships between the geometry of Banach spaces and the asymptotic properties of such martingales.

The contents of these chapters may be viewed as an effort to provide different generalizations of the following classical result due to Khinchine[1]:

Let (r_i) *be a sequence of Rademacher functions,i.e., independent, identically distributed real-valued random variables taking on values* ± 1 *with equal probabilities 1/2. Then, for any* $p, 0 < p < \infty$, *there exist constants* c_p, *and* C_p, *such that, for an arbitrary integer* n, *and any real numbers* $\alpha_1, \ldots, \alpha_n$,

$$c_p\Big(\sum_{I=1}^{n}\alpha_i^2\Big)^{1/2} \le \Big(\mathbf{E}\,\Big|\sum_{i=1}^{n}\alpha_i r_i\Big|^p\Big)^{1/p} \le C_p\Big(\sum_{I=1}^{n}\alpha_i^2\Big)^{1/2}.$$

Spaces that do not contain $\boldsymbol{c}_0$ are discussed in Chapter 4. Spaces of Rademacher cotype, and of Rademacher and stable types, are analyzed in Chapters 5, and 6. More details on spaces of type 2 are provided in Chapter 7, and Chapter 8 explains the concept of Beck convexity and its relationship to the laws of large numbers in Banach spaces. Finally, Chapter 9 provides the proof of the Macinkiewicz-Zygmund theorem in Banach spaces.

All the results related to the interplay between the geometry and martingales are provided with full proofs. In contrast, to keep the whole book of manageable length, the purely geometric and purely probabilistic results are provided without proofs which can be found in the cited literature. The reader is assumed to be familiar with the basic facts of functional analysis and probability theory.

To keep the size down several topics pertaining to the subject matter of this book have been left out. For instance, the theory of radonifying mappings, operators of type and cotype p, and various notions of orthogonality are not mentioned at all. These results

[1]For the best constants in the Khinchine inequality for real-valued random variables see S. Szarek (1976).

can be found in the bibliography listed at the end of the book. Results on extensions of our results to not necessarily locally convex linear metric spaces are also omitted.

The formulas, theorems, corollaries, propositions, definitions, and lemmas are numbered using three digits. So, formula (2.3.1) is the first formula in Section 3 of Chapter 2.

The author[2] is indebted to all the friends who, in the past, were influential in developing the theory: Alexandra Bellow, Michael Marcus, Czeslaw Ryll-Nardzewski, Joel Zinn, Patrice Assuad, Joe Diestel, Tadeusz Figiel, Stanislaw Kwapień and Gilles Pisier, and my first two graduate students, Jan Rosiński and Jerzy Szulga. Also, the benevolent mentorship of the author in his "salad days" by Kazimierz Urbanik, and Laurent Schwartz, is deeply appreciated.

[2]See his website http://sites.google.com/a/case.edu/waw for complete information about his work.

NOTATION:

$:=$	—	defined by
$\Longrightarrow$	—	implies
$\|A\|$	—	cardinality of the set A
A^c	—	complement of the set A
$a.s.$	—	almost surely (with probability 1)
$\mathcal{B}_X$	—	Borel sigma-field of the space $\boldsymbol{X}$
B_ϵ	—	ball of radius ϵ with center at 0
$B(\boldsymbol{x}, \epsilon)$	—	ball of radius ϵ with center at $\boldsymbol{x}$
$B(\boldsymbol{X}, \boldsymbol{Y})$	—	space of bounded operators from $\boldsymbol{X}$ to $\boldsymbol{Y}$
$\boldsymbol{C}$	—	set of complex numbers
$d(\boldsymbol{X}, \boldsymbol{Y})$	—	Banach-Mazur distance between spaces $\boldsymbol{X}$ and $\boldsymbol{Y}$
$(\boldsymbol{e}_i)$	—	canonical basis
$\mathbf{E}$	—	expectation
$\mathbf{E}(.\|\Sigma)$	—	conditional expectation with respect to σ-algebra Σ
(γ_i)	—	i.i.d. Gaussian random variables
$\boldsymbol{H}$	—	Hilbert space
I_A	—	indicator function of the set A
$i.i.d.$	—	independent and identically distributed
$J_\alpha(\mu)$	=	$\inf\{c : \mu\{\\|\boldsymbol{x}\\| > c\} \leq \alpha\}$
$\boldsymbol{l}^p$	—	space of sequences summable with p-th power
$\boldsymbol{L}^p$	—	space of functions integrable with p-th power
$\mathcal{L}(\mathcal{X})$	—	probability distribution of a random variable (or vector) X
$\Lambda_p(\Omega, \mathcal{F}, \mathbf{P}; \boldsymbol{X})$	—	Lorentz space of random vectors
$\mathbf{N}$	—	nonnegative integers
$(\Omega, \mathcal{F}, \mathbf{P})$	—	probability space
$\mathbf{P}(A)$	—	probability of set A
$\Pi_{p,q}(\boldsymbol{X}, \boldsymbol{Y})$	—	space of (p, q)-absolutely summing operators

$\pi_p(U)$	—	norm of the p-absolutely summing operator U
$\mathbf{R}$	—	set of real numbers
(r_i)	—	Rademacher i.i.d. symmetric random variables with values ± 1
$S_{\boldsymbol{X}}$	—	unit sphere of a normed space $\boldsymbol{X}$
$\boldsymbol{s}$	—	space of sequences with finitely many non-zero terms
$\sigma(X)$	—	σ-algebra spanned by the random vector X
$U \circ V$	—	superposition of operators u, and V
$\boldsymbol{x}, \boldsymbol{y}, \ldots$	—	elements of normed spaces
$\boldsymbol{X}, \boldsymbol{Y}, \ldots$	—	normed, or Banach spaces
$\boldsymbol{X}^*$	—	dual space of $\boldsymbol{X}$

Chapter 1

Preliminaries: Probability and geometry in Banach spaces

1.1 Random vectors in Banach spaces

Let $\boldsymbol{X}$ be a real separable Banach space with the dual space, $\boldsymbol{X}^*$, the unit ball $B_{\boldsymbol{X}}$, and the unit sphere $S_{\boldsymbol{X}}$. By definition, a random vector X with values in $\boldsymbol{X}$ is a strongly measurable map from the probability space $(\Omega, \Sigma, \mathbf{P})$ (always sufficiently rich) into $\boldsymbol{X}$ equipped with the Borel σ -algebra $\mathcal{B}_{\boldsymbol{X}}$. The set of all random vectors in $\boldsymbol{X}$ will be denoted $\boldsymbol{L}_0(\Omega, \Sigma, \mathbf{P}; \boldsymbol{X})$ or, simply, $\boldsymbol{L}_0(\boldsymbol{X})$, and will be equipped with the topology of convergence in probability which is determined by the family of gauges,

$$J_\alpha(\boldsymbol{X}, \mathrm{P}) := \inf\{c : \mathbf{P}(\|\boldsymbol{X}\| > c) \leq \alpha\}, \qquad \alpha \in (0,1). \quad (1.1.1)$$

Random vectors on product spaces $(\Omega_1 \times \Omega_2, \Sigma_1 \times \Sigma_2, \mathbf{P}_1 \times \mathbf{P}_2)$ satisfy the following *Fubini inequality*,[1]

$$J_\gamma(J_\delta(\boldsymbol{X}, \mathbf{P}_1), \mathbf{P}_2) \leq J_\alpha(J_\beta(\boldsymbol{X}, \mathbf{P}_2), \mathbf{P}_1), \quad (1.1.2)$$

whenever $\alpha + \beta \leq \gamma\delta$ (see, also, (1.3.1(b))).

[1] See L. Schwartz (1969/70).

By $\boldsymbol{L}_p(\Omega, \Sigma, \mathbf{P}; \boldsymbol{X})$, or, simply, $\boldsymbol{L}_p(\boldsymbol{X})$, $0 < p \le \infty$, we shall denote the space of random vectors X in $\boldsymbol{X}$ for which $\mathbf{E}\|X\|^p := \int_\Omega \|X(\omega)\|^p \mathbf{P}(d\omega) < \infty$, if $p < \infty$, and ess $\sup\|X\| < \infty$, if $p = \infty$, equipped with the corresponding topologies and quasi-norms. $\boldsymbol{l}_p$ will denote the analogous spaces on the set of positive integers $\mathbf{N}$, and $\boldsymbol{s}$ will denote the space of real sequences with finitely many non-zero terms.

If $A \subset \boldsymbol{L}_0(\Omega, \Sigma, \mathbf{P}; \boldsymbol{X})$ then

$$J_\alpha(A) := \sup\{J_\alpha(X, \mathbf{P}) : X \in A\}. \tag{1.1.3}$$

We shall also discuss the Lorentz spaces

$$\begin{aligned}\Lambda_p(\Omega, \Sigma, \mathbf{P}; \boldsymbol{X}) := \{X \in \boldsymbol{L}_0(\boldsymbol{X}) : \Lambda_p(X, \mathbf{P}) \\ := \sup_{c>0} c\mathbf{P}(\|X\| > c)^{1/p} < \infty\}.\end{aligned}$$

For a random vector X, $\mathcal{L}(X)$ denotes the distribution law of X, that is a Borel measure on $\boldsymbol{X}$ such that $\mathcal{L}(X)(A) = \mathbf{P}(X \in A)$, $A \in \mathcal{B}_{\boldsymbol{X}}$. For a general finite Borel measure μ on $\boldsymbol{X}$ we shall also introduce gauges,

$$J_\alpha(\mu) := \inf\{c > 0, \mu(\|\boldsymbol{x}\| > c) \le \alpha\}, \qquad 0 < \alpha < 1.$$

If $X \in \boldsymbol{L}_1(\boldsymbol{X})$, then $\mathbf{E}X$ will stand for the expectation of X in the sense of the Bochner integral. If Y is another random vector then $\mathbf{E}_Y X := \mathbf{E}(X|Y)$ stands for the conditional expectation of X with respect to the σ-algebra, $\sigma(Y)$, spanned by Y.

A sequence of random vectors (X_n) in $\boldsymbol{X}$ is said to be *exchangeable* if, for each $n \in \mathbf{N}$, each $r \le n$, and each injection

$$\{1, \dots, r\} \ni i \mapsto j_i \in \{1, \dots, n\}$$

the r-tuples $(X_1, \dots, X_r)$, and $(X_{j_1}, \dots, X_{j_r})$ are equally distributed.

In an analogous definition, a sequence (X_n) of random vectors is said to be *weakly exchangeable* if, for each $n \in \mathbf{N}$, each $r \le n$, and each order preserving injection

$$\{1, \dots, r\} \ni i \mapsto j_i \in \{1, \dots, n\}$$

the r-tuples $(X_1, \ldots, X_r)$, and $(X_{j_1}, \ldots, X_{j_r})$ are equally distributed.

Finally, a sequence (X_n) of random vectors is said to be *sign-invariant* if, for any $n \in \mathbf{N}$, and any choice of $\varepsilon_i = \pm 1$, the sequences $(X_1, \ldots, X_n)$, and $(\varepsilon_1 X_1, \ldots, \varepsilon_n X_n)$ are equally distributed.

Weakly exchangeable random vectors satisfy the following maximal inequality. For each $p \geq 1$ there exist constants $K_1, K_2 > 0$, such that, for any weakly exchangeable sequence $X_1, \ldots, X_n$ with $X_1 + \cdots + X_n = 0$,

$$K_1\Big(\mathbf{E}\sup_{1\le k\le n}\Big\|\sum_{i=1}^{k} X_i\Big\|^p\Big)^{1/p} \tag{1.1.4}$$

$$\le K_2 \inf\Big\{\Big(\mathbf{E}\Big\|\sum_{i=1}^{k} X_i\Big\|^p\Big)^{1/p} : |k - n/2| \le n^{1/2}]\Big\}$$

$$\le \Big(\mathbf{E}\Big\|\sum_{i=1}^{n} \rho_i X_i\Big\|^p\Big)^{1/p} \le \sup_{1\le k\le n}\Big(\mathbf{E}\Big\|\sum_{i=1}^{k} X_i\Big\|^p\Big)^{1/p}$$

$$\le \Big(\mathbf{E}\sup_{1\le k\le n}\Big\|\sum_{i=1}^{k} X_i\Big\|^p\Big)^{1/p},$$

where $\rho_i = (1 + r_i)/2, i \in \mathbf{N}$, are independent of (X_i), and (r_i) is the usual Rademacher sequence of independent symmetric random variables with values ± 1.

1.2 Random series in Banach spaces

The series $\sum_n X_n$ of independent random vectors with values in a general Banach space $\boldsymbol{X}$ enjoy a number of properties that will be repeatedly used in the following chapters.

Ito-Nisio Theorem.[2] *Let $\boldsymbol{X}$ be a Banach space, and let (X_n) be a sequence of independent random vectors with values in $\boldsymbol{X}$. Then the following three conditions are equivalent:*

[2]See K. Ito and S. Nisio (1968), but part of this result appeared in the earlier work of A. Tortrat.

(i) The series $\sum_n X_n$ *converges in probability;*

(ii) The series $\sum_n X_n$ *converges almost surely;*

(iii) The probability measures $\mathcal{L}(X_1+\cdots+X_n)$ *converge weakly, as* $n \to \infty$.

Under the additional condition that the random vectors (X_n) *are symmetric, the additional three conditions are equivalent to the conditions (i), (ii), and (iii):*

(iv) The measures $\mathcal{L}(X_1+\cdots+X_n), n \in \mathbf{N}$, *are uniformly tight;*

(v) There exists a random vector X *in* $\boldsymbol{X}$ *such that, for each* $\boldsymbol{x}^* \in \boldsymbol{X}^*$, *the sequence* $\boldsymbol{x}^*(X_1+\cdots+X_n) \to \boldsymbol{x}^*X$, *almost surely, as* $n \to \mathbf{N}$*;*

(vi) There exists a random vector X *in* $\boldsymbol{X}$ *such that the characteristic functionals*

$$\mathbf{E}\exp[i\boldsymbol{x}^*(X_1+\cdots+X_n)] \to \mathbf{E}\exp[i\boldsymbol{x}^*X], \qquad \boldsymbol{x}^* \in \boldsymbol{X}^*,$$

as $n \to \infty$.

As far as convergence in $\boldsymbol{L}_p$ is concerned we also have the following results:

Hoffman-Jorgensen Theorem.[3] *Let* (X_i) *be a sequence of independent random vectors in a Banach space* $\boldsymbol{X}$ *such that* $\sum_i X_i$ *converges almost surely. Then:*

(i) If $\mathbf{E}(\sup_i \|X_i\|)^p < \infty$, *then the series* $\sum_i X_i$ *converges in* $\boldsymbol{L}_q(\boldsymbol{X})$ *for every* $q \in [0, p)$*;*

(ii) If $\sum_i X_i \in \boldsymbol{L}_p(\boldsymbol{X})$ *for some* $p \in (0, \infty]$, *then*

$$\sup_n \|X_1+\cdots+X_n\| \in \boldsymbol{L}_p(\boldsymbol{X}),$$

and, if $p < \infty$, *then*

$$X_1+\cdots+X_n \to \sum_i X_i, \quad \text{in } \boldsymbol{L}_p(\boldsymbol{X}), \quad \text{as } n \to \infty.$$

The next theorem provides a condition on a pair of real random multiplier sequences guaranteeing that the almost sure convergence of one series implies the almost sure convergence of the other series.

[3]See J. Hoffmann-Jorgensen (1972/73).

Comparison Theorem.[4] *Let $(\boldsymbol{x}_n) \subset \boldsymbol{X}$, and let (ξ_n) be a sequence of uniformly nondegenerate real, symmetric, and independent random variables such that the series $\sum_n \xi_n \boldsymbol{x}_n$ converges almost surely. Then, if (η_n) is another sequence of real, symmetric, independent random variables such that, for some $\beta_0, 0 < \alpha \leq 1$, and all $\beta \geq \beta_0$,*

$$\mathbf{P}(|\xi_n| \geq \beta) \geq \alpha \mathbf{P}(|\eta_n| \geq \beta),$$

then the series $\sum \eta_n \boldsymbol{x}_n$ converges almost surely as well.

As a corollary to the above comparison result we shall cite the following,

Contraction Principle.[5] *Let $(\lambda_n) \subset [-1, 1]$. If $(x_n) \subset \boldsymbol{X}$, and the series $\sum_n r_n \boldsymbol{x}_n$ converges almost surely, then the series $\sum_n \lambda_n r_n \boldsymbol{x}_n$ converges almost surely as well.*

The following two results deal with the $\boldsymbol{L}_p$ behavior of the vector Rademacher series.

Kahane Theorem. *Let $\boldsymbol{X}$ be a Banach space. Then, for each $p, q > 0$, there exists a constant $K > 0$ such that*

$$\left(\mathbf{E}\left\|\sum_n r_n \boldsymbol{x}_n\right\|^p\right)^{1/p} \leq K\left(\mathbf{E}\left\|\sum_n r_n \boldsymbol{x}_n\right\|^q\right)^{1/q}.$$

Kwapień Theorem.[6] *Let $\boldsymbol{X}$ be a Banach space and $(\boldsymbol{x}_n) \subset \boldsymbol{X}$. If $\sum_i r_i \boldsymbol{x}_i$ converges almost surely then, for each $\alpha \in \mathbf{R}$,*

$$\mathbf{E} \exp \alpha \left\|\sum_i r_i \boldsymbol{x}_i\right\|^2 < \infty.$$

If the random multipliers θ_i in the series $\sum_i \theta_i \boldsymbol{x}_i$ are α-stable, $0 < \alpha \leq 2$, then we also have a number of inequalities and convergence results similar to those above that dealt with Rademacher

[4]See N.C. Jain and M.B. Marcus (1975). For extensions of this theorem see M.B. Marcus and W.A. Woyczyński (1979).

[5]See J.P. Kahane (1968).

[6]See S. Kwapień (1976).

multipliers. Let us recall that, in general, a random vector X in $\boldsymbol{X}$ (or its probability distribution $\mathcal{L}(X)$) is called *(symmetric) stable* if, for any $a, b > 0$, and any independent copies X_1, X_2, of X, there exists a $c > 0$ such that

$$\mathcal{L}(aX_1 + bX_2) = \mathcal{L}(cX).$$

In the above formula one can always take $c = (a^\alpha + b^\alpha)^{1/\alpha}$, for some $\alpha \in (0, 2]$, and then X (and $\mathcal{L}(X)$) are called *α-stable*, or stable with exponent α. Obviously, 2-stable random vectors are Gaussian. The next theorem gives a representation for the characteristic functional of an α-stable random vector.

Tortrat Theorem.[7] *If X is a symmetric α-stable random vector in $\boldsymbol{X}$, then there exists a finite Borel measure σ on the unit sphere $S_{\boldsymbol{X}}$ such that*

$$\mathbf{E}\exp[i\boldsymbol{x}^*X] = \exp\Big[-\int_{S_{\boldsymbol{X}}} |\boldsymbol{x}^*\boldsymbol{x}|^\alpha \sigma(d\boldsymbol{x})\Big].$$

A random vector Y (or its probability distribution $\mathcal{L}(Y)$) is said to have a Gaussian covariance if there exists a Gaussian random vector X in $\boldsymbol{X}$ such that

$$\mathbf{E}(\boldsymbol{x}^*X\boldsymbol{y}^*X) = \mathbf{E}(\boldsymbol{x}^*Y\boldsymbol{y}^*Y), \qquad \boldsymbol{x}^*, \boldsymbol{y}^* \in \boldsymbol{X}^*,$$

or, equivalently, if $\mathbf{E}\exp[i\boldsymbol{x}^*X] = \exp[-\mathbf{E}(\boldsymbol{x}^*Y)^2]$.

For any random vector Y such that, for each $\boldsymbol{x}^* \in \boldsymbol{X}^*$, the real random variable $\boldsymbol{x}^*Y \in \boldsymbol{L}_2(\mathbf{R})$, there exists an operator R_Y (or, $R_{\mathcal{L}(Y)}$) on $\boldsymbol{X}^*$ with values in $\boldsymbol{X}$ such that

$$\boldsymbol{x}^*R_Y\boldsymbol{y}^* = \mathbf{E}(\boldsymbol{x}^*Y\boldsymbol{y}^*Y).$$

The operator R_Y is called the *covariance operator* of Y. Any covariance operator admits the factorization,

$$\boldsymbol{X}^* \overset{A}{\mapsto} \mathbf{H} \overset{A^*}{\mapsto} \boldsymbol{X},$$

[7] See A. Tortrat (1976).

where $\mathbf{H}$ is a Hilbert space. The operator A is sometimes denoted by $R^{1/2}$.

Hoffmann-Jorgensen Theorem.[8] *Let $\boldsymbol{X}$ be a Banach space, and let (θ_n) be a sequence of independent and identically distributed (i.i.d.) α-stable random variables, $0 < \alpha \le 2$. Then, for each $p, q \in (0, \alpha)$, if $\alpha < 2$, and each $p, q \in (0, \infty)$, if $\alpha = 2$, there exists a constant $K > 0$ such that, for any finite sequence $(\boldsymbol{x}_n) \subset \boldsymbol{X}$,*

$$\left(\mathbf{E}\left\|\sum_n \theta_n \boldsymbol{x}_n\right\|^q\right)^{1/q} \le K\left(\mathbf{E}\left\|\sum_n \theta_n \boldsymbol{x}_n\right\|^p\right)^{1/p}.$$

The existence of quadratic exponential moments for Gaussian random series is guaranteed by the following result:

Landau-Shepp-Fernique Theorem.[9] *Let $\boldsymbol{X}$ be a Banach space, $(\boldsymbol{x}_n) \subset \boldsymbol{X}$, and let (γ_n) be a sequence of i.i.d. Gaussian random variables. If $\sum_n \gamma_n \boldsymbol{x}_n$ converges almost surely then there exists a constant $c > 0$ such that $\mathbf{E}\exp[c\|\sum_n \gamma_n \boldsymbol{x}_n\|^2] < \infty$.*

Finally, for α-stable random series with $0 < \alpha < 2$ we have the following results:

Schwartz Theorem.[10] *Let (θ_n) be a sequence of i.i.d. α-stable symmetric random variables with $0 < \alpha < 2$. Then:*

(i) For each $p \in (0, \alpha)$ there exists a constant $c > 0$ such that, for each $n \in \mathbf{N}$, and any $(a_1, \dots, a_n) \subset \mathbf{R}$,

$$\left(\mathbf{E}\left|\sum_{i=1}^n a_i\theta_i\right|^p\right)^{1/p} = c\left(\sum_{i=1}^n |a_i|^\alpha\right)^{1/\alpha};$$

(ii) For each $p \in (0, \alpha)$, and each $q \in (\alpha, \infty)$, there exist constants c, and C, such that, for each $n \in \mathbf{N}$, and any $(a_1, \dots, a_n) \subset \mathbf{R}$,

$$c\left(\sum_{i=1}^n |a_i|^\alpha\right)^{1/\alpha} \le \left(\mathbf{E}\left(\sum_{i=1}^n |a_i\theta_i|^q\right)^{p/q}\right)^{1/p} \le C\left(\sum_{i=1}^n |a_i|^\alpha\right)^{1/\alpha};$$

[8] See J. Hoffman-Jorgensen (1972/73, 1974).

[9] See H. J. Landau and L.A. Shepp (1970), and X. Fernique (1970).

[10] See L. Schwartz (1969/70).

(iii) For each $p \in (0,\alpha)$, there exist constants c, and C, such that, for each $n \in \mathbf{N}$, and any $(a_1,\dots,a_n) \subset \mathbf{R}$,

$$
\begin{aligned}
c \inf\Big\{a>0 : \sum_{i=1}^{n} \Phi_\alpha(a^{-1}|a_i|) \le 1\Big\} \\
\le \Big(\mathbf{E}\Big(\sum_{i=1}^{n} |a_i\theta_i|^\alpha\Big)^{p/\alpha}\Big)^{1/p} \\
\le C \inf\Big\{a>0 : \sum_{i=1}^{n} \Phi_\alpha(a^{-1}|a_i|) \le 1\Big\},
\end{aligned}
$$

where

$$
\Phi_\alpha(t) = \begin{cases} t^\alpha(1+\log(1/t)), & \text{for } t \in (0,1], \\ t, & \text{for } t \in (1,\infty); \end{cases}
$$

(iv) In particular, the series $\sum_i |a_i\theta_i|^\alpha$ converges almost surely if, and only if,

$$
\sum_i |a_i|^\alpha (1 + |\log(1/|a_i|)|) < \infty;
$$

(v) For each $p \in (0,\alpha)$, and each $q \in (0,\alpha)$, there exist constants c and C, such that for each $n \in \mathbf{N}$, and any $(a_1,\dots,a_n) \subset \mathbf{R}$,

$$
c\Big(\sum_i |a_i|^q\Big)^{1/q} \le \Big(\mathbf{E}\Big(\sum_i |a_i\theta_i|^q\Big)^{p/q}\Big)^{1/p} \le C\Big(\sum_i |a_i|^q\Big)^{1/q}.
$$

1.3 Basic geometry of Banach spaces

We shall start this section with a discussion of the *local properties* of Banach spaces. For two normed spaces, $\boldsymbol{X}, \boldsymbol{Y}$, the *Banach-Mazur distance* is is defined as follows:

$$
d(\boldsymbol{X}, \boldsymbol{Y}) := \inf \|I\| \cdot \|I^{-1}\|,
$$

where the infimum is taken over all ismorphisms $I : \boldsymbol{X} \mapsto \boldsymbol{Y}$.

A normed space $\boldsymbol{X}$ is said to be [crudely] *finitely representable* in $\boldsymbol{Y}$ if, [there exists $\lambda > 1$] for each $\lambda > 1$, and for each finite-dimensional $\boldsymbol{X}_1 \subset \boldsymbol{X}$, there exists a finite dimensional $\boldsymbol{Y}_1 \subset \boldsymbol{Y}$ such that $d(\boldsymbol{X}_1, \boldsymbol{Y}_1) \leq \lambda$. In other words, $\boldsymbol{X}$ is finitely representable in $\boldsymbol{Y}$ if, and only if, for each $\lambda > 1$, and for each finite-dimensional $\boldsymbol{X}_1 \subset \boldsymbol{X}$ there exists an isomorphism $I : \boldsymbol{X}_1 \mapsto \boldsymbol{Y}$ such that

$$\lambda^{-1}\|\boldsymbol{x}\|_{\boldsymbol{X}} \leq \|I\boldsymbol{x}\|_{\boldsymbol{Y}} \leq \lambda\|\boldsymbol{x}\|_{\boldsymbol{X}}, \qquad \boldsymbol{x} \in \boldsymbol{X}_1.$$

If P is a property of a normed space then we say that $\boldsymbol{X}$ has the property *super* P if each $\boldsymbol{Y}$ which is finitely representable in $\boldsymbol{X}$ has the property P. Note that, for each Banach space $\boldsymbol{X}$, the dual of the dual space, $\boldsymbol{X}^{**}$, is finitely representable in $\boldsymbol{X}$. This phenomenon is called the *Local Reflexivity Principle.*

Dvoretzky Theorem.[11] *For any $\lambda > 1$, and the integer $k \in$ $\mathbf{N}$, there exists an integer $N \in \mathbf{N}$ such that in every normed space $\boldsymbol{X}$ of dimension greater than N there exist linearly independent $\boldsymbol{x}_1, \ldots, \boldsymbol{x}_k \in \boldsymbol{X}$ such that, for any $a_1, \ldots, a_k \in \mathbf{R}$,*

$$\lambda^{-1}\Big(\sum_{i=1}^{k} |a_i|^2\Big)^{1/2} \leq \Big\|\sum_{i=1}^{k} a_i\boldsymbol{x}_i\Big\| \leq \lambda\Big(\sum_{i=1}^{k} |a_i|^2\Big)^{1/2}.$$

In particular, this implies that $\boldsymbol{l}_2$ is finitely representable in any infinite dimensional normed space.

Now, let $(\boldsymbol{e}_n)$ be a sequence of unit vectors spanning $\boldsymbol{X}$. The norm $\|.\|$ on $\boldsymbol{X}$ is said to be *invariant under spreading* for $(\boldsymbol{e}_n)$, if for each integer $n \in \mathbf{N}$, each $a_1, \ldots, a_n \in \mathbf{R}$, and each $k_1 < k_2 < \cdots < k_n \in \mathbf{N}$,

$$\Big\|\sum_{i=1}^{n} a_i\boldsymbol{e}_{k_i}\Big\| = \Big\|\sum_{i=1}^{n} a_i\boldsymbol{e}_i\Big\|.$$

The above concept permits us to give an example of the space in which the space $\boldsymbol{l}_q$ is also finitely representable.

[11]See A. Dvoretzky (1961). For shorter proofs, see T. Figiel (1974/75), and A. Szankowski (1974).

Krivine Theorem.[12] *Let* $(\boldsymbol{e}_n)$ *be an unconditional basis in a Banach space* $\boldsymbol{X}$ *equipped with the invariant under spreading norm* $\|.\|$. *Define,*

$$b := \inf\Big\{\lambda \le 1 : \lim_{n\to\infty} \lambda^n \Big\|\sum_{j=1}^{2^n} \boldsymbol{e}_j\Big\| = +\infty\Big\},$$

and

$$q = (\log_2 b^{-1})^{-1}.$$

Then, $\boldsymbol{l}_q$ *is finitely representable in* $\boldsymbol{X}$.

And operator $T : \boldsymbol{X} \mapsto \boldsymbol{Y}$ (of norm 1) is said to be [crudely] *finitely factorable* through a normed space $\boldsymbol{Z}$ if [there exists a λ] for each $\lambda > 1$, and each finite dimensional $\boldsymbol{X}_1 \subset \boldsymbol{X}$, there exists a finite dimensional $\boldsymbol{Z}_1 \subset \boldsymbol{Z}$, and an isomorphism $I : \boldsymbol{X}_1 \mapsto \boldsymbol{Z}_1$, such that

$$\lambda^{-1}\|T\boldsymbol{x}\|_{\boldsymbol{Y}} \le \|I\boldsymbol{x}\|_{\boldsymbol{Z}} \le \lambda\|\boldsymbol{x}\|_{\boldsymbol{X}}.$$

Clearly, $\boldsymbol{X}$ is finitely representable in $\boldsymbol{Y}$, if, and only if, the identity mapping of $\boldsymbol{X}$ is finitely factorable through $\boldsymbol{Y}$. In cases in which that is our main interest, as in the next theorem, we can state more explicitly:

The canonical embedding $\boldsymbol{l}_p \mapsto \boldsymbol{l}_q, 1 \le p \le q \le \infty$, is [crudely] finitely factorable through $\boldsymbol{X}$ if [there exists a $\lambda \in (0,1)$] , for each $\lambda \in (0,1)$, and each $n \in \mathbf{N}$, there exist $\boldsymbol{x}_1, \dots \boldsymbol{x}_n \in \boldsymbol{X}$ such that, for every $a_1, \dots, a_n \in \mathbf{R}$,

$$(1-\lambda)\Big(\sum_{i=1}^n |a_i|^q\Big)^{1/q} \le \Big\|\sum_{i=1}^n a_i \boldsymbol{x}_i\Big\| \le \Big(\sum_{i=1}^n |a_i|^p\Big)^{1/p}.$$

In the next theorem we prove that in some cases crude finite factorability implies finite factorability.

James Theorem.[13] *If* $1 \le p \le \infty$, *and the embedding* $\boldsymbol{l}_p \mapsto \boldsymbol{l}_\infty$ $(\boldsymbol{l}_1 \mapsto \boldsymbol{l}_p)$ *is crudely finitely factorable through* $\boldsymbol{X}$, *then it is finitely factorable through* $\boldsymbol{X}$.

[12] See J.L. Krivine ((1976).

[13] See R.C. James (1964), but the proof provided above is due to B. Maurey and G. Pisier (1976).

Proof. It is easy to notice that $\boldsymbol{l}_p \mapsto \boldsymbol{l}_\infty$ is finitely factorable through $\boldsymbol{X}$ if, and only if, for each $n \in \mathbf{N}$,

$$a_n := \inf\Big\{a : \forall \boldsymbol{x}_1, \ldots, \boldsymbol{x}_n \in \boldsymbol{X},$$

$$\inf_{1\le i\le n} \|\boldsymbol{x}_i\| \le a \sup\{\|\sum_{i=1}^{n} b_i \boldsymbol{x}_i\| : \sum_i |b_i|^p = 1\}\Big\} = 1.$$

On the other hand, $a_n \le 1$, and the sequence (a_n) is nonincreasing and submultiplicative, i.e.,

$$a_{nk} \le a_n a_k, \qquad n, k \in \mathbf{N}.$$

Indeed, let $\boldsymbol{x}_1, \ldots, \boldsymbol{x}_{nk} \in \boldsymbol{X}$. For each $i = 1, \ldots, n$, define

$$\boldsymbol{y}_i =:= \sum_{j=ik-k+1}^{ik} b_j \boldsymbol{x}_j,$$

so that

$$\|\boldsymbol{y}_i\| = \sup\Big\{\Big\|\sum_{j=ik-k+1}^{ik} b_j \boldsymbol{x}_j\Big\| : \sum_j |b_j|^p = 1\Big\}.$$

Then

$$\inf \|\boldsymbol{y}_i\| = a_n \sup\Big\{\Big\|\sum_{i=1}^{ik} b'_i \boldsymbol{y}_i\Big\| : \sum_i |b'_i|^p = 1\Big\}$$

$$\le a_n \sup\Big\{\Big\|\sum_{j=1}^{ik} b_j \boldsymbol{x}_j\Big\| : \sum_{j=1}^{nk} |b_j|^p = 1\Big\},$$

and, on the other hand, by construction,

$$\|\boldsymbol{y}_i\| > \frac{1}{a_k} \inf_{(i-1)k<j\le ik} \|\boldsymbol{x}_j\|, \quad i = 1, \ldots, n,$$

so that

$$\inf_{1\le j\le nk} \|\boldsymbol{x}_j\| \le a_n a_k \sup\Big\{\Big\|\sum_{j=1}^{nk} b_j \boldsymbol{x}_j\Big\| : \sum_{j=1}^{nk} |b_j|^p = 1\Big\}.$$

Now, if the embedding $\boldsymbol{l}_p \mapsto \boldsymbol{l}_\infty$ is crudely finitely factorable through $\boldsymbol{X}$, then $\liminf a_n > 0$. Then the submultiplicativity of (a_n) implies that $a_n = 1$ for each $n \in \mathbf{N}$, because if $a_{n_0} < 1$ then $a_{n_0^k} \to 0$, as $k \to \infty$. Therefore, the embedding $\boldsymbol{l}_p \mapsto \boldsymbol{l}_\infty$ is finitely factorable through $\boldsymbol{X}$.

For the finite factorability of the embedding $\boldsymbol{l}_1 \mapsto \boldsymbol{l}_p$, the proof is similar, but the sequence (a_n) should be defined as follows

$$a_n = \inf\Big\{a : \forall \boldsymbol{x}_1, \ldots, \boldsymbol{x}_n \in \boldsymbol{X},$$

$$\inf\{\|\sum_{i=1}^n b_i \boldsymbol{x}_i\| : \sum_i |b_i|^p = 1\} \leq a \sup_i \|x_i\|\Big\}. \qquad \text{QED}$$

Closed Interval Lemma. *Let* $\boldsymbol{X}$ *be a normed spacee. The set of q's for which the embedding* $\boldsymbol{l}_1 \mapsto \boldsymbol{l}_q$ *(* $\boldsymbol{l}_q \mapsto \boldsymbol{l}_\infty$*) is finitely factorable through* $\boldsymbol{X}$ *is a closed interval.*

The proof of the above statement is immediate from the definition of factorability, and the fact that if $q > p$, then $n^{(p-q)/pq}(\sum |a_i|^p)^{1/p} \leq (\sum |a_i|^q)^{1/q}$.

We conclude this section with the statement that for some spaces the finite factorability through any infinite dimensional space is always possible.

Dvoretzky-Rogers Lemma. *Let* $k \in \mathbf{N}$*. For any* $n \geq k$*, and any normed space* $\boldsymbol{X}$ *of dimension greater than* $4n^2$*, there exist* $\boldsymbol{x}_1, \ldots, \boldsymbol{x}_n \in \boldsymbol{X}$ *such that, for all* $a_1, \ldots, a_n \in \mathbf{R}$*,*

$$2^{-1} \sup_{1 \leq i \leq n} |a_i| \leq \Big\|\sum_{i=1}^n a_i \boldsymbol{x}_i\Big\| \leq \Big(\sum_{i=1}^n a_i^2\Big)^{1/2}.$$

In particular, the embedding $\boldsymbol{l}_2 \mapsto \boldsymbol{l}_\infty$ *is finitely factorable through any infinite dimensional Banach space.*

1.4 Spaces with invariant under spreading norms which are finitely representable in a given space

In this section we will outline a general procedure permitting us to produce spaces with invariant under spreading norms which are finitely representable in a given normed space $\boldsymbol{X}$. The procedure is based on the following combinatorial lemma:

Ramsey Lemma. *Let K be the set of k-tuples $\boldsymbol{n} = (n_1, \ldots, n_k)$ of different integers. If $K = K' \cup K''$ is a partition of K, then there exists a sequence (i_n) such that all $\boldsymbol{n}$ formed from members of (i_n) belong either to K', or to K''.*

Brunel-Sucheston Theorem.[14] *(i) If $(\boldsymbol{x}_n)$ is a bounded sequence in a Banach space $\boldsymbol{X}$ then, for some subsequence $(\boldsymbol{e}_n) \subset (\boldsymbol{x}_n)$, there exists a function*

$$S \ni \vec{a} = (a_1, \ldots, a_k) \mapsto L(\vec{a}) \in \mathbf{R}^+,$$

such that, for each $n_1 < n_2 < \cdots < n_k$, $(n_1, \ldots, n_k) \to \infty$

$$\Big\|\sum_i a_i \boldsymbol{e}_{n_i}\Big\| \to L(\vec{a}).$$

(ii) If $(\boldsymbol{e}_n)$ is not a Cauchy sequence then the mapping

$$\boldsymbol{X} \supset \Phi(S) \ni \Phi(\vec{a}) := \sum_i a_i \boldsymbol{e}_i \mapsto \Big|\sum_i a_i \boldsymbol{e}_i\Big| := L(\vec{a}) \in \mathbf{R}^+, \tag{1.4.1}$$

is an invariant under spreading norm on $\Phi(S) \subset \boldsymbol{X}$ (and we can assume that $|\boldsymbol{e}_i| = 1$.)

(iii) In the latter case the completion $\boldsymbol{Y}$ of $\Phi(S)$ under the norm $|.|$ is finitely representable in $\boldsymbol{X}$.

Sketch of the Proof. It is is sufficient to show the existence of $(\boldsymbol{e}_n) \subset (\boldsymbol{x}_n)$, and $L : S \mapsto \mathbf{R}^+$ such that, for every $\vec{a} \in S$, and every $\epsilon > 0$, there exists an $n_0 \in \mathbf{N}$ such that, for every $\ldots n_2 > n_2 > n_0$,

$$\Big|\Big\|\sum_i a_i \boldsymbol{e}_i\Big\| - L(\vec{a})\Big| \le \epsilon.$$

[14]See A. Brunel and L. Sucheston (1974, 1975, 1976).

Assume, without loss of generality, that $\|\boldsymbol{x}_n\| \leq 1, n \in \mathbf{N}$. Let $(a_1, \ldots, a_r)$ be a fixed r-tuple of rational numbers. Define

$$\Psi : K_r \ni \vec{n} = (n_1, \ldots, n_r) \mapsto \Psi(\vec{n}) = \left\| \sum_{i=1}^{r} a_i x_{n_i} \right\|.$$

The set $\{\Psi(\vec{n}) : \vec{n} \in K_r\}$ is bounded, say, by a constant $C > 0$. Now, consider the sets,

$$A_1 = \{\vec{n} : \Psi(\vec{n}) \leq C/2\}, \quad \text{and} \quad B_1 = \{\vec{n} : C/2 \leq \Psi(\vec{n}) \leq C\}.$$

By Ramsey lemma, either A_1, or B_1, contains all $\vec{n}$ formed from the terms of an infinite sequence, say $(i_n^{(1)}) \subset \mathbf{N}$. Assume that this is true for A_1 (the argument for B_1 is analogous), and consider

$$A_2 = \{\vec{n} : \Psi(\vec{n}) \leq C/4\}, \quad \text{and} \quad B_2 = \{\vec{n} : C/4 \leq \Psi(\vec{n}) \leq C/2\}.$$

There exists, again by Ramsey lemma, $(i_n^{(2)}) \subset (i_n^{(1)})$ such that all $\vec{n}$ formed from the terms of $(i_n^{(2)})$ are either in A_2, or in B_2. An so on, and so on. Let us denote by $L(a)$ the only point contained in $\bigcap_n A_n$. Then, for the diagonal sequence $(i_n^{(n)})$, we have $|\Psi(\vec{n}) - L(\vec{a})| < \epsilon$, for any choice of $n_1, \ldots, n_r$, sufficiently far in the sequence.

Now, let $a^{(1)}, a^{(2)}, \ldots,$ be the sequence of all $a \in S$ with rational coefficients. Utilizing again the diagonal argument one obtains a sequence of integers (i_n) such that, for every $(a^{(k)})$,

$$\lim \left\| \sum_i a_i^{(k)} \boldsymbol{x}_{n_i} \right\| = L(a^{(k)}),$$

where the limit is taken for $n_1 \to \infty$, $n_1 < n_2 < \cdots < n_k$, $(i_n) \supset (n_i)$.

There remains the case of an arbitrary $\vec{a} = (a_1, \ldots, a_r, 0, 0, \ldots) \in S$. Given an $\epsilon > 0$, and a rational $\vec{a}' = (a_1', \ldots, a_r', 0, 0, \ldots)$ with $\|\vec{a} - \vec{a}'\|_1 \leq \epsilon/2$, where $\|.\|_1$ denotes the $\boldsymbol{l}_1$ norm, we clearly have

$$\left\| \sum_{i=1}^{r} a_i \boldsymbol{x}_{n_i} \right\| - \left\| \sum_{i=1}^{r} a_i' \boldsymbol{x}_{n_i} \right\| \leq \|\vec{a} - \vec{a}'\|_1 \leq \epsilon/2.$$

Therefore, if n_1 is large enough, say $n_1 \geq N$,

$$\sup_{n_1>N} \Psi(\vec{n},\vec{a}) - \inf_{n_1>N} \Psi(\vec{n},\vec{a}') \leq \epsilon,$$

so that (i) holds true for an arbitrary $\vec{a}$.

(ii) The fact that (1.4.1) is invariant under spreading is obvious in view of (i). Also, the homogeneity and subadditivity of (1.4.1) follow directly from the definition. So, to complete the proof of (ii) it is sufficient to show that if $0 \neq \vec{a} \in S$, and $L(\vec{a}) = 0$, then the sequence $(\boldsymbol{e}_n)$ is Cauchy. Indeed, let q be the first integer such that $a_q \neq 0$. If $L(\vec{a}) = 0$, then, for each $\epsilon > 0$, there exists an $n_0 \in \mathbf{N}$ such that, for all $\cdots > n_2 > n_1 > p > n \geq n_0$,

$$\|a_q \boldsymbol{e}_q + a_{q+1}\boldsymbol{e}_{n_1} + a_{q+2}\boldsymbol{e}_{n_2} + \ldots \| \leq \epsilon(2|a_q|)^{-1},$$

and

$$\|a_q \boldsymbol{e}_p + a_{q+1}\boldsymbol{e}_{n_1} + a_{q+2}\boldsymbol{e}_{n_2} + \ldots \| \leq \epsilon(2|a_q|)^{-1}.$$

Thus, $n_0 \leq n \leq p$ implies $\|\boldsymbol{e}_n - \boldsymbol{e}_p\| \leq \epsilon$, and (ii) is proven.

(iii) Let $\boldsymbol{Y}_1$ be a finite dimensional subspace of $\boldsymbol{Y}$ with basis $\boldsymbol{y}_1, \ldots, \boldsymbol{y}_p$. To prove (iii) we have to find, for each $\epsilon > 0$, an invertible operator $V : \boldsymbol{Y}_1 \mapsto \boldsymbol{X}$, such that

$$\big|\|V\boldsymbol{x}\| - \|\boldsymbol{x}\|\,\big| < \epsilon\|\boldsymbol{x}\|, \qquad \boldsymbol{x} \in \boldsymbol{Y}_1.$$

Because $\boldsymbol{Y}$ is the completion of $\Phi(S)$ in the norm $|.|$, we can find $m \in \mathbf{N}$ and $\boldsymbol{x}_1, \ldots, \boldsymbol{x}_p \in \text{span}[\boldsymbol{e}_1, \ldots, \boldsymbol{e}_m] =: \boldsymbol{Y}_3$, such that the linear extension of the map $U\boldsymbol{y}_i = \boldsymbol{x}_i$ is an isomorphism of $\boldsymbol{Y}_1$ onto $\boldsymbol{Y}_2 := \text{span}[\boldsymbol{x}_1, \ldots, \boldsymbol{x}_p]$, and

$$\big|\,\|U\boldsymbol{x}'\| - \|\boldsymbol{x}'\|\,\big| \leq \epsilon\|\boldsymbol{x}'\|/4, \qquad \boldsymbol{x}' \in \boldsymbol{Y}_2.$$

On the other hand, the shift, $T(\sum a_i \boldsymbol{e}_i) = \sum a_i \boldsymbol{e}_{i+1}$, is a linear isometry acting from $\boldsymbol{Y}_3$ into $\Phi(S)$ equipped with $\|.\|$. Moreover.

$$\lim_{n\to\infty}\Big\|T^n\Big(\sum_{i=1}^m a_i\boldsymbol{e}_i\Big)\Big\| = \Big\|\sum_{i=1}^m a_i\boldsymbol{e}_i\Big\|,$$

and the convergence is uniform over the compact set of $\boldsymbol{x} \in \boldsymbol{Y}_3$, with $\|\boldsymbol{x}\| = 1$. Therefore, there exists a $q \in \mathbf{N}$ such that, for each $\boldsymbol{x} \in \boldsymbol{Y}_3$ with $\|\boldsymbol{x}\| = 1$, we have $\big|\|T^q\boldsymbol{x}\| - 1\big| < \epsilon/4$. Finally, for each $\boldsymbol{x} \in \boldsymbol{Y}_3$, we have $\big|\|T^q\boldsymbol{x}\| - 1\big| < \epsilon\|\boldsymbol{x}\|/4$, so that V can be taken to be equal to T^qU. QED

1.5 Absolutely summing operators and factorization results

Let $\boldsymbol{X}$ and $\boldsymbol{Y}$ be two normed spaces, and let $0 < q < p < \infty$. A linear operator $U : \boldsymbol{X} \mapsto \boldsymbol{Y}$ is said to be (p, q)*-absolutely summing* (in short, $U \in \Pi_{p,q}(\boldsymbol{X}, \boldsymbol{Y})$) if, there exists a constant $C > 0$ such that, for each $n \in \mathbf{N}$, and all $\boldsymbol{x}_1, \ldots, \boldsymbol{x}_n \in \boldsymbol{X}$,

$$\left(\sum_{i=1}^{n} \|U\boldsymbol{x}_i\|^p\right)^{1/p} \leq C \sup\left\{\left(\sum_{i=1}^{n} |\boldsymbol{x}^*\boldsymbol{x}_i|^q\right)^{1/q} : \boldsymbol{x}^* \in B(\boldsymbol{X}^*)\right\},$$

where $B(\boldsymbol{X}^*)$ is the unit ball of the dual space $\boldsymbol{X}^*$.

The smallest constant C for which the above inequality holds will be denoted by $\pi_{p,q}(U)$, and it is a complete norm on the linear space $\Pi_{p,q}(\boldsymbol{X}, \boldsymbol{Y})$. If $p = q$, then we shall simplify our notation by writing $\pi_{p,p}(U) = \pi_p(U)$, and $\Pi_{p,q} = \Pi_p$.

We shall say the $U : \boldsymbol{X} \mapsto \boldsymbol{Y}$ is 0*-absolutely summing* ($U \in \Pi_0(\boldsymbol{X}, \boldsymbol{Y})$) if, for each $\beta \in (0,1)$ there exists an $\alpha \in (0,1)$, and a constant $C > 0$, such that, for each probability measure μ on $\boldsymbol{X}$ with finite support,

$$J_\beta(U(\mu)) \leq \sup\{J_\beta(\boldsymbol{x}^*\mu) : \boldsymbol{x}^* \in B(\boldsymbol{X}^*)\},$$

where the gauge J_β has been defined in Section 1.1.

The notions of p-absolutely summing operators coincide for $0 < p < 1$ and, moreover, for each $0 \leq p < q < 1$, there exists a constant C such that, for all U, $\pi_q(U) \geq C\pi_p(U)$. On the other hand, $\pi_p(U)$ is a decreasing function of p, so that

$$\Pi_p(\boldsymbol{X}, \boldsymbol{Y}) \supset \Pi_q(\boldsymbol{X}, \boldsymbol{Y}), \qquad \text{if} \qquad q < p.$$

If $\boldsymbol{X}, \boldsymbol{Y}$ are Hilbert spaces then all classes $\Pi_p(\boldsymbol{X}, \boldsymbol{Y})$ coincide with the class of Hilbert-Schmidt operators.

Another characterization of p-absolutely summing operators is provided by the following result:

Pietsch Theorem.[15] *(i) Let $\boldsymbol{X}, \boldsymbol{Y}$ be normed spaces. A continuous linear operator $U : \boldsymbol{X} \mapsto \boldsymbol{Y}$ is p-absolutely summing*

[15] See A. Pietsch (1967).

if, and only if, there exists a Borel probability measure on $B(\boldsymbol{X}^*)$ such that

$$\|U\boldsymbol{x}\| \leq \pi_p(U)\Big(\int_{B(X^*)} |\boldsymbol{x}^*\boldsymbol{x}|^p \mu(d\boldsymbol{x}^*)\Big)^{1/p}.$$

(ii) In particular, a continuous linear operator U belongs to the class $\Pi_p(\boldsymbol{X}, \boldsymbol{L}_\infty[0,1])$ if, and only if, it can be factored thrugh a space $\boldsymbol{L}_p$ as follows:

$$U : \boldsymbol{X} \overset{V}{\mapsto} \boldsymbol{C}(B(\boldsymbol{X}^*)) \overset{I}{\mapsto} \boldsymbol{L}_p(B(\boldsymbol{X}^*), \mu) \overset{W}{\mapsto} \boldsymbol{L}_\infty[0,1],$$

where $\|V\| = 1$, $\|W\| = \pi_p(U)$, and I is the canonical embedding.

(iii) If $U \in \Pi_2(\boldsymbol{X}, \boldsymbol{Y})$, and $\boldsymbol{X}$, and $\boldsymbol{Y}$, are Banach spaces, then U can be factored through a Hilbert space $\boldsymbol{H}$ as follows:

$$U : \boldsymbol{X} \overset{V}{\mapsto} \boldsymbol{H} \overset{W}{\mapsto} \boldsymbol{Y},$$

where V, and W, are bounded operators. Moreover, if $\boldsymbol{X}$ is a Hilbert space, then $V : \boldsymbol{X} \mapsto \boldsymbol{H}$ is a Hilbert-Schmidt operator.

In the remainder of this section we shall discuss several results, including more factorization theorems and inequalities related to Lorentz norms which would be useful in the following chapters of the book.

Let $n \in \mathbf{N}$, and $1 < p < \infty$. Define a Lorentz-like gauge λ_p on $\mathbf{R}^n$ as follows:

$$\lambda_p((a_i)) = \big(\sup_{c>0} c^p |\{i : |a_i| > c\}|\big)^{1/p},$$

where $|A|$ denotes the cardinality of the set A. One can show that λ_p is equivalent to the quantity $\sup\{i^{1/p}|a_i|^* : 1 \leq i \leq n\}$, where $|a_i|^*$ denote the decreasing rearrangement of $|a_i|$'s. The relationship between the Lorentz norm Λ_p, and the $\boldsymbol{l}_q$ norm, is as follows:

$$\lambda_p((a_i)) \leq \Big(\sum_i |a_i|^p\Big)^{1/p},$$

and, if $p < q$, then

$$\Big(\sum_i |a_i|^q\Big)^{1/q} \leq \Big(\frac{q}{q-p}\Big)^{1/q} \lambda_p((a_i)).$$

Denote by S_n^p the set of those $(a_i) \in \mathbf{R}^n$ for which, for some $k > 0$, $|\{i : a_i \neq 0\}| = k$, and such that $a_i \neq 0 \implies |a_i| = k^{-1/p}$. Denote by σ_p the usual gauge of the convex envelope of S_n^p. Since, for each $(a_i) \in S_n^p$, $\sum_{i=1}^n |a_i|^p = 1$, we have, for $p \geq 1$, and $(a_i) \in \mathbf{R}^n$,

$$\Big(\sum_i |a_i|^p\Big)^{1/p} \leq \sigma_p((a_i)).$$

Now, if we denote by σ_p^* the polar gauge of the set $\{\boldsymbol{x}^* : |\boldsymbol{x}^*\boldsymbol{x}| \leq 1, \boldsymbol{x} \in S_n^p\}$, then one can check that

$$\lambda_p((a_i)) \leq \sigma_{p^*}^*((a_i)) \leq p^*\lambda_p((a_i)), \qquad (a_i) \in \mathbf{R}^n, \tag{1.5.1}$$

where $1/p + 1/p^* = 1$. By duality, for $1 < p < q < \infty$,

$$\sigma_p((a_i)) \leq \Big(\frac{p^*}{p^* - q^*}\Big)^{1/p^*} \Big(\sum_i |a_i|^p\Big)^{1/p}.$$

The next Lemma provides a relationship between the norms of operators from $\boldsymbol{l}_\infty^{(n)} \mapsto \boldsymbol{X}$, and $\boldsymbol{X} \mapsto \boldsymbol{l}_1^{(n)}$, and the gauges λ_p and σ_p.

Nikishin Lemma.[16] *Let $n \in \mathbf{N}^+$, $K > 0$, $1 \leq q < \infty$, and let $\boldsymbol{X}$ be a normed space.*

(i) If an operator $U : \boldsymbol{l}_\infty^{(n)} \mapsto \boldsymbol{X}$ satisfies the inequality,

$$\Big(\sum_{k=1}^n \|U(\alpha^k)\|^q\Big)^{1/q} \leq K \sup_{1\leq i\leq n} \sum_{k=1}^n |\alpha^k(i)|, \tag{1.5.2}$$

for any $\alpha^1, \ldots, \alpha^n \in \boldsymbol{l}_\infty^{(n)}$, $\alpha^k = (\alpha^k(1), \ldots, \alpha^k(n))$, then, for any $C \in \mathbf{R}^+$ there exists an $A_C \subset \{1, \ldots, n\}$, such that $|A_C| \geq n(1 - C^{-q})$, and, for any $(a_i) \in \mathbf{R}^n$,

$$\Big\|U\Big(\sum_{i\in A_C} a_i\boldsymbol{e}_i\Big)\Big\| \leq KC_n^{-1/q}\sigma_q((a_i)).$$

(ii) Dually, if an operator $U : \boldsymbol{X} \mapsto \boldsymbol{l}_1^{(n)}$ satisfies the inequality

$$\sum_{i=1}^n \sup_{1\leq k\leq n} |U(\boldsymbol{x}_k)(i)| \leq K\Big(\sum_{k=1}^n \|\boldsymbol{x}_k\|^q\Big)^{1/q}, \tag{1.5.3}$$

[16] This proof of Nikishin's Lemma is due to B. Maurey amd G. Pisier (1976).

for all $\boldsymbol{x}_1, \ldots, \boldsymbol{x}_n \in \boldsymbol{X}$, *then, for each* $C \in \mathbf{R}^+$, *there exists an* $A_C \subset \{1, \ldots, n\}$ *such that* $|A_C| \geq n(1 - C^{-q})$, *and such that*

$$\lambda_q\Big(\big(U(\boldsymbol{x})(i)\big)_{i \in A_C}\Big) \leq KCn^{-1/q}\|\boldsymbol{x}\|, \qquad \boldsymbol{x} \in \boldsymbol{X},$$

Proof. We shall only show Part (i). Part (ii) then follows by a duality argument. So, let $C \in \mathbf{R}^+$. Call the set $A \subset \{1, \ldots, n\}$ an *N-subset* if there exist $\varepsilon_1, \ldots, \varepsilon_n = \pm 1$ such that

$$\Big\|U\Big(\sum_{i \in A} \varepsilon_i \boldsymbol{e}_i\Big)\Big\| > KCn^{-1/q}|A|^{1/q}.$$

Clearly, all N–subsets are non-empty. If there are no N-subsets, then put $A_C = \{1, \ldots, , n\}$ and proceed as follows.

Assume that there exists at least one N-subset. Denote by $A_1, \ldots, A_m$, a maximal system of N-subsets of $\{1, \ldots, n\}$ that are pairwise disjoint. By definition of an N-subset, for each $j = 1, \ldots, m$, there exist $\varepsilon_1^j, \ldots, \varepsilon_m^j = \pm 1$, such that

$$\Big\|U\Big(\sum_{i \in A_j} \varepsilon_i^j \boldsymbol{e}_i\Big)\Big\| > KCn^{-1/q}|A_j|^{1/q}.$$

Since $m \leq n$, on can apply (1.5.2), and get the inequality,

$$\Big(\sum_{j=1}^{m}\Big\|U\Big(\sum_{i \in A_j} \varepsilon_i^j \boldsymbol{e}_i\Big)\Big\|^q\Big)^{1/q} \leq K \sup_{1 \leq i \leq n} \sum_{1 \leq j \leq m} (I_{A_j})(i),$$

from which it follows that

$$KCn^{1/q}\Big(\sum_{j=1}^{m} |A_j|\Big)^{1/q} \leq K,$$

so that

$$\Big|\bigcup_{j=1}^{m} A_j\Big| \leq nC^{-q}.$$

Put $A_C = \{1, \ldots, n\} \setminus \{\cup_{j=1}^m A_j\}$, and let $B \subset A_C$. Because of the maximality of $A_1, \ldots, A_m$, for each $\varepsilon_1, \ldots, \varepsilon_n = \pm 1$,

$$\Big\|U\Big(\sum_{i \in B} \varepsilon_i \boldsymbol{e}_i\Big)\Big\| \leq KCn^{-1//q}|B|^{1/q},$$

so that

$$\Big\|U\Big(\sum_{i\in B}|B|^{-1/q}\varepsilon_i\boldsymbol{e}_i\Big)\Big\|\le KCn^{-1/q},$$

and, by convexity, for each $(a_i)\in\mathbf{R}^n$,

$$\Big\|U\Big(\sum_{i\in A_C}a_i\boldsymbol{e}_i\Big)\Big\|\le KCn^{-1/q}\sigma_q\big((a_i)\big). \qquad \text{QED}$$

Nikishin Theorem.[17] *Let K be a convex subset of $\boldsymbol{L}_0(\Omega,\Sigma,\mathbf{P})$ formed from all positive functions. Then, for each $\epsilon>0$, there exists a subset $\Omega_\epsilon\subset\Omega,\Omega_\epsilon\in\Sigma$, with $\mathbf{P}(\Omega\setminus\Omega_\epsilon)\le 2\epsilon$, such that, for all $f\in K$,*

$$\int_{\Omega_\epsilon}f\,d\mu\le 2J_\epsilon(K,P).$$

Maurey Theorem.[18] *Let $\boldsymbol{X}$ be a normed space, $0<p\le q\le\infty$, and let (T,Σ,μ) be a measure space. A continuous operator $U:\boldsymbol{X}\mapsto\boldsymbol{L}_p(\mu)$ admits a factorization*

$$U:\boldsymbol{X}\overset{V}{\mapsto}\boldsymbol{L}_q(\mu)\overset{T_g}{\mapsto}\boldsymbol{L}_p(\mu),$$

where V is bounded, and T_g is the opearator of multiplication by a function $g\in\boldsymbol{L}_r(\mu),1/p=1/q+1/r$, if, and only if, for each $(\boldsymbol{x}_i)\in\boldsymbol{l}_q(\boldsymbol{X})$,

$$\int_T\Big(\sum_i\|U\boldsymbol{x}_i\|\Big)^{p/q}d\mu<\infty.$$

Maurey-Rosenthal Theorem.[19] *Let $p>1$, $\boldsymbol{X}$ be a normed space, and assume that $U:\boldsymbol{X}\mapsto\boldsymbol{L}_1$ does not factor through the space $\boldsymbol{L}_p$. Then, for each $n\in\mathbf{N}$, there exist $\boldsymbol{x}_1^*,\dots,\boldsymbol{x}_n^*\in S_{\boldsymbol{X}^*}$ such that, for each $a_1,\dots,a_n\in\mathbf{R}$,*

$$\Big\|\sum_{i=1}^n a_i\boldsymbol{x}_i^*\Big\|\le 2\Big(\sum_{i=1}^n|a_i|^q\Big)^{1/q},\qquad 1/p+1/q=1.$$

---

[17] See B. Maurey (1974).
[18] See B. Maurey (1974).
[19] See B. Maurey (1974).

Finally, we provide a result about unconditional basic sequences that will be used in the analysis of Banach spaces of cotype q in Chapter 5.

Johnson Lemma.[20] *Let $k_0 \in \mathbf{N}, k > 1$, and $\epsilon, \delta > 0$. For each $n \in \mathbf{N}$, there exists an $N = N(k_0, \delta, \epsilon, n) \in \mathbf{N}$ such that, for each Banach space $\boldsymbol{X}$, and all $\boldsymbol{x}_1, \ldots, \boldsymbol{x}_N \in B_{\boldsymbol{X}}$ satisfying the property*

$$(\mathrm{P}_{k_0,\delta}): \qquad \forall A \subset \{1, \ldots, N\}, |A| \geq k_0, \qquad \sup_{(i,j) \in A \times A} \|\boldsymbol{x}_i - \boldsymbol{x}_j\| \geq \delta,$$

there exists a subsequence $\boldsymbol{x}_{i_1}, \ldots, \boldsymbol{x}_{i_{2n}}$ such that the sequence of increments $(\boldsymbol{x}_{i_{2j}} - \boldsymbol{x}_{i_{2j-1}})$ is an unconditional basic sequence with the constant[21] *at most $2 + \epsilon$, and satisfying the condition $\|\boldsymbol{x}_{i_{2j}} - \boldsymbol{x}_{i_{2j-1}}\| \geq \delta/2, j = 1, \ldots, n$.*

Proof. Assume, to the contrary, that for given k_0, ϵ, δ, there exists an $n \in \mathbf{N}$ such that, for all $N \in \mathbf{N}$, there exists a Banach space $\boldsymbol{X}_N$ and $\boldsymbol{x}_1^N, \ldots, \boldsymbol{x}_N^N \in B_{\boldsymbol{X}_N}$ which satisfy the condition $(\mathrm{P}_{k_0,\delta})$, but do not satisfy the assertion of the Lemma.

Now, let $\mathcal{U}$ be a non-trivial ultrafilter on $\mathbf{N}$. Define a seminorm on the set $\mathcal{S} \ni a = (a_i)$ via the formula

$$|||a||| = \lim_{\mathcal{U}} \Big\| \sum_i a_i \boldsymbol{x}_i^N \Big\|_{\boldsymbol{X}_N},$$

and define $\boldsymbol{X} = \mathcal{S}/|||.|||$. Let $(\boldsymbol{e}_n)$ be the canonical basis in $\mathcal{S}$, and $(\bar{\boldsymbol{e}}_n)$ its image in $\boldsymbol{X}$. The sequence $(\bar{\boldsymbol{e}}_n)$ is bounded in $\boldsymbol{X}$, and, by the Brunel-Sucheston Theorem, there exists a subsequence $(\bar{\boldsymbol{e}}_k') \subset (\bar{\boldsymbol{e}}_k)$ such that, for every $(a_j) \in \mathcal{S}$,

$$L((a_j)) := \lim_{k_1 < k_2 < \cdots < k_j < \ldots, k_1 \to \infty} ||| \sum_j a_j \bar{\boldsymbol{e}}_{k_j}' |||$$

exists, and, because $(\bar{\boldsymbol{e}}_k')$ is not Cauchy in view of the condition $(\mathrm{P}_{k_0,\delta})$, it is a norm on $\mathcal{S}$, for which the sequence $(\boldsymbol{e}_{2k} - \boldsymbol{e}_{2k-1})$ is unconditional with constant ≤ 2.

[20]See B. Maurey and G. Pisier (1976)

[21]Recall that C is called the unconditional constant of a basis $(\boldsymbol{e}_i)$, if it is the smallest constant such that $\| \sum_i \varepsilon_i a_i \boldsymbol{e}_i \| \leq C \| \sum_i a_i \boldsymbol{e}_i \|$, for any $\varepsilon_i = \pm 1$.

Consider the space $\boldsymbol{Y}$ spanned by $\boldsymbol{e}_1, \ldots, \boldsymbol{e}_{2n}$ in $\mathcal{S}$ equipped with the norm L. For each $i, N \in \mathbf{N}$ define the linear operator,

$$U_{i,N} : \boldsymbol{Y} \ni \boldsymbol{e}_j \mapsto U_{i,n}(\boldsymbol{e}_j) = \boldsymbol{x}^N_{m_{i+j}} \in \boldsymbol{X}_N, \qquad j = 1, \ldots, 2n.$$

In view of the above construction, for each $\boldsymbol{y} \in \boldsymbol{Y}$,

$$\|\boldsymbol{y}\| = \lim_{i \to \infty} \lim_{\mathcal{U}} \|U_{i,N}\boldsymbol{y}\|_{\boldsymbol{X}_N}. \tag{1.5.4}$$

On the other hand, the functions $\boldsymbol{Y} \ni \boldsymbol{y} \mapsto \|U_{i,N}\boldsymbol{y}\| \in \mathbf{R}$ are equicontinuous, so that the limit in (1.5.4) is uniform over the unit ball of $\boldsymbol{Y}$. Therefore, for i, N sufficiently large, and all $\boldsymbol{y} \in \boldsymbol{Y}$,

$$\big(1 + \min(\epsilon, 1)/2\big)^{-1/2} \|\boldsymbol{y}\| \le \|U_{i,N}\boldsymbol{y}\| \le \big(1 + \min(\epsilon, 1)/2\big)^{1/2} \|\boldsymbol{y}\|,$$

from which it follows that the sequence

$$\big(\boldsymbol{x}^N_{m_{i+2j}} - \boldsymbol{x}^N_{m_{i+2j_1}}\big), \quad j = 1, \ldots, n,$$

is unconditional with constants less than $2 + \epsilon$, and of the norm uniformly larger than $\delta/2$. A contradiction. QED

Chapter 2

Dentability, Radon-Nikodym Theorem, and Martingale Convergence Theorem

2.1 Dentability

The following concept of a *dentable*[1] subset of a Banach space $\boldsymbol{X}$ is essential for the study of the Radon-Nikodym Theorem and the Martingale Convergence Theorem in Banach spaces:

Definition 2.1.1. A set $A \subset \boldsymbol{X}$ is said to be *dentable* if, for every $\epsilon > 0$, there exists an $x \in A$ such that $\boldsymbol{x} \notin \overline{\text{conv}}(A \setminus B_\epsilon(x))$, where $\overline{\text{conv}}(A)$ denotes the closed convex hull of A, and $B_\epsilon(x)$ stands for the ball in $\boldsymbol{X}$ with center at $\boldsymbol{x}$, and radius ϵ.

This concept leads in a natural way to the definition of a *dentable Banach space* itself:

Definition 2.1.2. The Banach space $\boldsymbol{X}$ is said to be *dentable* if every bounded subset of $\boldsymbol{X}$ is dentable.

It is clear that the unit balls of the Banach spaces, $\boldsymbol{c}_0$, $\boldsymbol{L}_1[0,1]$, and $\boldsymbol{C}[0,1]$, are not dentable. On the other hand all reflexive

[1]See M.A. Rieffel (1967).

Banach spaces are dentable. This fact follows from Theorem 1.1. in Section 2, and the classical results on Radon-Nikodym Theorem in Banach spaces due to Dunford, Morse, Pettis and Phillips.

It follows directly from the definition that the property of dentability is invariant under equivalent renorming of the Banach space, and for dentability of $\boldsymbol{X}$ it is sufficient to verify that each closed convex subset of $\boldsymbol{X}$ is dentable. Indeed, we have the following

Proposition 2.1.1.[2] *If $A \subset \boldsymbol{X}$, and $\overline{\text{conv}}(A)$ is dentable, then A is dentable as well.*

Proof. For a given $\epsilon > 0$, one can find $\boldsymbol{y} \in \overline{\text{conv}}(A)$ such that,

$$\boldsymbol{y} \notin Q := \overline{\text{conv}}\big[\overline{\text{conv}}(A) \setminus B_{\epsilon/2}\big].$$

Now, let $\boldsymbol{x} \in A \setminus Q$. Such an $\boldsymbol{x}$ does exist because $\boldsymbol{y} \in \overline{\text{conv}}(A) \setminus Q$, Q is closed and convex, so that Q cannot contain A. Obviously, $\boldsymbol{x} \notin \overline{\text{conv}}(A \setminus B_\epsilon(\boldsymbol{x}))$ because $\boldsymbol{x} \in B_{\epsilon/2}(y)$, and $A \setminus B_\epsilon(\boldsymbol{x})) \subset Q$. QED

Moreover, one can prove[3] that X is dentable if and only if after an arbitrary equivalent renorming the new unit ball is dentable. Also, X is dentable if and only if every bounded closed set in X has an extreme point in its convexification[4]. The property is also obviously related to the so called Krein-Milman Property: Every bounded, closed and convex subset of X possesses an extreme point. Recall that an extreme point of a convex set A in a real vector space is a point in A which does not lie in any open line segment joining two points of A.

Example 2.1.1. (a) A simple example of a dentable set is a set A which has a fixed denting point, that is a point $\boldsymbol{x}$ such that, for any $\epsilon > 0$, we have $\boldsymbol{x} \notin \overline{\text{conv}}(A \setminus B_\epsilon(\boldsymbol{x}))$.

(b) Also, a *strongly exposed point* of A is always a denting point of A. Recall that a strongly exposed point is a point $\boldsymbol{x}$ such that

---

[2]See M.A. Rieffel (1967).

[3]See W.J. Davis and R.R. Phelps (1974).

[4]See R.E. Huff and P.D. Morris (1976).

if $\boldsymbol{x}^* \in \boldsymbol{X}^*$, and $\alpha \in \mathbf{R}$, are such that

$$\{\boldsymbol{y} : \boldsymbol{x}^*\boldsymbol{y} = \alpha\} \cap A = \{\boldsymbol{x}\},$$

then the condition

$$\{\boldsymbol{y}_n\} \subset A, \quad \boldsymbol{x}^*\boldsymbol{y}_n \to \alpha, \quad \text{implies} \quad \|\boldsymbol{y}_n - \boldsymbol{x}\| \to 0.$$

(c) On the other hand, every denting point of A is an extreme point of $\overline{\text{conv}}A$.

As a matter of fact, one can show that the above examples reflect the general situation and prove[5] that each closed, convex, bounded and non-empty set A in a dentable space has always a denting point.

Another characterization of dentable sets uses the concept of slices of the set. More precisely, slice of the the set $A \subset \boldsymbol{X}$ is a set of the form $\{\boldsymbol{x} : \boldsymbol{x}^*\boldsymbol{x} \geq \beta\} \cap A$, with non-empty $\{\boldsymbol{x} : \boldsymbol{x}^*\boldsymbol{x} > \beta\} \cap A$, with $\beta \in \mathbf{R}$ and $\boldsymbol{x}^* \in \boldsymbol{X}^*$. The following result is also due to Davis (1973-74).

Proposition 2.1.2. *A closed, bounded, convex, and non-empty set $A \subset \boldsymbol{X}$ is dentable if, and only if, for any $\epsilon > 0$, there exists a slice of A with diameter less than ϵ.*

Proof. If A is dentable then, for arbitrary $\epsilon > 0$ we can find an $\boldsymbol{x} \in A$ such that $\boldsymbol{x} \notin \overline{\text{conv}}(A \setminus B_\epsilon(\boldsymbol{x}))$, so that, for $\boldsymbol{x}^* \in \boldsymbol{X}^*$ and $\alpha \in \mathbf{R}$,

$$\boldsymbol{x}^*\boldsymbol{x} > \alpha > \sup\Big\{\boldsymbol{x}^*\boldsymbol{x} : \boldsymbol{x} \in \overline{\text{conv}}(A \setminus B_\epsilon(\boldsymbol{x}))\Big\}.$$

In particular,

$$\{\boldsymbol{x} : \boldsymbol{x}^*\boldsymbol{x} \geq \alpha\} \cap A \subset B_\epsilon(\boldsymbol{x}),$$

which completes the proof of one implication. The other implication is straightforward. QED

A more subtle result[6], using the Ryll-Nardzewski's Fixed Point Theorem, facilitates checking the dentability of many sets and

[5] See W.J. Davis (1973-74).

[6] Due to E. Asplund and I. Namioka (1967).

spaces. In particular, it implies that all weakly compactly generated Banach spaces (thus all separable duals) are dentable.

Proposition 2.1.3. *Weakly compact convex sets in a separable Banach space are dentable.*

Proof. It is sufficient to show that, for any $\epsilon > 0$, and arbitrary weakly compact and convex $K \subset \boldsymbol{X}$, one can find a closed and convex $C \subset K$ with diameter of $K \setminus C$ less than ϵ.

Denote by P the weak closure of the set of extreme points in K, and let

$$P \subset \bigcup_{i=1}^{\infty} B_{\epsilon/4}(\boldsymbol{x}_i), \qquad \boldsymbol{x}_i \in P.$$

Because P is weakly compact, and thus second category, there is an $\boldsymbol{x} \in P$, and a weakly open neighborhood W of $\boldsymbol{x}$, such that

$$P \cap B_{\epsilon/4} \supset W \cap P \neq \emptyset.$$

Denote

$$K_1 = \overline{\text{conv}}(P \setminus W), \qquad K_2 - \overline{\text{conv}}(W \setminus P).$$

Evidently, by Krein-Milman Theorem, K is the convex hull of weakly compact sets K_1, and K_2, and, furthermore, $K_1 \neq K$ because the extreme points of K_1 lie within $P \setminus W$[7]. Now, define

$$C_r = \left\{t\boldsymbol{k}_1 + (1-t)\boldsymbol{k}_2 : \boldsymbol{k}_1 \in K_1, \boldsymbol{k}_2 \in K_2, r \leq t \leq 1\right\}, \quad 0 \leq r \leq 1.$$

Clearly, C_r's are weakly compact, convex, and increase as $r \to 0+$, with $C_0 = K$, and $C_1 = K_1$. Finally, note $C_r \neq K$, for all $0 < r < 1$, because if $C_r = K$, then each extreme point $\boldsymbol{z}$ of K has the form,

$$\boldsymbol{z} = \lambda \boldsymbol{x}_1 + (1-\lambda)\boldsymbol{x}_2, \qquad xx_1 \in K_1, \boldsymbol{x}_2 \in K_2, \lambda \in [r, 1].$$

Hence, $\boldsymbol{z} = \boldsymbol{x}_1 \in K_1$, contradicting the statement that $K_1 \neq K$. Notice that, if $\boldsymbol{y} \in K \setminus C_r$, then $\boldsymbol{y}$ is of the form,

$$\boldsymbol{y} = \lambda \boldsymbol{x}_1 + (1-\lambda)\boldsymbol{x}_2, \qquad \boldsymbol{x}_1 \in K_1, \boldsymbol{x}_2 \in K_2, \lambda \in [0, r],$$

[7]See, e.g., J. Kelley and I. Namioka (1963), 15.2.

and

$$\|\boldsymbol{y}-\boldsymbol{x}_2\| = \|\lambda\boldsymbol{x}_1+(1-\lambda)\boldsymbol{x}_2-\boldsymbol{x}_2\| = |\lambda|\|\boldsymbol{x}_1-\boldsymbol{x}_2\| \leq r\|\boldsymbol{x}_1-\boldsymbol{x}_2\|,$$

so that $\boldsymbol{y} \in K\setminus C_r$ lies within the distance $r\cdot\mathrm{diam}(K_2)$ of K_2. But $\mathrm{diam}(K_2) \leq \epsilon/2$, thus, as $r \to 0+$, C_r has the desired property $\mathrm{diam}(K - C_r) < \epsilon$. QED

An apparently weaker concept of σ-dentability[8] employs the following concept of σ-convexification of a set:

$$\sigma - \mathrm{conv}(B) := \Big\{\sum_{i=1}^{\infty}\lambda_i\boldsymbol{b}_i : \lambda_{\geq}0, \sum_i \lambda_i = 1, \boldsymbol{b}_i \in B\Big\}, \qquad B \subset \boldsymbol{X}.$$

Definition 2.1.3. The set $A \subset \boldsymbol{X}$ is said to be *σ-dentable* if, for each $\epsilon > 0$, there exists an $\boldsymbol{x} \in A$ such that $\boldsymbol{x} \notin \sigma-\mathrm{conv}(A\setminus B_\epsilon(x))$.

However, we have the following

Proposition 2.1.4.[9] *The Banach space* $\boldsymbol{X}$ *is dentable if, and only if, it is σ-dentable.*

Proof. Obviously, it is sufficent to show that if $\boldsymbol{X}$ is σ-dentable then it is also dentable. Assume to the contrary that that there exists a set $A \subset \boldsymbol{X}$ which is bounded and not dentable. Take $\boldsymbol{x} \in \boldsymbol{X}$ so that $(\boldsymbol{x} + A) \cap (-\boldsymbol{x} - A) = \emptyset$. Then $B = \mathrm{conv}((\boldsymbol{x} + A) \cap (-\boldsymbol{x} - A))$ is a closed, convex, symmetric, and also non-dentable by Proposition 1.1, so that we may claim that the unit ball $B_1(0) \subset \boldsymbol{X}$ is not dentable. Indeed, were it dentable, the closed convex body $\overline{B + B_1(0)} \subset \boldsymbol{X}$ which generates on $\boldsymbol{X}$ the norm equivalent to the original norm, would also be dentable by Proposition 2.1.1. But it is not since if B is not dentable then one can find $\epsilon > 0$ such that for all $\boldsymbol{x} \in B, \boldsymbol{x} \in \overline{\mathrm{conv}}(B - B_\epsilon(\boldsymbol{x}))$, so that if $\boldsymbol{x} + \boldsymbol{y} \in B + B_1(0)$ then also

$$\boldsymbol{x}+\boldsymbol{y} \in \overline{\mathrm{conv}}\big((\boldsymbol{y}+B)-B_\epsilon(\boldsymbol{y}+\boldsymbol{x})\big) \subset \overline{\mathrm{conv}}\big((B_1(0)+B)-B_\epsilon(\boldsymbol{y}+\boldsymbol{x})\big).$$

Now, we shall show that non-dentability of $B_1(0)$ implies non-σ-dentability of $\mathrm{int}B_1(0)$ what, in turn, would contradict the σ-dentability of $\boldsymbol{X}$.

[8]Introduced by H.B. Maynard (1973).

[9]See W.J. Davis and R.R. Phelps (1974).

Take $\epsilon > 0$ such that, for each $\boldsymbol{x} \in B_1(0)$, and $\boldsymbol{x} \in \overline{\text{conv}}(B_1(0) - B_\epsilon(\boldsymbol{x}))$. If $\|\boldsymbol{x}\| < 1 - \epsilon/4$, then, for some $\lambda > 0$ such that $\|\lambda\boldsymbol{x}\| < 1$, we have $\|\boldsymbol{x} - \lambda\boldsymbol{x}\| > \epsilon/4$, and $\|\boldsymbol{x} + \lambda\boldsymbol{x}\| > \epsilon/4$. Thus $\boldsymbol{x} \in \text{conv}(B_1(0) - B_{\epsilon/4}(\boldsymbol{x}))$. If $1 > \|\boldsymbol{x}\| > 1 - \epsilon/4$ then $B_{\epsilon/4}(\boldsymbol{x}) \subset B_\epsilon(\boldsymbol{x}/\|\boldsymbol{x}\|)$, so that

$$\boldsymbol{x}/\|\boldsymbol{x}\| \in \overline{\text{conv}}(B_1(0) - B_{\epsilon/4}(\boldsymbol{x}).$$

For small ϵ, the origin 0 is an interior point of $\overline{\text{conv}}(B_1(0) - B_{\epsilon/4}(\boldsymbol{x})$, so that the entire segment $[0, \boldsymbol{x}/\|\boldsymbol{x}\|)$ is in the interior of that set. In particular,

$$\boldsymbol{x} \in \text{conv}(\text{int} B_1(0) - B_{\epsilon/4}(\boldsymbol{x}),$$

so that $\text{int} B_1(0)$ is not dentable . QED

The equivalence proven in the above Proposition shows that it is sufficient to know what separable Banach spaces are dentable in order to know what are all dentable Banach spaces. In particular, $\boldsymbol{X}$ is dentable if, and only if, each closed separable subspace of $\boldsymbol{X}$ is dentable. More precisely, we have the following result:

Proposition 2.1.5.[10] *Dentability is a separably determined property, i.e., $A \subset \boldsymbol{X}$ is σ-dentable if, and only if, each countable subset of A is σ-dentable.*

Proof. Assume, to the contrary, that A is not σ-dentable. Then we can find an $\epsilon > 0$ such that, for each $\boldsymbol{x} \in A$, also $\boldsymbol{x} \in \sigma - \text{conv}(A \setminus B_\epsilon(\boldsymbol{x}))$. Thus, choosing an arbitrary $\boldsymbol{y} \in A$, we can find a sequence $\{\boldsymbol{y}_n\} \subset A$ such that

$$\boldsymbol{y} = \sum_n \lambda_n \boldsymbol{y}_n, \quad \text{and} \quad \|\boldsymbol{y} - \boldsymbol{y}_n\| \geq \epsilon.$$

Now, apply to each $\boldsymbol{y}_n$ the same denial of the σ-dentability of A. Reiterating this procedure one gets an infinite tree in A which is countable and, by definition, not σ-dentable. QED

[10]Due to J. Diestel (1973).

2.2 Dentability versus Radon-Nikodym property, and martingale convergence

It turns out that the dentability property of a Banach space is closely related to certain measure-theoretic and martingale-theoretic theorems in such Banach spaces.

Definition 2.2.1. We say that the Banach space $\boldsymbol{X}$ has the *Radon-Nikodym Property (RNP)* if every $\boldsymbol{X}$-valued measure m on (Ω, Σ), for which the total variation measure

$$|m|(E) := \sup\{\sum_i \|m(E_i)\| : E_i \in \Sigma \text{ are disjoint with } \bigcup_i E_i = E \in \Sigma\},$$

is finite, and which is absolutely continuous with respect to a finite positive measure μ, admits with respect to μ a Bochner integrable density.

The following illuminating example shows a Banach space without the Radon-Nikodym property.

Example 2.2.1.[11] The Banach space $\boldsymbol{X} = \boldsymbol{c}_0$ of real sequences convergent to 0 does not have the Radon-Nikodym property. Indeed, let (Ω, Σ, μ) be a finite positive measure space containing no atoms. We'll construct a measure $m : \Sigma \to \boldsymbol{c}_0$ (the elements of $\boldsymbol{c}_0$ will be denoted $(a_{n,i}), n = 1, 2, \ldots, 2^n \leq i < 2^{n+1}$) as follows:

$$m : \Sigma \ni E \to m(E) = (\mu(E \cap E_{n,i})) \in \boldsymbol{c}_0,$$

where $(E_{n,i}), n = 1, 2, \ldots, 2^n \leq i < 2^{n+1}$, is a sequence of measurable sets such that $\mu(E_{n,i}) = 2^{-n}\mu(\Omega)$, and such that $E_{n,i}$ is the disjoint union of $E_{n+1,2i}$ and $E_{n+1,2i+1}$. Such a sequence $(E_{n,i})$ does exist in view of non-atomicity of μ.

Evidently, $\|m(E)\| \leq \mu(E)$ so that m is absolutely continuous with respect to μ and has finite total variation. However, m does not have a Bochner integrable density with respect to μ. Indeed, if such a density, say $f : \Omega \mapsto \boldsymbol{c}_0$, existed then, denoting by $(e_{n,i})$

[11] Due to D.R. Lewis (1972).

the standard basis in $\boldsymbol{l}_1$, we would have, for each $i = 2, 3, \ldots$, and $E \in \Sigma$, the equality,

$$\int_E f(\omega)e_{n,i}\mu(d\omega) = \mu(E \cap E_{n,i}).$$

Hence, for each i, there would exist a μ-null set $C_i \subset \Omega$ such that $f(\omega)e_{n,i} = I_{E_{n,i}}(\omega)$, for $\omega \notin C_i$. Choose, $\omega_0 \in \Omega \setminus \bigcup_i C_i$. By the very construction of $(E_{n,i})$, we have $I_{E_{n,i,}}(\omega) = 1$ for infinitely many indices i, so that $\lim_i f(\omega)e_{n,i} \neq 0$, which gives the desired contradiction.

Now let's turn to the issue of martingale convergence in Banach spaces and recall that, given a probability space $(\Omega, \Sigma, \mathbf{P})$, and a sequence of increasing sub-σ-algebras $\Sigma_1 \subset \Sigma_2 \subset \cdots \subset \Sigma$, an $\boldsymbol{X}$-valued (Σ_n)-martingale is a sequence $\{M_n\}$ of strongly measurable functions such that $M_n$meas Σ_n, and

$$\mathbf{E}(M_{n+1}|\Sigma_n) = M_n.$$

Definition 2.2.2. Let $1 \leq p < \infty$. We say that the Banach space $\boldsymbol{X}$ has the L_p *-Martingale Convergence Property* ($\boldsymbol{X} \in (MCP_p)$), if for each $\boldsymbol{X}$-valued martingale (M_n, Σ_n), such that $\sup_n \mathbf{E}\|M_n\|^p < \infty$, there exists an $M_\infty \in L^p(\Omega, \Sigma, \mathbf{P}; \boldsymbol{X})$ such that $M_n \to M_\infty$, almost surely in norm.

Not all Banach spaces have the Martingale Convergence Property, and here is the counterexample:

Example 2.2.2.[12] The Banach space $\boldsymbol{X} = \mathbf{L}^1(0,1)$ does not have (MCP_1). Indeed, let Σ_n be the binary Borel algebra in (0,1) generated by the intervals $(m/2^n, (m+1)/2^n), 0 \leq m \leq 2^{n-1}, n = 1, 2, \ldots,$. Define

$$M_n(\omega) = 2^n \left(I_{(0,(m+1)/2^n]} - I_{(0,m/2^n]} \right),$$

if $\omega \in (m/2^n, (m+1)/2^n$, and 0 elsewhere. It is easy to check that $(M_n, \Sigma_n, n \geq 1)$ is a martingale with values in $\mathbf{L}^1(0,1)$ and moreover,

$$\|M_n(\omega)\| \equiv 1, \quad a.e,$$

[12] Due to S.D. Chatterji (1960).

$$\mathbf{E}\|M_n(\omega)\|^p = 1, \quad n \geq 1,$$
$$\mathbf{E}\big(\sup_n \|M_n(\omega) - M_{n-1}(\omega)\|\big) = 1, \quad M_0 = 0.$$

But, if $\omega \neq p/2^q$, then $X_n(\omega)$ does not converge to any limit, either weakly, or strongly.

The following Theorem is the main result of this chapter:

Theorem 2.2.1.[13] *For a Banach space $\boldsymbol{X}$ the following properties are equivalent:*
(D) $\boldsymbol{X}$ is dentable,
(RNP) $\boldsymbol{X}$ has the Radon-Nikodym Property,
(MCP_p) $\boldsymbol{X}$ has the $\mathbf{L}^p$-Martingale Convergence Property, for any $1 \leq p < \infty$.

Proof. We shall prove the theorem in the circular fashion,

$$(D) \implies (RNP) \implies (MCP_p) \implies (D).$$

$(D) \implies (RNP)$. Assume that $\boldsymbol{X}$ is dentable, and m is an $\boldsymbol{X}$-valued measure on (Ω, Σ) with finite total variation which is absolutely continuous with respect to a finite positive measure μ[14]. The task is to find a Bochner Σ-measurable $f : \Omega \mapsto \boldsymbol{X}$ such that, for each $E \in \Sigma$,

$$m(E) = \int_E f(\omega)\mu(d\omega).$$

Our **First Observation** is that under the above assumption m has a *locally almost dentable average range* with respect to μ, that is for each $E \in \Sigma$, and arbitrary $\epsilon > 0$, there exists an $F \subset E$ such that $\mu(E \setminus F) < \epsilon$, and the set

$$AR(F) := \Big\{\frac{m(F')}{\mu(F')} : F' \subset F, \mu(F') > 0\Big\} \subset \boldsymbol{X}$$

[13]The equivalence of the first and second conditions follows from the work of M.A. Rieffel (1967), H.B. Maynard (1973), and W.J. Davis and R.R. Phelps (1974). A short and direct proof of this equivalence can also be found in R.E. Huff (1974). The equivalence of the second and third conditions has been proved by S.D. Chatterji (1969).

[14]The proof is the same in the case of a σ-finite μ.

is dentable. Indeed, because of deniability of $\boldsymbol{X}$ it is sufficient to prove that $AR(F)$ is bounded, and the last statement can be verified as follows:

The total variation $|m|$ is a finite positive measure on Σ which is also absolutely continuous with respect to μ and, thus by the real-valued Radon-Nikodym Theorem, for some $\varphi \in \boldsymbol{L}^1(\Omega, \Sigma, \mu; \mathbf{R})$,

$$|m|(E) = \int_E \varphi(\omega)\mu(d\omega), \qquad E \in \Sigma,$$

so that, given $\epsilon > 0$, there exists a constant K such that $\varphi(\omega) < K$ on $E_0 \in \Sigma$, with $\mu(\Omega \setminus E_0) < \epsilon$. Thus, given $E \in \Sigma$, and $\epsilon > 0$, we take $F = E \cap E_0$. Then $\mu(E \setminus F) \leq \mu(\Omega \setminus E_0) < \epsilon$, and for each $F' \subset F = E \cap E_0$, with $\mu(F') > 0$, we shall have

$$\left\| \frac{m(F')}{\mu(F')} \right\| \leq \frac{|m|(F')}{\mu(F')} \leq K.$$

Our **Second Observation** is that, for any $\epsilon > 0$, and $E \in \Sigma$ with $\mu(E) > 0$, one can find an $F \subset E$ with $\mu(F) > 0$, and an $\boldsymbol{x} \in \boldsymbol{X}$ such that , for all $F' \subset F$,

$$\|m(F') - \boldsymbol{x}\mu(F')\| < \epsilon\mu(F').$$

Indeed, by first observation, there is an $E_d \subset E$ of positive measure μ such that $AR(E_d)$ is dentable, and, in particular, for the ϵ given above one can find

$$\boldsymbol{x} = \frac{m(F_0)}{\mu(F_0)} \in AR(E_d), \qquad F_0 \subset E_d,\ \mu(F_0) > 0,$$

such that

$$\boldsymbol{x} \notin Q := \overline{\text{conv}}\big(AR(E_d) \setminus B_\epsilon(\boldsymbol{x})\big).$$

Now, either F_0 can be taken as the desired F, or not. If it can, we are done, and if not, we can find $E_1 \subset F_0$ such that $\mu(E_1) \geq 1/k_1$ (and let k_1 be the smallest integer for which such an E_1 exists), and

$$\frac{m(E_1)}{\mu(E_1)} \in Q.$$

Next, either $F_1 = F_0 \setminus E_1$ is the looked for F, or not. If yes, then we are done again. If not, we can find $E_2 \subset F_1$ as above. So, either we find our F in a finite number of steps, or, by induction, we choose a sequence (E_i) of pairwise disjoint subsets of F_0, and a sequence of (minimal in the above sense) integers $k_i \uparrow \infty$(because $\mu(F_0) < \infty$), such that

$$\mu(E_i) \geq \frac{1}{k_i}, \quad \text{and} \quad \frac{m(E_i)}{\mu(E_i)} \in Q,$$

and such that if

$$E' \subset F_0 \setminus \bigcup_{i=1}^{n} E_i, \quad \text{and} \quad \frac{m(E')}{\mu(E')} \in Q,$$

then $\mu(E') < 1/(k_n - 1)$ (remember the minimality of k_n !!). But, in this case we can surely take

$$F = F_0 \setminus \bigcup_{i=1}^{\infty} E_i \subset F_0 \setminus \bigcup_{i=1}^{n} E_i, \qquad n = 1, 2, \ldots,$$

because, if for any $F' \subset F$ we have $\mu(F') > 0, m(F')/\mu(F') \in Q$, then for each $n = 1, 2, \ldots,$ we would also have $\mu(F') < 1/(k_n - 1)$ and that would imply $\mu(F') = 0$, a contradiction.

If F were of μ measure 0, then also $m(F) = 0$ and, by a convexity argument, we would have that

$$\boldsymbol{x} = \frac{m(F_0)}{\mu(F_0)} = \frac{m(\bigcup E_i)}{\mu(\bigcup E_i)} = \sum_{i=1}^{\infty} \frac{m(E_i)}{\mu(E_i)} \cdot \frac{\mu(E_i)}{\mu(\bigcup E_i)} \in Q,$$

yielding another contradiction.

Third Observation: For each $\epsilon > 0$, one can find a sequence $(\boldsymbol{x}_n) \subset \boldsymbol{X}$, and an (at most countable) partition $(E_i) \subset \Sigma$ of Ω such that

$$F \subset E_i, \ \mu(F) > 0 \quad \Longrightarrow \|m(F) - \boldsymbol{x}_i \mu(F)\| \leq \epsilon \mu(F). \qquad (*)$$

Indeed, using repeatedly the Second Observation either we exhaust Ω in a finite number of steps or else we find a sequence

(E_i) of pairwise disjoint subsets of Ω, a sequence $\boldsymbol{x}_i \subset \boldsymbol{X}$, and a nondecreasing sequence $k_i \uparrow \infty$ of (minimal, as in the Second Observation) integers such that, for each $i = 1, 2, \ldots$, (*) holds, $\mu(E_i) \geq 1/k_i$, and if, for some n,

$$E \subset \Omega \setminus \bigcup_{i=1}^{n} E_i$$

is such that, for some $\boldsymbol{x} \in \boldsymbol{X}$, and all $F \subset E, \mu(F) > 0$, and $\|m(F) - \boldsymbol{x}\mu(F)\| < \epsilon\mu(F)$, then $\mu(E) < 1/(k_n - 1)$., But, in the latter case, $\mu(\Omega \setminus \bigcup E_i) = 0$ because, otherwise, using the Second Observation, we could find $E \subset \Omega \setminus \bigcup E_i$, $\mu(E) > 0$, as above, and that would mean that, for each n, $\mu(E) < 1/(k_n - 1)$, i.e., $\mu(E) = 0$, a contradiction.

Now, we are ready to complete the proof of the implication $(D) \implies (RNP)$ and construct the density f as follows:

Let $\Pi(\ni \pi)$ be the directed family of finite partitions of Ω into sets of positive measure, and put

$$f_\pi = \sum_{E \in \pi} \frac{m(E)}{\mu(E)} I_E.$$

If we manage to show that $\lim_\pi f_\pi$ exists in $L^1(\Omega, \Sigma, \mu; \boldsymbol{X})$ then, clearly, $f = \lim_\pi f_\pi$ is the desired density because then, in particular, for each $E \in \Sigma, \mu(E) > 0$,

$$\int_E f d\mu = \lim_{\pi \in \Pi} \int_E f_\pi d\mu,$$

and because, for $\pi \geq \{E, \Omega \setminus E\}$,

$$\int_E f_\pi d\mu = m(E).$$

So the proof will be complete as soon as we show that (f_π) satisfy the Cauchy condition in $L^1(\Omega, \Sigma, \mu; \boldsymbol{X})$.

Take $\epsilon > 0$. Because $|m|$ is a finite positive measure, absolutely continuous with respect to μ, we can choose $E \subset \Omega$ such that

$|m|(\Omega \setminus E) < \epsilon/6$, and a $\delta > 0$ such that $|m|(F) < \epsilon/6$ whenever $\mu(F) < \delta$.

Now, employing the Third Observation, in the decomposition $E = \bigcup E_i$ (and using $\epsilon/6$ instead of ϵ) take n such that

$$\mu\Big(E \setminus \bigcup_{i=1}^{n} E_i\Big) < \delta,$$

and

$$\pi_0 = \Big\{E_1, \dots, E_n, E \setminus \bigcup_{i=1}^{n} E_i, \Omega \setminus E\Big\}.$$

Then, for arbitrary $\pi \geq \pi_0$, which is evidently of the form,

$$\pi = \Big\{F_{ij} : 1 \leq i \leq n, 1 \leq j \leq m\Big\} \cup \Big\{G_1, \dots, G_l\Big\} \cup \Big\{F_1, \dots, F_k\Big\},$$

with

$$\bigcup_{j=1}^{k} F_{ij} = E_i, \qquad 1 \leq j \leq n,$$

we have

$$\int \|f_\pi(\omega) - f_{\pi_0}(\omega)\| \, \mu(d\omega) = \sum_{i=1}^{n} \sum_{j=1}^{k_i} \Big\| \frac{m(F_{ij})}{\mu(F_{ij})} - \frac{m(E_j)}{\mu(E_j)} \Big\| \, \mu(Fij)$$

$$+ \sum_{i=1}^{l} \Big\| \frac{m(G_i)}{\mu(G_i} - \frac{m\big(E - \bigcup_{i=1}^{n} E_i\big)}{\mu\big(E - \bigcup_{i=1}^{n} E_i\big)} \Big\| \, \mu(G_i)$$

$$+ \sum_{i=1}^{k} \Big\| \frac{m(F_i)}{\mu(F_i} - \frac{m(\Omega \setminus E)}{\mu(\Omega \setminus E)} \Big\| \, \mu(F_i)$$

$$\leq \sum_{i=1}^{n} \sum_{j=1}^{k_i} \|m(F_{ij}) - \boldsymbol{x}_i \mu(F_{ij})\| + \sum_{i=1}^{n} \|m(E_i) - \boldsymbol{x}_i \mu(E_i)\|$$

$$+2 \sum_{I=1}^{l} \|m(G_i)\| + 2 \sum_{i=1}^{k} \|m(f_i)\|$$

$$\leq \sum_{i=1}^{n}\sum_{J=1}^{k_i} \frac{\epsilon}{\sigma\mu(E)}\mu(F_{ij}) + \sum_{i=1}^{n} \frac{\epsilon}{\sigma\mu(E)}\mu(E_i)$$

$$+2|m|\Big(E \setminus \bigcup_{i=1}^{n} E_i\Big) + 2|m|(\Omega \setminus E)$$

$$\leq \frac{\epsilon}{6} + \frac{\epsilon}{6} + 2\frac{\epsilon}{6} + 2\frac{\epsilon}{6} = \epsilon,$$

which completes the proof of the implication $(D) \implies (RNP)$.

$(RNP) \implies (MCP_p)$. Assume that $\boldsymbol{X}$ has the Radon-Nikodym property, $p > 1$ and that (M_n, Σ_n) is an $\boldsymbol{X}$-valued $\boldsymbol{L}^p$-bounded martingale on (Ω, Σ, μ), where μ is a probability measure.

Now, the martingale property implies that $\boldsymbol{X}$-valued, countably additive measures

$$m_n(E) := \int_E M_n(\omega)\,\mu(d\omega), \qquad E \in \Sigma_n,\ n = 1, 2, \ldots,$$

extend to a finitely additive set function, m, on the algebra $\bigcup_{i=1}^{\infty} \Sigma_i$. The total variation of m on any set $E \in \Sigma_i$, say $E \in \Sigma_n$,

$$|m|(E) = \int_E \|M_n\|\,d\mu \leq [\mu(E)]^{\frac{1}{q}} \cdot \sup_n \Big(\int_\Omega \|M_n\|^p\Big)^{\frac{1}{p}} < \infty,$$

$$q = \frac{p}{p-1} < \infty.$$

Moreover, the above inequality shows that $|m|$ is countably additive on $\bigcup \Sigma_n$, and thus extends to a measure on $\Sigma_\infty = \sigma(\bigcup \Sigma_n)$ which is absolutely continuous with respect to μ. And, since $\|m(E)\| \leq |m|(E), E \in \Sigma_n$, m is also countably additive on Σ_n, extends to Σ_∞, and is absolutely continuous with respect to μ.[15] Thus, by the RNP, one can find a function $g \in \boldsymbol{L}^1(\Omega, \Sigma_\infty, \mu; \boldsymbol{X})$ such that

$$m(E) = \int_E g\,d\mu, \qquad E \in \Sigma_\infty.$$

[15] For all the basic properties of vector set functions and their variation, see N. Dunford and J.T. Schwartz (1958), Chapter III, and Chapter IV.10.

Evidently, for $n = 1, 2, \dots$, $\mathbf{E}(g|\Sigma_n) = M_n$, because , for each $E \in \Sigma_n$,

$$\int_E g \, d\mu = m(E) = m_n(E) = \int_E M_n \, d\mu,$$

and, by the general theorem on conditional expectations,[16] as $n \to \infty$,

$$M_n = \mathbf{E}(g|\Sigma_n) \to \mathbf{E}(g|\Sigma_\infty) = g, \qquad \text{a.s., and in} \quad \boldsymbol{L}^p((\Omega, \Sigma, \mu; \boldsymbol{X}),$$

which completes the proof of the implication $(RNP) \implies (MCP_p)$.

$(MCP_p) \implies (D)$. Suppose the space in not dentable, i.e., by Proposition 2.1.4, it is not σ-dentable. Then there exists a bounded, closed and convex set $A \subset \boldsymbol{X}$ which is not σ-dentable, that is, for some $\epsilon > 0$, and each $\boldsymbol{x} \in A$, there are positive numbers $\alpha_i(\boldsymbol{x}), \sum_i \alpha_i(\boldsymbol{x}) = 1$, and $\boldsymbol{a}_i(\boldsymbol{x}) \in A$, $i = 1, 2, \dots$, such that

$$\inf_i \|\boldsymbol{x} - \boldsymbol{a}_i(\boldsymbol{x})\| > \epsilon,$$

and

$$\boldsymbol{x} = \sum \alpha_i(\boldsymbol{x}) \cdot \boldsymbol{a}_i(\boldsymbol{x}).$$

Now, we shall construct a bounded martingale M_n with values in A which diverges almost surely (a.s.), thus contradicting (MCP_p). We proceed by induction, and construct a tree as in the the scheme pictured below:

$$M_0 = \boldsymbol{x}$$

$$M_1(0,1), \qquad M_1(02), \qquad M_1(0,3), \qquad \dots$$

$$M_2(0,1,1), M_2(0,1,2) \dots M_2(0,2,1) M_2(0,2,2) \dots M_2(0,3,1),$$
$$M_2(0,3,2) \dots$$

$$\dots \quad \dots \qquad \dots \quad \dots \qquad \dots \quad \dots$$

Take as the probability space

$$\Omega = \mathbf{N} \times \mathbf{N} \times \dots, \qquad \Sigma = \sigma(2^{\mathbf{N}} \times 2^{\mathbf{N}} \times \dots), \qquad \mathbf{P} = \lim \mathbf{P}_n,$$

[16] See J. Neveu (1972), Prop. V-2-6.

where $\mathbf{P}_n$ are probability measures on the increasing sequence of σ-algebras

$$\Sigma_n = \underbrace{2^{\mathbf{N}} \times 2^{\mathbf{N}} \times \cdots \times 2^{\mathbf{N}}}_{n-\text{times}} \times \mathbf{N} \times \mathbf{N} \times \ldots,$$

and define

$$M_n(\omega) = M_n(i_0, i_1, i_2, \ldots) = M_n(i_0, \ldots, i_n),$$

where M_0 is any point $\boldsymbol{x} \in A$ a.s., and inductively, given M_n, and $\mathbf{P}_n$, we define

$$M_{n+1}(i_0, i_1, \ldots, i_n, i_{n+1}) = \boldsymbol{a}_{i_{n+1}}(M_n(i_0, \ldots, i_n)),$$

and

$$\begin{aligned} \mathbf{P}_{n+1}(\{i_0, i_1, \ldots, i_{n+1}\}) = \mathbf{P}_{n+1}(\{i_0, i_1, \ldots, i_n\}) \\ \cdot \alpha_{i_{n+1}}(M_n(\{i_0, i_1, \ldots, i_n\})). \end{aligned}$$

The construction implies that (M_n) is a martingale which has values in the bounded set $A \subset \boldsymbol{X}$, and which is evidently divergent since, for any $\omega = (i_0, i_1, \ldots) \in \Omega$,

$$\|M_{n+1}(\omega) - M_n(\omega)\| > \epsilon, \qquad n = 1, 2, \ldots.$$

This completes the proof of the implication $(MCP_p) \implies (D)$, and of the Theorem 2.2.1 itself. QED

2.3 Dentability and submartingales in Banach lattices and lattice bounded operators

In this section[17] we will discuss behavior of submartingales with values in Banach lattices, that is, Banach spaces $(\boldsymbol{X}, \|.\|)$ endowed with a partial order $\leq$ satisfying the following properties:

[17]The submartingale results of this section are due to J. Szulga and W.A. Woyczyński (1974).

(i) Translation invariance; $\boldsymbol{x} \leq \boldsymbol{y}$ implies $\boldsymbol{x} + \boldsymbol{z} \leq \boldsymbol{y} + \boldsymbol{z}$;

(ii) Positive homogeneity: For any scalar $\alpha \geq 0$, $\boldsymbol{x} \leq \boldsymbol{y}$ implies $\alpha\boldsymbol{x} \leq \alpha\boldsymbol{y}$;

(iii) For any pair of vectors $\boldsymbol{x}, \boldsymbol{y} \in \boldsymbol{X}$ there exists a supremum, $\boldsymbol{x} \vee \boldsymbol{y} \in \boldsymbol{X}$ with respect to partial order $\leq$;

(iv) The norm is monotone, that is $|\boldsymbol{x}| \leq |\boldsymbol{y}|$ implies $\|\boldsymbol{x} \leq \|\boldsymbol{y}\|$.

In the above notation $|\boldsymbol{x}| := \boldsymbol{x}^+ + \boldsymbol{x}^-$, where $\boldsymbol{x}^+ := \boldsymbol{x} \vee 0, \boldsymbol{x}^- := -\boldsymbol{x} \vee 0$. The norm dual $\boldsymbol{X}^*$ of $\boldsymbol{X}$ is also a Banach lattice with the natural ordering, and by $\boldsymbol{X}_+$, and $\boldsymbol{X}^*_+$ we denote the non-negative cones in $\boldsymbol{X}$, and $\boldsymbol{X}^*$, respectively.

Definition 2.3.1. Let (Ω, Σ, P) be a probability space, and let $\Sigma_1 \subset \Sigma_2 \subset \cdots \subset \Sigma$ be a sequence of sub-σ-algebras. The sequence $(X_n, n \in \mathbf{N})$ of Banach lattice $\boldsymbol{X}$-valued random vectors with $X_n \in \boldsymbol{L}^1(\Omega, \Sigma_n, P; \boldsymbol{X})$ is said to be a *sub-martingale* if

$$\mathbf{E}(X_{n+1}|\Sigma_n) \geq X_n, \quad n = 1, 2, \ldots, a.s.$$

For real-valued submartingales the following results is classical[18]:

Proposition 2.3.1. *If* $(X_n, n \in \mathbf{N})$ *is a real-valued submartingale, and* $\sup_n \mathbf{E}X_n^+ < \infty$, *then there exists an* $X_\infty \in \boldsymbol{L}^1$ *such that* $X_n \to X_\infty$, *a.s.*

The results presented below show how the above Proposition (which can also be dually formulated for supermartingales) carries over to the case of submartingales with values in Banach lattices.[19]

Definition 2.3.2. The set $A \subset \boldsymbol{X}$ is said to be *order bounded* if there exists an $\boldsymbol{x}_0 \in \boldsymbol{X}$ such that, for all $\boldsymbol{y} \in A$, $|\boldsymbol{y}| \leq \boldsymbol{x}_0$. The linear operator T from a Banach space $\boldsymbol{Y}$ into a Banach lattice $\boldsymbol{X}$ is said to be *lattice bounded* if it maps the unit ball of $\boldsymbol{Y}$ into an order bounded subset of $\boldsymbol{X}$.

---

[18]See J. Neveu (1972).

[19]The initial work was motivated by L. Schwartz's (1973) extensive work on supermartingales that have measures as their values. The results had an elegant application to the problems of disintegration of measures. His model fits into our general framework.

The following Proposition is well known[20]:

Proposition 2.3.2. (a) *Let $\boldsymbol{C}$ be the Banach space of real continuous functions on the unit interval, and let $\boldsymbol{X}$ be a separable Banach lattice with an order continuous norm. Then $T : \boldsymbol{C} \to \boldsymbol{X}$ is lattice bounded if, and only if there exists a function $g : [0,1] \to \boldsymbol{X}$ of bounded 0-variation,*

$$\text{ess var}\, g(t) := \sup \sum |g(t_{i+1}) - g(t_i)| \in \boldsymbol{X},$$

such that

$$Tf = \int_0^1 f(t)dg(t), \qquad f \in \boldsymbol{C},$$

where the integral is understood as an order limit of Stieltjes sums;

(b) The operator $T : \boldsymbol{l}^q \mapsto \boldsymbol{l}^p$ $(q > 1, p \geq 1)$ is lattice bounded if, and only if it is of the form,

$$T\boldsymbol{y} = \Big(\sum_{k=1}^{\infty} a_{ik} y_k\Big)_{i \in \mathbf{N}}, \qquad \boldsymbol{y} = (y_k) \in \boldsymbol{l}^q,$$

where

$$\sum_{i=1}^{\infty} \Big[\sum_{k=1}^{\infty} |a_{ik}|^{q/(q-1)}\Big]^{(q-1)p/q} < \infty.$$

(c) The operator $T : \boldsymbol{Y} \mapsto \boldsymbol{L}^p[0,1], p \geq 1$, where $\boldsymbol{Y}$ is a separable Banach space, is lattice bounded if, and only if, it is of the form,

$$(T\boldsymbol{y})(t) = f^*(t)\boldsymbol{y},$$

where $\boldsymbol{y} \in \boldsymbol{Y}$, $t \in [0,1]$, and $f^* : [0,1] \mapsto \boldsymbol{Y}$ is *- weakly measurable and such that $\|f^*\| \in \boldsymbol{L}^p[0,1]$.

Now we shall formulate a result on dentable Banach lattices that will be used later on.

Proposition 2.3.3. *If $\boldsymbol{X}$ is a dentable Banach lattice and the sequence $\boldsymbol{x}_0 \leq \boldsymbol{x}_1 \leq \boldsymbol{x}_3 \leq \dots$ in $\boldsymbol{X}$ is norm bounded then it is convergent.*

[20]See, e.g., L.V. Kantorovich, B.Z. Vulikh and A.G. Pinsker (1950).

Proof. Because $\boldsymbol{X}$ is dentable it does not contain isomorphic copies of $\boldsymbol{c}_0$ (see, Example 2.1.2, and Theorem 2.1.1), and in every such Banach lattice monotone norm-bounded sequences are convergent.[21] QED

Now we turn to the investigation of analogues of Proposition 2.3.1 for $\boldsymbol{X}$-valued submartingales. Note, that the classical Doob's condition, $\sup_n \mathbf{E}X_n^+ < \infty$, for real-valued random variables has two different versions for Banach-lattice-valued random vectors, namely, order boundedness of $\mathbf{E}X_n^+, n \in \mathbf{N}$, and norm boundedness, $\sup \|X_n^+\| < \infty$. Both boil down to the Doob's condition in the real case. However, as we shall see below, in general, neither is sufficient to assure a.s. convergence of a submartingale $(X_n), n \in \mathbf{N}$.

It is not difficult to check that for both real- and vector-valued sub-martingales, the set $(\mathbf{E}X_n^+, n \in \mathbf{N})$ is order bounded if, and only if, $(\mathbf{E}|X_n|, n \in \mathbf{N})$ is such. However, even for vector-valued martingales it might happen that $\sup_n \mathbf{E}\|X_n^+\| < \infty$, and still $\sup_n \mathbf{E}\|X_n^-\| = \infty$, so that it will not be surprising to see that the condition $\sup_n \mathbf{E}\|X_n^+\| < \infty$ does not imply, in general, the a.s. convergence of the submartingale (X_n) even in dentable Banach lattices. On the other hand, the condition $\sup_n \mathbf{E}\|X_n^+\| < \infty$ is stronger than order boundedness of $(\mathbf{E}X_n^+, n \in \mathbf{N})$ for any sequence (X_n) of random vectors with values in the Banach lattice $\boldsymbol{X}$ because, for each $\boldsymbol{x}^* \in \boldsymbol{X}_+^*$, $\sup \boldsymbol{x}^*\mathbf{E}(X_n^*) \leq \sup_n \mathbf{E}\|X_n^+\|\|\boldsymbol{x}^*\| < \infty$, and because the set $A \subset \boldsymbol{X}$ is order bounded if, and only if, for each $\boldsymbol{x}^* \in \boldsymbol{X}_+^*$, the set $\boldsymbol{x}^*A$ is bounded on the real line. The last statement follows from the fact that in a Banach lattice $\boldsymbol{X}$, $\boldsymbol{x} \geq 0$ if, and only if, for each $\boldsymbol{x}^* \in \boldsymbol{X}_+^*, \boldsymbol{x}^*\boldsymbol{x} \geq 0$.

It is not hard to see that if $\boldsymbol{X}$ is a dentable Banach lattice then in order to produce examples of,

(1) a martingale (M_n) with values in $\boldsymbol{X}$ such that $\sup_n \mathbf{E}\|M_n^+\| < \infty$ and, at the same time, $\sup_n \mathbf{E}\|m_n^-\| = \infty$, and $\sup_n \mathbf{E}\|M_n\| = \infty$,

and

(2) a submartingale (X_n) with values in $\boldsymbol{X}$ such that

[21] See, e.g., Theorem 14 in L. Tzafriri (1972).

$\sup_n \mathbf{E}\|X_n^+\| < \infty$, and X_n diverges a.s.,

it is sufficient to find

(3) an a.s. divergent sequence (Y_n) of non-negative independent random vectors in $\boldsymbol{L}^1(\Omega, \Sigma, P; \boldsymbol{X})$ such that both, $\sup_n \mathbf{E}\|Y_n\| < \infty$, and $\sup_n \|\sum_{i=0}^n \mathbf{E}Y_i\| < \infty$.

Indeed, given such a sequence (Y_n) it is enough to take $\Sigma_n = \sigma(Y_0, \ldots, Y_n)$, $Z_0 = 0$,

$$Z_n = \sum_{i=0}^{n-1} Y_i,$$

$$M_n = -Z_{n+1} + \mathbf{E}Z_{n+1},$$

and

$$X_n = M_n + Z_n = \mathbf{E}Z_{n+1} - Y_n, \qquad n \geq 1.$$

Now, (M_n) defined in such a way is a zero-mean martingale such that $\sup_n \mathbf{E}\|M_n^+\| \leq \sup_n \|\mathbf{E}Z_{n+1}\| < \infty$ (because $M_n^+ \leq \mathbf{E}Z_{n+1}$) but, at the same time,

$$\mathbf{E}\|M_n\| = \mathbf{E}\|Z_{n+1} - \mathbf{E}Z_{n+1}\| \geq \mathbf{E}\|Z_{n+1}\| - \|\mathbf{E}Z_{n+1}\|$$

is unbounded. Also, (X_n) is evidently a submartingale that is divergent a.s., and for which

$$\sup_n \mathbf{E}\|X_n^+\| \leq \sup_n \|\mathbf{E}Z_{n+1}\| < \infty.$$

Below, we provide an example of sequences (Y_n) of random vectors with values in certain classical Banach lattices that satisfy the condition (3).

Example 2.3.1. Let $\boldsymbol{X} = \boldsymbol{l}^p, p > 1$ (the reason why $p = 1$ is excluded is provided in the Corollary at the end of this Section), $\Omega_i = [0, 1)$, Σ_i be the family of all Borel subsets of $[0, 1)$, and λ_i be the Lebesgue measure on $\Omega_i, i \in \mathbf{N}$. Define

$$\Omega = \prod_{i \in \mathbf{N}} \Omega_i, \qquad \Sigma = \prod_{i \in \mathbf{N}} \Sigma_i, \qquad P = \prod_{i \in \mathbf{N}} \lambda_i,$$

$Y_0(\omega_0, \omega_1, \ldots) = 0$, and

$$Y_{2^{n-1}+k}(\omega_0, \omega_1, \ldots) = I_{[k/2^{n-1}, (k+1)/2^{n-1})}(\omega_{2^{n-1}+k})\boldsymbol{e}_{2^{n-1}+k},$$

where $n = 1, 2, \ldots,\ k = 0, 1, \ldots, 2^{n-1} - 1$, I_A is the indicator function of the set A, and $(\boldsymbol{e}_n, n \in \mathbf{N})$ is the standard basis in $\boldsymbol{l}^p$.

By definition $Y_n, n \in \mathbf{N}$, are independent, non-negative and in $\boldsymbol{L}^1(\Omega, \Sigma, P; \boldsymbol{X})$,

$$\sup_{n\in\mathbf{N}} \mathbf{E}\|Y_n\| \leq 1, \qquad \text{and} \qquad \sup_{n\in\mathbf{N}} \Big\| \sum_{i=1}^{n} \mathbf{E}Y_i \Big\| < \infty,$$

because

$$\mathbf{E}Y_0 = 0, \qquad \mathbf{E}Y_{2^{n-1}+k} = 2^{1-n}\boldsymbol{e}_{2^{n-1}+k},$$

for $n = 1, 2, \ldots,\ k = 0, 1, \ldots 2^{n-1} - 1$, so that

$$\sup_{n\in\mathbf{N}} \Big\| \sum_{i=0}^{n} \mathbf{E}Y_i \Big\| = \Big\| \sum_{i=0}^{\infty} \mathbf{E}Y_i \Big\| = \Big(\sum_{i=0}^{\infty} 2^{i(1-p)} \Big)^{1/p},$$

but, at the same time, (Y_n) is divergent for each $\omega \in \Omega$ because, for each $\omega \in \Omega$ there exist sequences $(n_i), (n_i') \subset \mathbf{N}$ such that $\|Y_{n_i}(\omega)\| = 1$, and $\|Y_{n_i'}(\omega)\| = 0$.

Now, in the next two theorems[22] we discuss positive results concerning convergence of sub-martingales in Banach lattices.

We start with the observation that the Doob's decomposition of real martingales survives in Banach lattices. Namely, if $(X_n, \Sigma_n, n \in \mathbf{N})$ is a submartingale with values in a Banach lattice $\boldsymbol{X}$, then

$$X_n = M_n + Z_n,$$

where (M_n, Σ_n) is a martingale, and the sequence $(Z_n, n \in \mathbf{N})$ is predictable, i.e., $Z_n \in \boldsymbol{L}^1(\Omega, \Sigma_{n-1}, P; \boldsymbol{X})$. As in the real case, to prove the above statement it is sufficient to define, $Z_0 = 0, M_0 = X_0$, and

$$M_n = X_0 + \sum_{i=1}^{n} \big[X_i - \mathbf{E}(X_i | \Sigma_{i-1}) \big],$$

and

$$Z_n = \sum_{i=1}^{n} \mathbf{E}(X_i - X_{i-1} | \Sigma_{i-1}), \qquad n = 1, 2, \ldots .$$

[22]Due to J. Szulga and W.A. Woyczyński (1974).

The above Example 2.3.1 shows that, in general, the condition $\sup_n \mathbf{E}\|X_n^+\| < \infty$ for the submartingales $X_n = M_n + Z_n$ does not imply its a.s. convergence. However, we have the following result:

Theorem 2.3.1. *For a separable Banach lattice* $\boldsymbol{X}$ *the following conditions are equivalent:*

(i) $\boldsymbol{X}$ *is dentable;*

(ii) For each $\boldsymbol{X}$*-valued submartingale* $(X_n = M_n + Z_n, n \in \mathbf{N})$ *satisfying the conditions,*

$$\sup_n \mathbf{E}\|X_n^+\| < \infty, \quad \text{and} \quad \sup_n \mathbf{E}\|X_n^-\| < \infty,$$

there exists an $X_\infty \in \boldsymbol{L}^1(\Omega, \Sigma, P; \boldsymbol{X})$ *such that* $X_n \to X_\infty$*, a.s.;*

(iii) For each $\boldsymbol{X}$*-valued submartingale* $(X_n = M_n + Z_n, n \in \mathbf{N})$ *satisfying the conditions,*

$$\sup_n \mathbf{E}\|X_n^+\|^p < \infty, \quad \text{and} \quad \sup_n \mathbf{E}\|X_n^-\|^p < \infty,$$

for some $p \in (1, \infty)$*, there exists an* $X_\infty \in \boldsymbol{L}^p(\Omega, \Sigma, P; \boldsymbol{X})$ *such that* $X_n \to X_\infty$*, a.s., and in* $\boldsymbol{L}^p$*.*

Proof. *(i)* $\implies$ *(ii)*[*(i)* $\implies$ *(iii)*]. Since $X_n = M_n + Z_n \geq M_n$, a.s., we also have that $X_n^+ \geq M_N^+$, a.s., and the monotonicity of the norm implies that

$$\sup_{n\in\mathbf{N}} \mathbf{E}\|M_n\|^p \leq 2^p(\sup_{n\in\mathbf{N}} \mathbf{E}\|X_n^+\|^p + \sup_{n\in\mathbf{N}} \mathbf{E}\|M_n^-\|^p).$$

Hence, by Theorem 2.1.1, there exists $M_\infty \in \boldsymbol{L}^1$ [$M_\infty \in \boldsymbol{L}^p$], such that $M_n \to M_\infty$, a.s. [$M_n \to M_\infty$, a.s., and in $\boldsymbol{L}^p$] . Because $Z_n = X_n - M_n$, we have that $Z_n \leq X_n^+ + M_n^-$, so that $\sup_{n\in\mathbf{N}} \mathbf{E}\|Z_n\|^p < \infty, 1 \leq p < \infty$. Utilizing again the monotonicity of the norm, and the Lebesgue Monotone Convergence Theorem, we get that $\mathbf{E}\sup_{n\in\mathbf{N}} \|Z_n\|^p < \infty$, so that $\sup_{n\in\mathbf{N}} \|Z_n\| < \infty$, a.s. However, because of Proposition 2.3.3, there exists a random vector Z_∞ such that $Z_n \to Z_\infty$, a.s. The Fatou Lemma yields that

$$\mathbf{E}\|Z_\infty\|^p \leq \liminf_{n\in\mathbf{N}} \mathbf{E}\|Z_n\|^P \leq \sup_{n\in\mathbf{N}} \mathbf{E}\|Z_n\|^p < \infty,$$

so that $Z_\infty \in \boldsymbol{L}^p, 1 \leq p < \infty$, and letting

$$X_\infty = M_\infty + Z_\infty,$$

completes the proof of the first two implications.

$(ii) \implies (i)[(iii) \implies (i)]$. If $X_n = M_n$, then the conditions in $(ii)[(iii)]$ boil down to $\sup_{n\in\mathbf{N}} \mathbf{E}\|M_n\| < \infty$ $[\sup_{n\in\mathbf{N}} \mathbf{E}\|M_n\|^p < \infty]$, and Theorem 2.1.1. gives the dentability of $\boldsymbol{X}$. QED

In the next theorem we impose weaker assumptions on the submartingale (X_n), but then the convergence takes place only for a transformed submartingale.

Theorem 2.3.2. *Let $\boldsymbol{X}$ be a separable Banach lattice, $\boldsymbol{Y}$ a dentable separable Banach lattice, and $T : \boldsymbol{X} \mapsto \boldsymbol{Y}$ be a linear bounded positive operator such that its transpose $T^* : \boldsymbol{Y}^* \mapsto \boldsymbol{X}^*$ is lattice bounded. If $(X_n, n \in \mathbf{N})$ is a submartingale with values in $\boldsymbol{X}$ such that $(\mathbf{E}(X_n^+), n \in \mathbf{N})$ is order bounded, then there exists a $Y_\infty \in \boldsymbol{L}^1(\boldsymbol{Y})$ such that the submartingale $TX_n \to Y_\infty$, a.s., as $n \to \infty$.*

Proof. Let $X_n = M_n + Z_n$, as before, and let $\mathbf{E}(X_n^+) \leq \boldsymbol{x}_0 \in \boldsymbol{X}_+$, for all $n \in \mathbf{N}$. We will show that under the above assumptions,

$$\sup_{n\in\mathbf{N}} \mathbf{E}\|(TX_n)^+\| < \infty,$$

and

$$\sup_{n\in\mathbf{N}} \mathbf{E}\|(TX_n)^-\| < \infty,$$

what, in view of Theorem 2.3.1, would give the desired result because the Doob's decomposition for the sub-martingale TX_n is $TM_n + TZ_n$.

Indeed,

$$\sup_{n\in\mathbf{N}} \mathbf{E}\|(TX_n)^+\| \leq \sup_{n\in\mathbf{N}} \mathbf{E}\|T(X_n^+)\|$$

$$= \sup_{n\in\mathbf{N}} \mathbf{E} \sup\{\boldsymbol{y}^* T(X_n) : \|\boldsymbol{y}^*\| \leq 1, 0 \leq \boldsymbol{y}^* \in \boldsymbol{Y}^*\}$$

$$= \sup_{n\in\mathbf{N}} \mathbf{E} \sup\{(T^*\boldsymbol{y}^*)X_n^+ : \|\boldsymbol{y}^*\| \leq 1, 0 \leq \boldsymbol{y}^* \in \boldsymbol{Y}^*\}.$$

However, the transpose T^* of a positive T is also positive, and thus the fact that T^* is lattice bounded implies the existence of an $\boldsymbol{x}_0^* \in \boldsymbol{X}_+^*$ such that, for each $\boldsymbol{y}^*$ with $\|\boldsymbol{y}^*\| \leq 1$, we have $|T^*\boldsymbol{y}^*| \leq x_0^*$. Thus, we get that

$$\sup_{n\in\mathbf{N}} \mathbf{E}\|TX_n)^+\| \leq \mathbf{E}(\boldsymbol{x}_0^* X_n^+) = \boldsymbol{x}_0^*\mathbf{E}(X_n^+) \leq \boldsymbol{x}_0^*\boldsymbol{x} < \infty.$$

Proceeding as above, and utilizing the inequality

$$\mathbf{E}(TM_n)^- = \mathbf{E}(TM_n)^+ - \mathbf{E}(TM_n)$$

$$= \mathbf{E}(TM_n)^+ - \mathbf{E}(TM_0) \leq \mathbf{E}(TX_n)^+ - \mathbf{E}(TM_0)$$

we get that

$$\sup_{n\in\mathbf{N}} \mathbf{E}\|(TM_n)^-\| \leq \boldsymbol{x}_0^*\boldsymbol{x}_0 + |\boldsymbol{x}_0^*\mathbf{E}(TM_0)| < \infty,$$

because $M_0 \in \boldsymbol{L}^1$. This ends the proof. QED

Because $\boldsymbol{l}^1$ is a dentable Banach lattice (see, Proposition 2.1.3, and the preceding comments), and because the operator $[\mathrm{Id}(\boldsymbol{l}^1, \boldsymbol{l}^1)]^*$ is lattice bounded in $\boldsymbol{l}^\infty$, we also obtain the following,

Corollary 2.3.1. *If* $(X_n, n \in \mathbf{N})$ *is a sub-martingale with values in* $\boldsymbol{l}^1$ *such that* $(\mathbf{E}(X_n^+), n \in \mathbf{N})$ *is order bounded, then there exists an* $X_\infty \in \boldsymbol{L}^1(\Omega, \Sigma, P; \boldsymbol{l}^1)$ *such that* $X_n \to X_\infty$, *a.s. as* $n \to \infty$.

Chapter 3

Uniform Convexity and Uniform Smoothness

3.1 Basic concepts

We begin this chapter by describing the concept of uniform convexity of a Banach space.[1]

Definition 3.1.1. Let $\boldsymbol{X}$ be a Banach space of dimension ≥ 2. The *modulus of convexity* of $\boldsymbol{X}$ is defined by the formula:

$$\delta_{\boldsymbol{X}}(\epsilon) := \inf\{1-\|(\boldsymbol{x}+\boldsymbol{y})/2\| : \|\boldsymbol{x}\| = \|\boldsymbol{y}\| = 1, \|\boldsymbol{x}-\boldsymbol{y}\| = \epsilon\}, \quad 0 \leq \epsilon \leq 2.$$

$\boldsymbol{X}$ is said to be *uniformly* convex if $\delta_{\boldsymbol{X}}(\epsilon) > 0$, for $\epsilon > 0$. $\boldsymbol{X}$ is said to be *q-uniformly convex* if there exists a constant C such that $\delta_{\boldsymbol{X}}(\epsilon) \geq C\epsilon^q$, $q \geq 2$.

Example 3.1.1. The $\boldsymbol{L}^P$ space, $p \geq 1$, is $\max(p, 2)$-uniformly convex.[2]

Definition 3.1.2. Let $\boldsymbol{X}$ be a Banach space of dimension of at

[1]Introduced by J.A. Clarkson (1936).

[2]Consult V.D. Milman (1972) for a detailed exposition of metric geometry of Banach spaces. More information about moduli of convexity and smoothness of Orlicz and other spaces can be found in T. Figiel (1976) and M.M. Day (1973).

least 2. The *modulus of smoothness* of $\boldsymbol{X}$ is defined by the formula:

$$\rho_{\boldsymbol{X}}(\tau) := \sup\{\|(\boldsymbol{x}+\boldsymbol{y})/2\|+\|(\boldsymbol{x}-\boldsymbol{y})/2\|-1 : \|\boldsymbol{x}\| = 1, \|\boldsymbol{y}\| = \tau\}.$$

$\boldsymbol{X}$ is said to be *uniformly smooth* if $\rho_{\boldsymbol{X}}(\tau) = o(\tau), \tau \to 0$. $\boldsymbol{X}$ is said to be *p-uniformly smooth*, $1 < p \leq 2$, if there exists a constant C such that $\rho_{\boldsymbol{X}}(\tau) \leq C\tau^p$.

It is evident that the above notions are not invariant under equivalent renorming of the Banach space. So, we say that the Banach space $(\boldsymbol{X}, \|.\|)$ is *uniformly convexifiable (q-uniformly convexifiable, uniformly smoothable, p-uniformly smoothable)* if it admits a norm equivalent to $\|.\|$ that is uniformly convex (q-uniformly convex, uniformly smooth, p-uniformly smooth).

The Banach space $\boldsymbol{X}$ is uniformly convex if, and only if, the dual space $\boldsymbol{X}^*$ is uniformly smooth, and the relationship between their moduli of convexity and smoothness is given by the following result:[3]

Proposition 3.1.1 *For any Banach space $\boldsymbol{X}$ the modulus of smoothness of the dual space $\boldsymbol{X}^*$ is the function conjugate in the sense of Young to the modulus of convexity of $\boldsymbol{X}$, that is,*

$$\rho_{\boldsymbol{X}^*}(\tau) = \sup\{\frac{\tau\epsilon}{2} - \delta_{\boldsymbol{X}}(\epsilon) : 0 \leq \epsilon \leq 2\}, \qquad \tau > 0.$$

Proof. First, note that for every positive ϵ, and τ,

$$\delta_{\boldsymbol{X}}(\epsilon) + \rho_{\boldsymbol{X}^*}(\tau) \geq \frac{\tau\epsilon}{2}. \tag{3.1.1}$$

Indeed, if $\boldsymbol{x}, \boldsymbol{y} \in \boldsymbol{X}$, $\|\boldsymbol{x}\| = \|\boldsymbol{y}\| = 1$, $\|\boldsymbol{x}-\boldsymbol{y}\| = \epsilon$, and $\boldsymbol{x}^*, \boldsymbol{y}^* \in \boldsymbol{X}^*$ are such that $\|\boldsymbol{x}^*\| = \|\boldsymbol{y}^*\| = 1$, and $\boldsymbol{x}^*(\boldsymbol{x}+\boldsymbol{y}) = \|\boldsymbol{x}+\boldsymbol{y}\|$, $\boldsymbol{y}^*(\boldsymbol{x}-\boldsymbol{y}) = \|\boldsymbol{x}-\boldsymbol{y}\|$, then

$$2\rho_{\boldsymbol{X}^*}(\tau) \geq \|\boldsymbol{x}^* + \tau\boldsymbol{y}^*\| + \|\boldsymbol{x}^* - \tau\boldsymbol{y}^*\| - 2$$

$$\geq \boldsymbol{x}^*\boldsymbol{x} + \tau\boldsymbol{y}^*\boldsymbol{x} + \boldsymbol{x}^*\boldsymbol{y} - \tau\boldsymbol{y}^*\boldsymbol{y} - 2$$

[3]Due to J. Lindenstrauss (1963).

$$= \boldsymbol{x}^*(\boldsymbol{x}+\boldsymbol{y}) + \tau \boldsymbol{y}^*(\boldsymbol{x}-\boldsymbol{y}) - 2 = \|\boldsymbol{x}+\boldsymbol{y}\| + \tau\epsilon - 2$$

so that

$$2 - \|\boldsymbol{x}+\boldsymbol{y}\| \geq \tau\epsilon - 2\rho_{\boldsymbol{X}^*}(\tau),$$

which gives (3.1.1).

Now, let $\boldsymbol{x}^*, \boldsymbol{y}^* \in \boldsymbol{X}^*$ satisfy the conditions, $\|\boldsymbol{x}^*\| = 1, \|\boldsymbol{y}^*\| = \tau$, and let $\alpha > 0$. Then, there exist $\boldsymbol{x}, \boldsymbol{y} \in \boldsymbol{X}$ such that $\|\boldsymbol{x}\| = \|\boldsymbol{y}\| = 1$, and

$$\boldsymbol{x}^*\boldsymbol{x} + \boldsymbol{y}^*\boldsymbol{x} \geq \|\boldsymbol{x}^* + \boldsymbol{y}^*\| - \alpha, \qquad \boldsymbol{x}^*\boldsymbol{y} - \boldsymbol{y}^*\boldsymbol{y} \geq \|\boldsymbol{x}^* - \boldsymbol{y}^*\| - \alpha.$$

Therefore,

$$\|\boldsymbol{x}^* + \boldsymbol{y}^*\| + \|\boldsymbol{x}^* - \boldsymbol{y}^*\| \leq \boldsymbol{x}^*\boldsymbol{x} + \boldsymbol{y}^*\boldsymbol{x} + \boldsymbol{x}^*\boldsymbol{y} - \boldsymbol{y}^*\boldsymbol{y} + 2\alpha$$

$$= \boldsymbol{x}^*(\boldsymbol{x}+\boldsymbol{y}) + \boldsymbol{y}^*(\boldsymbol{x}-\boldsymbol{y}) + 2\alpha \leq \|\boldsymbol{x}+\boldsymbol{y}\| + \tau\|\boldsymbol{x}-\boldsymbol{y}\| + 2\alpha$$

$$\leq 2 + 2\sup\{\epsilon\tau/2 - \delta_{\boldsymbol{X}}(\epsilon) : 0 \leq \epsilon \leq 2\} + 2\alpha.$$

In view of the arbitrariness of α, we get our Proposition. QED

Corollary 3.1.1. *For every Banach space $\boldsymbol{X}$ the modulus of smoothness satisfies the following inequality:*

$$\rho_{\boldsymbol{X}}(\tau) \geq (1+\tau^2)^{1/2} - 1, \qquad \tau > 0,$$

where the right-hand side of the inequality is the modulus of smoothness of the Hilbert space. Its asymptotics at 0 is, obviously, of the order τ^2.

Example 3.1.2. It is easy to see that the space $\boldsymbol{L}^p$, $p > 1$ is $\min(p, 2)$ uniformly smooth.

The next result[4] shows that p-uniform smoothness is equivalent to being on "one side of the parallelogram equality".

Proposition 3.1.2. *A Banach space $\boldsymbol{X}$ is p-uniformly smooth if, and only if, there exists a constant $C > 0$ such that, for all $\boldsymbol{x}, \boldsymbol{y} \in \boldsymbol{X}$,*

$$\|\boldsymbol{x}+\boldsymbol{y}\|^p + \|\boldsymbol{x}-\boldsymbol{y}\|^p \leq 2\|\boldsymbol{x}\|^p + C\|\boldsymbol{y}\|^p. \qquad (3.1.2)$$

[4]Due to P. Assuad (1974), and J. Hoffmann-Jorgensen (1974).

Proof. Let us start with the observation that $\boldsymbol{X}$ is p-uniformly smooth if, and only if, there exists a constant K such that, for all $\boldsymbol{x}, \boldsymbol{y} \in \boldsymbol{X}$,

$$\left\|\frac{\boldsymbol{x}+\boldsymbol{y}}{2}\right\| + \left\|\frac{\boldsymbol{x}-\boldsymbol{y}}{2}\right\| \leq \|\boldsymbol{x}\| \left(1 + K\frac{\|\boldsymbol{y}\|^p}{\|\boldsymbol{x}\|^p}\right).$$

Now, assume that $\boldsymbol{X}$ is p-uniformly smooth. Then,

$$\frac{1}{2}\left[\|\boldsymbol{x}+\boldsymbol{y}\| - (\|\boldsymbol{x}\| + \|\boldsymbol{y}\|) + \|\boldsymbol{x}-\boldsymbol{y}\| - (\|\boldsymbol{x}\| - \|\boldsymbol{y}\|)\right] \leq K\frac{\|\boldsymbol{y}\|^p}{\|\boldsymbol{x}\|^p}.$$

If $\|y\| \leq \|\boldsymbol{x}\|$, then $\max(\|\boldsymbol{x}+\boldsymbol{y}\|, \|\boldsymbol{x}-\boldsymbol{y}\|) \leq 2\|\boldsymbol{x}\|$. Given the following two elementary inequalities for real numbers u, and v,

$$u^p - v^p \leq pu^{p-1}(u-v), \qquad u, v \geq 0,\ p \geq 1,$$

$$\frac{1}{2}\left(|u+v|^p + |u-v|^p\right) \leq |u|^p + |v|^p, \qquad u, v \in \mathbf{R},\ 1 \leq p \leq 2,$$

we obtain the following estimate:

$$\frac{1}{2}\left(\|\boldsymbol{x}+\boldsymbol{y}\|^p + \|\boldsymbol{x}-\boldsymbol{y}\|^p\right.$$

$$\leq \frac{1}{2}\left[(\|\boldsymbol{x}\| + \|\boldsymbol{y}\|)^p + (\|\boldsymbol{x}\| - \|\boldsymbol{y}\|)^p\right] + p\big(2\|\boldsymbol{x}\|\big)^{p-1}\|\boldsymbol{x}\|K\frac{\|\boldsymbol{y}\|^p}{\|\boldsymbol{x}\|^p}$$
$$\leq \|\boldsymbol{x}\|^p + \|\boldsymbol{y}\|^p + p2^{p-1}K\|\boldsymbol{y}\|^p = \|\boldsymbol{x}\|^p + (1 + p2^{p-1}K)\|\boldsymbol{y}\|^p.$$

On the other hand, if $\|\boldsymbol{y}\| \geq \|\boldsymbol{x}\|$, then $\max(\|\boldsymbol{x}+\boldsymbol{y}\|, \|\boldsymbol{x}-\boldsymbol{y}\|) \leq 2\|\boldsymbol{y}\|$, so that

$$\frac{1}{2}\left(\|\boldsymbol{x}+\boldsymbol{y}\|^p + \|\boldsymbol{x}-\boldsymbol{y}\|^p\right) \leq \|\boldsymbol{x}\|^p + 2^p\|y\|^p,$$

and consequently we get the desired inequality with

$$C = \max\big(2^p, [1 + p2^{p-1}K]\big).$$

Conversely, if the inequality (3.1.2) is satisfied, then $\boldsymbol{X}$ is p-uniformly smooth because

$$\rho_{\boldsymbol{X}}(\tau) := \sup\{\|(\boldsymbol{x}+\boldsymbol{y})/2\| + \|(\boldsymbol{x}-\boldsymbol{y})/2\| - 1 : \|\boldsymbol{x}\| = 1, \|\boldsymbol{y}\| = \tau\}$$

$$\le \sup\{(\|(\boldsymbol{x}+\boldsymbol{y})/2\| + \|(\boldsymbol{x}-\boldsymbol{y})/2\|)^p - 1 : \|\boldsymbol{x}\| = 1, \|\boldsymbol{y}\| = \tau\}$$

$$\le \sup\Big\{\frac{1}{2}\Big(\|(\boldsymbol{x}+\boldsymbol{y})\|^p + \|(\boldsymbol{x}-\boldsymbol{y})\|^p\Big) - 1 : \|\boldsymbol{x}\| = 1, \|\boldsymbol{y}\| = \tau\Big\} \le c\tau^p,$$

which completes the proof of the proposition. QED

Remark 3.1.1. Another way to look at uniform smoothness is related to the concept of the (G_α)-type of Banach space[5] which is important to the study of random series in Banach spaces, and is defined as follows: A Banach space $\boldsymbol{X}$ is said to be of (G_α)-type, $\boldsymbol{X} \in (G_\alpha)$, $0 < \alpha \le 1$, if there exists a mapping

$$G : \{\boldsymbol{x} \in \boldsymbol{X} : \|\boldsymbol{x}\| = 1\} \mapsto \{\boldsymbol{x}^* \in \boldsymbol{X} : \|\boldsymbol{x}^*\| = 1\},$$

such that

$$\|G(\boldsymbol{x})\| = \|\boldsymbol{x}\|^\alpha, \qquad G(\boldsymbol{x})\boldsymbol{x} = \|\boldsymbol{x}\|^{1+\alpha},$$

and there exists a constant $C > 0$, such that, for all $\boldsymbol{x}$, and $\boldsymbol{y}$, of norm one,

$$\|G(\boldsymbol{x}) - G(\boldsymbol{y})\|| \le C\|\boldsymbol{x}-\boldsymbol{y}\|^\alpha.$$

It turns out[6] that $\boldsymbol{X} \in (G_\alpha)$ if, and only if, $\boldsymbol{X}$ is $(1+\alpha)$- uniformly smooth which provides an alternative descriptions of p-uniform smoothness.

3.2 Martingales in uniformly smooth and uniformly convex spaces

Uniform smoothness. Now, we turn to the investigations of interrelations between uniform smoothness, and uniform convexity of Banach spaces, and the behavior of martingales in such spaces.

Theorem 3.2.1.[7] *A Banach space $\boldsymbol{X}$ is p-uniformly smooth if, and only if, there exists a positive constant K such that, for*

[5]See W.A. Woyczyński (1973).
[6]See J. Hoffmann-Jorgensen (1974).
[7]Due to P. Assuad (1974).

any two random vectors $X_1, X_2 \in \boldsymbol{L}^p(\Omega, \Sigma, P; \boldsymbol{X})$, *such that*

$$X_1 \text{ mes } \Sigma_1, \quad X_2 \text{ mes } \Sigma_2, \qquad \Sigma_1 \subset \Sigma_2 \subset \Sigma, \tag{3.2.1}$$

and

$$\|2X_1 - \mathbf{E}(X_2|\Sigma_1)\| \geq \|X_1\|, \quad a.s.,$$

we have the inequality

$$\mathbf{E}\big(\|X_2\|^p - \|X_1\|^p \mid \Sigma_1\big) \leq K\mathbf{E}\big(\|X_2 - X_1\|^p \mid \Sigma_1\big). \tag{3.2.2}$$

Proof. Assume that $\boldsymbol{X}$ is p-uniformly smooth. Then, by Jensen's Inequality and (3.2.1),

$$\mathbf{E}\big(\|2X_1 - X_2\|^p \mid \Sigma_1\big) \geq \|2X_1 - \mathbf{E}(X_2|\Sigma_1)\|^p \geq \|X_1\|^p,$$

so that, by Proposition 3.1.2, substituting $\boldsymbol{x} = X_1, \boldsymbol{y} = X_2 - X_1$, and averaging conditionally given Σ_1, we see that

$$\begin{aligned}\mathbf{E}\big(\|X_2\|^p + \|X_1\|^p \mid \Sigma_1\big) &\leq \mathbf{E}\big(\|X_2\|^p + \|2X_1 - X_2\|^p \mid \Sigma_1\big)\\ &\leq 2\mathbf{E}\big(\|X_1\|^p|\Sigma_1\big) + C\mathbf{E}\big(\|X_2 - X_1\|^p \mid \Sigma_1\big),\end{aligned}$$

which yields (3.2.2).

Now, conversely, assume that (3.2.2) holds true for any X_1, X_2, defined above. So, in particular, given $\boldsymbol{x}, \boldsymbol{y} \in \boldsymbol{X}$, we can take $X_1 = \boldsymbol{x}, X_2 = \boldsymbol{x} + \varepsilon\boldsymbol{y}$, where ε is a Bernoulli random variable, and Σ_1 is a trivial σ-algebra. Then (3.2.1) is clearly satisfied because $\mathbf{E}(X_1|\Sigma_1) = X_1$, and for such X_1, X_2, the formula (3.2.2) becomes the inequality,

$$\mathbf{E}(\|\boldsymbol{x} + \varepsilon\boldsymbol{y}\|^p - \|\boldsymbol{x}\|^p) \leq K\mathbf{E}\|\varepsilon\boldsymbol{y}\|^p, \qquad \boldsymbol{x}, \boldsymbol{y} \in \boldsymbol{X},$$

which, in view of Proposition 3.1.2, means that $\boldsymbol{X}$ is p-uniformly smooth. QED

Definition 3.2.1. A (Σ_n)-adapted sequence of random vectors in $\boldsymbol{X}$, X_n mes $\Sigma_n, \Sigma_n \subset \Sigma_{n+1} \subset \cdots \subset \Sigma,$, $n = 1, 2, \ldots$, is said to be a *norm supermartingale* if , for each $n = 1, 2, \ldots$,

$$\|2X_n - \mathbf{E}(X_{n+1}|\Sigma_n)\| \geq \|X_n\|.$$

Example 3.2.1. Obviously, each martingale with values in any Banach space is a norm supermartingale. Also, if (M_n, Σ_n) is a martingale with values in any Banach space, and (X_n) is a sequence of positive, decreasing, and predictable (X_{n+1} mes Σ_n) real-valued random variables, then the sequencee $(X_n M_n)$ is a norm supermartingale (an application of this fact will be given later on in this section). Finally, any positive real-valued supermartingale is also a norm supermartingale, and some "rapidly" growing submartingales are norm supermartingales as well.

The following result is a direct consequence of Theorem 3.2.1.

Corollary 3.2.1. *The Banach space* $\boldsymbol{X}$ *is* p*-uniformly smooth if, and only if, there exists a constant* $K > 0$ *such that, for any* $\boldsymbol{X}$*-valued norm supermartingale* $(X_n, \Sigma_n), n = 0, 1, 2, \ldots,$ *in* $\boldsymbol{L}^p(\boldsymbol{X})$,

$$\sup_n \mathbf{E}\|X_n\|^p \leq \mathbf{E}\|X_0\|^p + K \sum_{n=1}^{\infty} \mathbf{E}\|X_{n+1} - X_n\|^p.$$

The above Corollary, Theorem 3.1.1, and the fact that every p-uniformly smooth space is reflexive (and thus dentable[8]), yield the following

Theorem 3.2.2. *If* $\boldsymbol{X}$ *is a* p*-uniformly smooth Banach space,* $1 < p \leq 2$, *and* (M_n) *is an* $\boldsymbol{X}$*-valued martingale such that* $\sum_n \mathbf{E}\|M_{n+1} - M_n\|^p < \infty$, *then there exists a random vector* $M_0 \in \boldsymbol{L}^p(\Omega, \Sigma_\infty, P; \boldsymbol{X})$ *such that* $M_n \to M_\infty$ *a.s. and in* $\boldsymbol{L}^p$.

Applying the above theorem and the Kronecker Lemma to the martingale

$$M_n' = \sum_{i=1}^{n} \frac{M_i - M_{i-1}}{i},$$

we get the following *Strong Law of Large Numbers for Martingales* in Banach spaces:

[8]See, e.g., V.D. Milman (1971), p 79.

Corollary 3.2.2. *If (M_n, Σ_n) is a martingale with values in a p-uniformly smooth Banach space such that $M_0 = 0$, and*

$$\sum_{i=1}^{\infty} \frac{\mathbf{E}\|M_i - M_{i-1}\|^p}{i^p} < \infty,$$

then

$$\lim_{n\to\infty} \frac{M_n}{n} = 0, \qquad \text{a.s., and in } \boldsymbol{L}^p.$$

Remark 3.2.1. Theorem 3.2.2 and Corollary 3.2.2 generalize results for sums of independent Banach space valued random vectors.[9] It is worth noticing that p-uniform smoothness is not uniquely characterized by the behavior of general martingales. Indeed, from the proof of Theorem 3.2.1, and Corollary 3.2.1, we immediately get the following characterization:

Banach space $\boldsymbol{X}$ is p-uniformly smooth if, and only if, there exists a positive constant C such that, for any $n = 1, 2, \ldots$, any $\boldsymbol{x} \in \boldsymbol{X}$, and any independent, zero-mean random vectors, $X_1, \ldots, X_n$, with values in $\boldsymbol{X}$,

$$\mathbf{E}\Big\|\boldsymbol{x} + \sum_{i=1}^{n} X_i\Big\|^p \le \|\boldsymbol{x}\|^p + C\sum_{i=1}^{n} \|X_i\|^p.$$

The behavior of Banach space valued martingales with independent increments, that is sums of independent random vectors, will be discussed at length in later chapters of this book in connection with p-type Banach spaces, that is spaces for which there exists a positive constant C such that, for any $n = 1, 2, \ldots$, and any independent zero-mean random vectors $X_1, \ldots, X_n$ with values in $\boldsymbol{X}$,

$$\mathbf{E}\Big\|\sum_{i=1}^{n} X_i\Big\|^p \le C\sum_{i=1}^{n} \|X_i\|^p.$$

Despite the formal similarity of the concepts of p-uniform smoothness and p-type, they do not coincide. Of course, the above discussion shows that every p-uniformly smooth Banach space is also

[9]See W.A. Woyczyński (1973,1974).

of p-type. However, the reverse implication is not true (even after renorming). There exists an example[10] of a Beck-convex Banach space (thus of p-type for some $p > 1$) which is non-reflexive, so that it cannot be p-uniformly smooth for any $p > 1$.

Uniform convexity. Now, we turn to the relationship between the concept of uniform convexity of a Banach space and the behavior of martingales with values in such spaces.

Theorem 3.2.3.[11] *For a Banach space $\boldsymbol{X}$ the following four conditions are equivalent:*

(i) $\boldsymbol{X}$ is q-uniformly convex;

(ii) There exists a constant $L > 0$ such that, for any $\boldsymbol{x}, \boldsymbol{y} \in \boldsymbol{X}$,

$$\|\boldsymbol{x}+\boldsymbol{y}\|^q + \|\boldsymbol{x}-\boldsymbol{y}\|^q \geq 2\|\boldsymbol{x}\| + L\|\boldsymbol{y}\|^q,$$

(that is, we are on the other side of the parallelogram equality as compared with p-uniform smoothness);

(iii) There exists a constant $K > 0$ such that, for any $\boldsymbol{X}$-valued martingale $(M_n, n = 0, 1, \dots) \subset \boldsymbol{L}^q$,

$$\sup_n \mathbf{E}\|M_n\|^q \geq \mathbf{E}\|M_0\|^q + K\sum_{n=1}^{\infty} \mathbf{E}\|M_{n+1} - M_n\|^q;$$

(iv) There exists a constant $K > 0$ such that, for any two $\boldsymbol{X}$-valued random vectors in $\boldsymbol{L}^q$, X_1 mes Σ_1, and X_2 mes Σ_2, with $\Sigma_1 \subset \Sigma_2$, and such that

$$\|2X_1 - \mathbf{E}(X_2|\Sigma_1)\| = \|X_1\|,$$

we have the inequality,

$$\mathbf{E}\big(\|X_2\|^q - \|X_1\|^q|\Sigma_1\big) \geq K\mathbf{E}\big(\|X_2 - X_1\|^q|\Sigma_1\big).$$

Proof. We shall prove the following implications :

$$(i) \implies (iv) \implies (iii) \implies (ii) \implies (i).$$

[10]Due to R.C. James (1972).

[11]Due to P. Assuad (1974).

$(i) \implies (iv)$. Assume that $\boldsymbol{X}$ is q-uniformly convex. Hence, it is also reflexive[12], and so is $\boldsymbol{X}^*$, which, additionally, is p-uniformly smooth (with $1/p + 1/q = 1$) in view of Corollary 3.2.1.

Now, let Y_1^*, and Y_2^*, be two random vectors in $\boldsymbol{L}^p(\Omega, \Sigma, \mathbf{P}; \boldsymbol{X}^*)$. Then, the random vectors

$$X_1^* = \mathbf{E}(Y_1^*|\Sigma_1), \qquad \text{and} \qquad X_2^* - X_1^* = Y_2^* - \mathbf{E}(Y_2^*|\Sigma_1),$$

give rise to a martingale X_1^*, X_2^*, with values in $\boldsymbol{X}^*$. In view of Theorem 3.2.1,

$$\mathbf{E}\big(Y_1^* X_1 + Y_2^*(X_2 - X_1)\big)$$

$$= \mathbf{E}\Big[X_1^* X_1 + X_1^*(X_2 - X_1) + (X_2^* - X_1^*)X_1 + \big(Y_2^* - \mathbf{E}(Y_2^*|\Sigma_1)\big)(X_2 - X_1)\Big]$$

$$= \mathbf{E}\Big[X_1^* X_1 + X_1^*(X_2 - X_1) + (X_2^* - X_1^*)X_1 + (X_2^* - X_1^*)(X_2 - X_1)\Big]$$

$$= \mathbf{E}X_2^* X_2 \le \mathbf{E}\|X_2^*\|\|X_2\| \le \frac{1}{p}\mathbf{E}\|X_2^*\|^p + \frac{1}{q}\mathbf{E}\|X_2\|^q$$

$$\le \frac{1}{p}\mathbf{E}\|X_1^*\|^p + \frac{1}{p}K\mathbf{E}\|X_2^* - X_1^*\|^p + \frac{1}{q}\mathbf{E}\|X_2\|^q$$

$$\le \mathbf{E}\Big(\frac{1}{p}\|Y_1^*\|^P + \frac{1}{p}2^{p+1}K\|Y_2^*\|^p + \frac{1}{q}\|X_2\|^q\Big).$$

Finally, because of the arbitrariness of Y_1^*, and Y_2^*, we get that

$$\mathbf{E}\big(\|X_1^*\|^q + \frac{1}{c}\|X_2 - X_1\|^q\big) \le \mathbf{E}\|X_2\|^q,$$

where the last inequality is motivated by the inequality

$$\mathbf{E}(Y_1^* X_1) \le \mathbf{E}\|Y_1^*\|\|X_1\| \le \frac{1}{p}\mathbf{E}\|Y_1^*\|^p + \frac{1}{q}\mathbf{E}\|X_1\|^q,$$

a similar inequality for $\mathbf{E}(Y_2^*(X_2 - X_1))$, and the fact that if, for given reals c, b, the inequality, $ab \le a^p/p + c$, is satisfied for all real a, then $c \le b^q/q$. This completes the proof of the implication$(i) \implies (iv)$.

---

[12]See Theorem 1.3 in V.D. Milman (1971).

(iv) $\Longrightarrow$ *(iii)*. This implication follows directly by the operations of summing up and averaging.

(iii) $\Longrightarrow$ *(ii)*. This implication follows by setting $M_0 = \boldsymbol{x}$, $M_1 = \boldsymbol{x} + \varepsilon \boldsymbol{y}$, and $M_n = M_1$, for $n = 2, 3, \ldots$. Here, ε is a Bernoulli random variable.

(ii) $\Longrightarrow$ *(i)*. Let us apply *(ii)* to the vectors

$$\boldsymbol{x} = \|\boldsymbol{u}\|\boldsymbol{v} + \|\boldsymbol{v}\|\boldsymbol{u}, \qquad \text{and} \qquad \boldsymbol{y} = \|\boldsymbol{u}\|\boldsymbol{v} - \|\boldsymbol{v}\|\boldsymbol{u},$$

thus obtaining the inequality,

$$\big\|2\|\boldsymbol{u}\|\boldsymbol{v}\big\|^q + \big\|2\|\boldsymbol{v}\|\boldsymbol{u}\big\|^q \geq 2\big\|\|\boldsymbol{u}\|\boldsymbol{v} + \|\boldsymbol{v}\|\boldsymbol{u}\big\|^q + L\big\|\|\boldsymbol{u}\|\boldsymbol{v} - \|\boldsymbol{v}\|\boldsymbol{u}\big\|^q,$$

wherefrom it follows that,

$$2^{q+1}\|\boldsymbol{u}\|^q\|\boldsymbol{v}\|^q \geq 2\big\|\|\boldsymbol{u}\|\boldsymbol{v} + \|\boldsymbol{v}\|\boldsymbol{u}\big\|^q + L\big\|\|\boldsymbol{u}\|\boldsymbol{v} - \|\boldsymbol{v}\|\boldsymbol{u}\big\|^q,$$

so that, for any $\boldsymbol{u}, \boldsymbol{v}$, such that $\|\boldsymbol{u}\| = \|\boldsymbol{v}\| = 1$, and $\|\boldsymbol{u} - \boldsymbol{v}\| = \epsilon$, we get that

$$1 - \left\|\frac{\boldsymbol{u} + \boldsymbol{v}}{2}\right\| \geq \frac{L}{2^{q+1}}\epsilon^q.$$

Finally, because, for any real number $\alpha, 0 \leq \alpha \leq 1$, $1 - \alpha^q \leq q(1 - \alpha)$, we get that

$$\delta_{\boldsymbol{X}}(\epsilon) = \inf\left\{1 - \left\|\frac{\boldsymbol{u} + \boldsymbol{v}}{2}\right\| : \|\boldsymbol{u}\| = \|\boldsymbol{v}\| = 1, \|\boldsymbol{u} - \boldsymbol{v}\| = \epsilon\right\}$$

$$\geq \frac{1}{q}\inf\left\{1 - \left\|\frac{\boldsymbol{u} + \boldsymbol{v}}{2}\right\|^q : \|\boldsymbol{u}\| = \|\boldsymbol{v}\| = 1, \|\boldsymbol{u} - \boldsymbol{v}\| = \epsilon\right\} \geq \frac{L}{q2^{q+1}}\epsilon^q,$$

which implies the q-unifrom convexity of $\boldsymbol{X}$ and completes the proof of the theorem. QED

Paley-Walsh Martingales. The results of the previous subsection can be refined using the concept of the *Paley-Walsh martingale*:

Definition 3.2.2. Paley-Walsh martingale is the dyadic martingale on $\Omega = \{+1, -1\} \times \{+1, 1\} \times \ldots$, equipped with the Bernoulli product probability, and adapted to the natural σ-field filtration, Σ_n, spanned by the first n factors of the product space Ω.

In this context, the following result characterises p-smooth and q-convex spaces:

Theorem 3.2.4.[13] *(i) A Banach space $\boldsymbol{X}$ is q-uniformly convexifiable if, and only if, there exists a constant $C > 0$ such that, for any $\boldsymbol{X}$-valued Paley-Walsh martingale (M_n),*

$$C \sup_n \mathbf{E}\|M_n\|^q \geq \mathbf{E}\|M_0\|^q + \sum_{n=0}^{\infty} \mathbf{E}\|M_{n+1} - M_n\|^q, \qquad (3.2.3)$$

(ii) A Banach space $\boldsymbol{X}$ is p-uniformly smoothable if, and only if, there exists a constant $C > 0$ such that, for any $\boldsymbol{X}$-valued Paley-Walsh martingale (M_n),

$$\sup_n \mathbf{E}\|M_n\|^p \leq C\Big[\mathbf{E}\|M_0\|^p + \sum_{n=0}^{\infty} \mathbf{E}\|M_{n+1} - M_n\|^q,\Big] \qquad (3.2.4)$$

Proof. The "only if" parts of statements (i), and (ii), automatically follow from the results of the previous subsection (Corollary 3.2.1 and Theorem 3.2.1). So, next, we will prove the "if" part of the statement (i) (The proof of (ii) is completely analogous and we will omit it).

Assume that (3.2.3) holds true and define the new norm on $\boldsymbol{X}$ by means of the formula

$$\sharp \boldsymbol{x} \sharp := \inf\Big[C \sup_n \mathbf{E}\|M_n\|^q - \sum_{n=0}^{\infty} \mathbf{E}\|M_{n+1} - M_n\|^q\Big]^{1/q}.$$

where the infimum is taken over all Paley-Walsh martingales (M_n) with values in $\boldsymbol{X}$, such that $M_0 = \mathbf{E}M_n = \boldsymbol{x}$, and satisfying the condition $\sup \mathbf{E}\|M_n\|^q < \infty$.

Evidently, the norm $\sharp.\sharp$ is positively homogeneous, and additionally it satisfied the inequality,

$$\|\boldsymbol{x}\| \leq \sharp \boldsymbol{x} \sharp \leq C^{1/q}\|\boldsymbol{x}\|,$$

[13] Due to G. Pisier (1974).

so it is equivalent to the original norm $\|.\|$. The first inequality follows directly from (3.2.3), and the second from the definition of $\sharp.\sharp$, if one takes $M_n = \boldsymbol{x}, n = 0, 1, 2, \ldots$. So, in view of the above inequality and Theorem 3.2.3, it is sufficient to show that, for all $\boldsymbol{x}, \boldsymbol{y} \in \boldsymbol{X}$,

$$\sharp(\boldsymbol{x}+\boldsymbol{y})/2)\sharp^q + \|(\boldsymbol{x}-\boldsymbol{y})/2\|^q \leq \left(\sharp\boldsymbol{x}\sharp^q + \sharp\boldsymbol{y}\sharp^q\right)/2,$$

which also will show the convexity of the set $\{\boldsymbol{x} : \sharp\boldsymbol{x}\sharp \leq 1\}$, that is the triangle inequality for the norm $\sharp.\sharp$.

Now, let $\boldsymbol{x}, \boldsymbol{y} \in \boldsymbol{X}$. By the definition of the new norm, for any $\alpha > 0$, there exist Paley-Walsh martingales, (X_n), and (Y_n), with values in $\boldsymbol{X}$ such that

$$X_0 = \boldsymbol{x}, \qquad \sup_n \mathbf{E}\|X_n\|^q < \infty,$$

$$Y_0 = \boldsymbol{y}, \qquad \sup_n \mathbf{E}\|Y_n\|^q < \infty,$$

and

$$C \sup_n \mathbf{E}\|X_n\|^q - \sum_{n=0}^{\infty} \mathbf{E}\|X_{n+1} - X_n\|^q \leq \sharp\boldsymbol{x}\sharp^q + \alpha,$$

$$C \sup_n \mathbf{E}\|Y_n\|^q - \sum_{n=0}^{\infty} \mathbf{E}\|Y_{n+1} - Y_n\|^q \leq \sharp\boldsymbol{y}\sharp^q + \alpha.$$

At this point we will construct a new Paley-Walsh martingale (Z_n) via the following formulas:

$$Z_0 = \frac{\boldsymbol{x}+\boldsymbol{y}}{2},$$

$$Z_n = \frac{1+\varepsilon_1}{2} X_{n-1}(\varepsilon_2, \varepsilon_3, \ldots) + \frac{1-\varepsilon_1}{2} Y_{n-1}(\varepsilon_2, \varepsilon_3, \ldots).$$

Evidently, $\sup_n \mathbf{E}\|Z_n\|^q < \infty$, and

$$\sharp(\boldsymbol{x}+\boldsymbol{y})/2\sharp^q \leq C \sup_n \mathbf{E}\|Z_n\|^q - \sum_{n=0}^{\infty} \mathbf{E}\|Z_{n+1} - Z_n\|^q$$

$$\leq C \sup_n \frac{\mathbf{E}\|X_n\|^q + \mathbf{E}\|Y_n\|^q}{2} - \left\|\frac{\boldsymbol{x}-\boldsymbol{y}}{2}\right\|^q$$

$$-\sum_{n=0}^{\infty}\frac{\mathbf{E}\|X_{n+1}-X_n\|^q+\mathbf{E}\|Y_{n+1}-Y_n\|^q}{2}$$

$$\leq \frac{\sharp\boldsymbol{x}\sharp^q+\sharp\boldsymbol{y}\sharp^q}{2}+\alpha-\left\|\frac{\boldsymbol{x}-\boldsymbol{y}}{2}\right\|^q,$$

which proves the theorem in view of the arbitrariness of α. QED

Convergence of rearranged nonrandom series in p-uniformly smooth Banach spaces. The classical Steinitz Theorem states that if a subsequence of the partial sums of a real series converges then one can finid a rearrangement of the series which converges.[14] The result below shows its validity for general p- uniformly smooth Banach spaces .

Theorem 3.2.5.[15] *Let $\boldsymbol{x}$ be a p-uniformly smooth Banach space. If the sequence $(\boldsymbol{x}_i)\subset\boldsymbol{X}$ is such that $\sum_i\|\boldsymbol{x}_i\|^p<\infty$, and*

$$\sum_{i=1}^{n_k}\boldsymbol{x}_i\to\boldsymbol{x}\in\boldsymbol{X},\qquad\text{as}\qquad k\to\infty,$$

for a certain subsequence $(n_k)\subset\boldsymbol{N}$, then there exists a rearrangement σ of positive integers such that

$$\sum_{i=1}^{\infty}\boldsymbol{x}_{\sigma(i)}=\boldsymbol{x}.$$

Proof. Let us start with an observation that if $\boldsymbol{X}$ is p-uniformly smooth then there exists a constant $K>0$ such that, for any $n\in\boldsymbol{N}$, and any $\boldsymbol{x}_1,\ldots,\boldsymbol{x}_n\in\boldsymbol{X}$ such that $\boldsymbol{x}_1+\cdots+\boldsymbol{x}_n=0$, one can find a permutation σ of the set $(1,2,\ldots,n)$ such that, for each $k, 1\leq k\leq n$,

$$\left\|\sum_{i=1}^{k}\boldsymbol{x}_{\sigma(i)}\right\|^p\leq K\sum_{i=1}^{k}\|\boldsymbol{x}_{\sigma(i)}\|^p. \tag{3.2.5}$$

[14]An analogue of this theorem for $\boldsymbol{L}^p$ has been demonstrated by M.I. Kadec (1954).

[15]Due to P. Assuad (1974).

Indeed, to prove the above statement it is sufficient to note that given such $\boldsymbol{x}_1, \ldots, \boldsymbol{x}_n \in \boldsymbol{X}$, and taking as the probability space Ω the set of all permutations σ of the set $\{1, \ldots, n\}$ equipped with the uniform probability distribution, $\boldsymbol{x}_{\sigma(1)}, \ldots, \boldsymbol{x}_{\sigma(n)}$, become exchangeable random variables so that

$$Y_k(\sigma) = \frac{X_{\sigma(1)} + \cdots + X_{\sigma(n-k)}}{n-k}, \qquad k = 0, 1, \ldots, n-1,$$

becomes a martingale with values in $\boldsymbol{X}$ (relative to the natural σ-algebras). Since, for any real-valued random variables, $\xi_1, \ldots, \xi_n$, such that $\mathbf{E}(\xi_{i+1}|\Sigma_i) \geq 0$, for $i = 0, \ldots, n-1$, one can find an $\omega \in \Omega$ such that, for every $i = 1, \ldots, n$, $\xi_i(\omega) \geq 0$, applying Theorem 3.2.1 to the norm supermartingale,

$$Y_k(n-k) = \boldsymbol{x}_{\sigma(1)} + \cdots + \boldsymbol{x}_{\sigma(n-k)},$$

we obtain the existence of σ such that (3.2.5) is satisfied.

An obvious calculation shows that, for any $\boldsymbol{x}_1, \ldots, \boldsymbol{x}_n \in \boldsymbol{X}$, with $\boldsymbol{x}_1 + \cdots + \boldsymbol{x}_n = \boldsymbol{y}$, one can find a permutation σ of $\{1, \ldots, n\}$ such that, for each $k, 1 \leq k \leq n$,

$$\Big\|\sum_{i=1}^{k} \boldsymbol{x}_{\sigma(i)}\Big\|^p \leq L\Big(\|\boldsymbol{y}\|^p + \sum_{i=1}^{k} \|x_{\sigma(i)}\|^p\Big), \tag{3.2.6}$$

Thus, the above inequality implies the statement of the theorem after rearranging the series $\sum_i \boldsymbol{x}_i$ in blocks $(n_i, n_{i+1} - 1)$ according to the permutation guaranteeing (3.2.6). QED

Remark 3.2.2. It is also worth recalling the theorem[16] which states that every uniformly smooth (uniformly convex) Banach space may be renormed so that it becomes p-uniformly smooth (q-uniformly convex) for some $p > 1$ ($q < \infty$).

3.3 General concept of super-property

Definition 3.3.1. Let $\mathcal{P}$ be a property of a Banach space. We say that the Banach space $\boldsymbol{X}$ has the property *super* $\mathcal{P}$ if every Banach space $\boldsymbol{Y}$ that is finitely representable in $\boldsymbol{X}$ has the property $\mathcal{P}$.

[16]Due to G. Pisier (1974), see also Pisier's monograph (2016).

Recall that $\boldsymbol{Y}$ is said to be *finitely representable* in $\boldsymbol{X}$ if, for any $\epsilon > 0$, and any finite-dimensional $\boldsymbol{Y}^{fd} \subset \boldsymbol{Y}$, one can find a finite-dimensional subspace $\boldsymbol{X}^{fd} \subset \boldsymbol{X}$ such that

$$\text{dist}(\boldsymbol{X}^{fd}, \boldsymbol{Y}^{fd}) := \inf\{\|T\|\,\|T^{-1}\| : \text{isomorphism } T : \boldsymbol{X}^{fd} \mapsto \boldsymbol{Y}^{fd}\} \le 1 + \epsilon.$$

The following implications are evident:

$$\text{super } \mathcal{P} \implies \mathcal{P},$$
$$\text{super (super } \mathcal{P}) \implies \text{super } \mathcal{P},$$
$$(\mathcal{P} \implies \mathcal{Q}) \implies (\text{super } \mathcal{P} \implies \text{super } \mathcal{Q}).$$

Roughly speaking, $\boldsymbol{X} \in$ super-$\mathcal{P}$ if any Banach space $\boldsymbol{Y}$ the finite dimensional subspaces thereof are "similar" to those of $\boldsymbol{X}$ has $\mathcal{P}$.

Below, we collect some purely geometric characterizations of of the property of *super-reflexivity* that will be needed in further sections. Let us begin by introducing the concept of the *finite tree property.*[17]

Definiton 3.3.2. Banach space $\boldsymbol{X}$ is said to have the *finite tree property* if there exists an $\epsilon > 0$ such that, for any $n \in \mathbf{N}$, one can find a binary tree

$$\{\boldsymbol{x}(\varepsilon_1, \dots, \varepsilon_k) : 1 \le k \le n, \varepsilon_i = \pm 1\},$$

contained in the unit ball of $\boldsymbol{X}$, and such that

$$\boldsymbol{x}(\varepsilon_1, \dots, \varepsilon_k) = \frac{1}{2}\big(\boldsymbol{x}(\varepsilon_1, \dots, \varepsilon_k, 1) + \boldsymbol{x}(\varepsilon_1, \dots, \varepsilon_k, -1)\big)$$

and

$$\|\boldsymbol{x}(\varepsilon_1, \dots, \varepsilon_k) - \boldsymbol{x}(\varepsilon_1, \dots, \varepsilon_k, \varepsilon_{k+1})\| \ge \frac{\epsilon}{2}, \qquad k = 1, \dots n - 1.$$

Note that the above tree is essentially a finite Paley-Walsh martingale with uniformly big increments.

[17]The concepts of super-reflexivity, and finite tree property, were introduced by R.C. James (1972).

Definition 3.3.3. A sequence $(\boldsymbol{x}_n) \subset \boldsymbol{X}$ is said to be *basic* with constant δ if, for any scalar sequence $(\alpha_n) \subset \mathbf{R}$, and all integers n, m,

$$\delta \Big\| \sum_{i=1}^{n} \alpha_i \boldsymbol{x}_i \Big\| \leq \Big\| \sum_{i=1}^{n+m} \alpha_i \boldsymbol{x}_i \Big\|.$$

In the following theorem we cite several characterizations of super-reflexivity without proofs.

Theorem 3.3.1.[18] *The following properties of a Banach space $\boldsymbol{X}$ are equivalent:*

(i) $\boldsymbol{X}$ is super-reflexive;

(ii) There exists an $n \in \mathbf{N}$, and $\epsilon > 0$, such that, for all $\boldsymbol{x}_1, \ldots, \boldsymbol{x}_n \in \boldsymbol{X}$,

$$\inf_{1 \leq k \leq n} \Big\| \sum_{i=1}^{k} \boldsymbol{x}_i - \sum_{i=k+1}^{n} \boldsymbol{x}_i \Big\| \leq n(1-\epsilon) \sup_{1 \leq i \leq n} \|\boldsymbol{x}_i\|;$$

(iii) $\boldsymbol{X}$ does not possess the finite tree property;

(iv) For each $\delta > 0$, there exists a $p > 1$, and a constant $C > 0$, such that, for any finite basic sequence $(\boldsymbol{x}_n) \subset \boldsymbol{X}$ with constant δ,

$$\Big\| \sum \boldsymbol{x}_n \Big\| \leq C \Big(\sum \|\boldsymbol{x}_n\|^p \Big)^{1/p};$$

(v) For each $\delta > 0$, there exists a $q < \infty$, and a constant $C > 0$, such that, for any finite basic sequence $(\boldsymbol{x}_n) \subset \boldsymbol{X}$ with constant δ,

$$\Big\| \sum \boldsymbol{x}_n \Big\| \geq \frac{1}{C} \Big(\sum \|\boldsymbol{x}_n\|^q \Big)^{1/q};$$

(vi) $\boldsymbol{L}^2(\Omega, \mu; \boldsymbol{X}), \mu(\Omega) > 0$, is super-reflexive.

3.4 Martingales in super-reflexive Banach spaces

The following theorem provides a characterization of super-reflexivity of $\boldsymbol{X}$ in terms of convergence properties of $\boldsymbol{x}$-valued

[18] For proofs, see J.J. Schäffer and K. Sundaresan (1970) ((a) $\Leftrightarrow$ (b)), R.C. James (1972, 1974) ((a) $\Leftrightarrow$(c) $\Leftrightarrow$(d) $\Leftrightarrow$(e)) , and G. Pisier (1974)((a) $\Leftrightarrow$ (f)).

martingales. For the definitions of the Radon-Nikodym Property (RNP), and the $\boldsymbol{L}^p$ Martingale Convergence Theorem (MCT_p), see Chapter 2.

Theorem 3.4.1.[19] *For the Banach space $\boldsymbol{X}$ the following properties are equivalent:*

(i) $\boldsymbol{X}$ is super-reflexive;

(ii) $\boldsymbol{X}$ has the super-MCT_p property;

(iii) $\boldsymbol{X}$ has the super-RNP property;

(iv) There exists a constant $C > 0$, and $p > 1$, such that, for each $\boldsymbol{X}$-valued martingale $(M_n) \subset \boldsymbol{L}^2$,

$$\sup_n \|M_n\|_2 \le C\Big[\|M_0\|_2^p + \sum_{n=0}^{\infty} \|M_{n+1} - M_n\|_2^p\Big]^{1/p}, \tag{3.4.1}$$

where

$$\|M\|_2 := (\mathbf{E}\|M\|^2)^{1/2};$$

(v) There exists a constant $C > 0$, and $p > 1$, such that, for each $\boldsymbol{X}$-valued Paley-Walsh martingale (M_n) the inequality (3.3.1) holds true;

(vi) There exists a constant $C > 0$, and $q < \infty$, such that, for each $\boldsymbol{X}$-valued martingale $(M_n) \subset \boldsymbol{L}^2$,

$$\Big[\|M_0\|_2^q + \sum_{n=0}^{\infty} \|M_{n+1} - M_n\|_2^q\Big]^{1/q} \le C \sup_n \|M_n\|_2; \tag{3.4.2}$$

(vii) There exists a constant $C > 0$, and $q < \infty$, such that for each $\boldsymbol{X}$-valued Paley-Walsh martingale (M_n) the inequality (3.4.2) holds true.

Proof. We will prove the following implications: $(i) \Longrightarrow (ii) \Longrightarrow (i)$, $(ii) \Leftrightarrow (iii)$, $(i) \Longrightarrow (vi) \Longrightarrow (vii) \Longrightarrow (i)$, $(i) \Longrightarrow (iv) \Longrightarrow (v) \Longrightarrow (vii)$.

$(i) \Longrightarrow (ii)$ This implication follows from the fact that super-reflexive spaces are reflexive, and that reflexive spaces satisfy (MCT_p) (see Chapter 2).

[19] Due to G. Pisier (1974).

$(ii) \implies (i)$ Assume that $\boldsymbol{X}$ is not super-reflexive, i.e., there exists a Banach space $\boldsymbol{Y}$ which is finitely representable in $\boldsymbol{X}$, and which has the infinite ϵ-tree in its unit a ball, for some $\epsilon > 0$. But the same tree can be viewed as a bounded Paley-Walsh martingale with values in $\boldsymbol{X}$ which does not converge a.s. This evidently contradicts (ii).

$(ii) \implies (iii)$ This follows directly from Theorem 2.2.1 in Chapter 2.

$(i) \implies (iv)$, and $(i) \implies (vi)$, follow directly from Theorem 3.3.1 because the increments of a square integrable martingale form a basic sequence in $\boldsymbol{L}(\Omega, \Sigma, P; \boldsymbol{X})$ with constant 1. Indeed, for any $(\alpha_i) \subset \mathbf{R}$,

$$\Big(\mathbf{E}\Big\|\sum_{i=1}^{n+m} \alpha_i(M_i - M_{i-1})\Big\|^2\Big)^{1/2} = \Big[\mathbf{E}\Big(\mathbf{E}\Big(\Big\|\sum_{i=1}^{n+m} \alpha_i(M_i - M_{i-1})\Big\|^2 \Big| \Sigma_n\Big)\Big)\Big]^{1/2}$$

$$\geq \Big[\mathbf{E}\Big\|\mathbf{E}\Big(\sum_{i=1}^{n+m} \alpha_i(M_i - M_{i-1}) \Big| \Sigma_n\Big)\Big\|^2\Big]^{1/2} = \Big[\mathbf{E}\Big\|\sum_{i=1}^{n+m} \alpha_i(M_i - M_{i-1})\Big\|^2\Big]^{1/2}$$

$(iv) \implies (v)$, and $(vi) \implies (vii)$ are obvious implications.

$(vii) \implies (i)$ Assume that $\boldsymbol{X}$ is not super-reflexive. Then, by Theorem 3.1.1, it has the finite tree property, that is there exists an $\epsilon > 0$ such that, for all $n \in \mathbf{N}$, there exists a Paley-Walsh martingale (M_n) of length n, with values in the unit ball of $\boldsymbol{X}$, such that

$$\|M_{n+1} - M_n\| \geq \epsilon/2.$$

Thus, were the inequality (3.4.2) satisfied, we would have that

$$1 \geq (\mathbf{E}\|M_n\|^2)^{1/2} \geq \frac{1}{C}\Big[\sum_{k=1}^{n} \|M_n - M_{n-1}\|_2^q\Big]^{1/q} \geq \frac{\epsilon}{2C} n^{1/q},$$

a contradiction.

$(v) \implies (vii)$ We will prove this implication for $\boldsymbol{X}^*$ but that suffices because, by the implication $(vii) \implies (i)$, we would have that $\boldsymbol{X}^*$ is super-reflexive, and so is $\boldsymbol{X}$, since super-reflexivity is a self-dual property.[20]

[20]See R.C. James (1972).

So, let (m_N^*) be a Walsh-Paley martingale with values in $\boldsymbol{X}^*$. Then

$$\Big(\|M_0^*\|_2^q + \sum_{k=1}^{n} \|M_k^* - M_{k-1}^*\|_2^q\Big)^{1/q}$$

$$= \sup\Big\{\mathbf{E}M_0^* X_0 + \sum_{k=1}^{n} \mathbf{E}(M_K^* - M_{k-1})X_k : \|X_0\|_2^p$$

$$+ \sum_{k=1}^{n} \|X_k\|_2^p \le 1, X_i \in \boldsymbol{L}^p(\boldsymbol{X})\Big\}$$

$$= \sup\Big\{\mathbf{E}[M_n^*\mathbf{E}(X_0|\Sigma_0)] + \sum_{k=1}^{n} \mathbf{E}\big[M_n^*(\mathbf{E}(X_k|\Sigma_k) - \mathbf{E}(X_k|\Sigma_{k-1})\big]\Big\}$$

$$\le \|M_n^*\|_2 \sup\Big\{\Big\|\mathbf{E}(X_0|\Sigma_0) + \sum_{k=1}^{n}\big[\mathbf{E}(X_k|\Sigma_k) - \mathbf{E}(X_k|\Sigma_{k-1})\big]\Big\|_2\Big\}$$

$$\le C\|M_n^*\|_2 \sup\Big\{\|\mathbf{E}(X_0|\Sigma_0)\|_2^p + \sum_{k=1}^{n} \|\mathbf{E}(X_k|\Sigma_k) - \mathbf{E}(X_k|\Sigma_{k-1})\|_2^p\Big\}^{1/p}$$

$$\le 2C\|M_n^*\|_2,$$

which completes the proof of the Theorem. QED

Remark 3.4.1. One can show[21] that the super-reflexivity is equivalent to uniform convexifiability, and this property, in turn, is equivalent to uniform smoothability. Actually, in this case the space may be equipped with an equivalent norm which is at the same time uniformly smooth and uniformly convex. Using the martingale techniques developed above, one can demonstrate,[22] that uniform convexifiability is equivalent to q-uniform convexifiability for some $q < \infty$.

---

[21] See P. Enflo (1972).

[22] See G. Pisier (1974).

Chapter 4

Spaces that do not contain $\boldsymbol{c}_0$

4.1 Boundedness and convergence of random series

Let us begin by discussing the boundedness of the vector-valued Rademacher series, and its relationship to the Banach space $\boldsymbol{c}_0$ of all real sequences convergent to 0..

Definition 4.1.1. The Rademacher vector sums $\sum_{i=1}^n r_i \boldsymbol{x}_i$, $\mathbf{P}(r_i = \pm 1) = 1/2$, $(\boldsymbol{x}_i) \subset \boldsymbol{X}$, are said to be *almost surely bounded* if

$$\mathbf{P}\Big(\sup_n \Big\|\sum_{i=1}^n r_i \boldsymbol{x}_i\Big\| = \infty\Big) = 0.$$

It is clear that $\sum_{i=1}^n r_i \boldsymbol{x}_i$ are a.s. bounded if, and only if, for each $\epsilon \in (0,1)$, the gauges $J_\epsilon(\sum_{i=1}^n r_i \boldsymbol{x}_i; \mathbf{P})$ are bounded.[1]

Theorem 4.1.1.[2] *The following properties of the Banach space $\boldsymbol{X}$ are equivalent:*

(i) $\boldsymbol{X}$ does not contain a subspace isomorphic to $\boldsymbol{c}_0$;

[1]See Chapter 1 for the definition of gauges J_ϵ.

[2]Due to S. Kwapień (1974).

(ii) For each sequence $(\boldsymbol{x}_n) \subset \boldsymbol{X}$ such that $\sum_{i=1}^n r_i \boldsymbol{x}_i$ is a.s. bounded, we have $\boldsymbol{x}_n \to 0$, as $n \to \infty$.

Proof. $(ii) \implies (i)$. Assume $\boldsymbol{X} \supset \boldsymbol{c}_0$, and let $(\boldsymbol{e}_i)$ be the canonical basis in $\boldsymbol{c}_0$. Then, $\| \sum_{i=1}^n r_i \boldsymbol{e}_i \| = 1$, but still $\boldsymbol{e}_n$'s do not converge to 0.

$(i) \implies (ii)$. Suppose (ii) is not satisfied. Then there exists a sequence $(\boldsymbol{x}_i) \subset \boldsymbol{X}$, with $\inf_i \|\boldsymbol{x}_i\| > 0$, and a constant $M < \infty$, such that

$$\mathbf{P}\Big(\sup_n \Big\| \sum_{i=1}^n r_i \boldsymbol{x}_i \Big\| < M\Big) > \frac{1}{2}.$$

Utilizing the fact that for each set A in the σ-algebra spanned by $(r_i, i \in \mathbf{N})$,

$$\lim_i \mathbf{P}(A \cap (r_i = 1)) = \lim_i \mathbf{P}(A \cap (r_i = -1)) = \mathbf{P}(A)/2,$$

we can find by induction a sequence $n_i \uparrow \infty$ of integers such that for every sequence of signs, $\varepsilon_i = \pm 1$,

$$\mathbf{P}\Big(\sup_n \Big\| \sum_{i=1}^n r_i \boldsymbol{x}_i \Big\| < M, \varepsilon_i r_{n_i} = 1, i = 1, \ldots, k\Big) \tag{4.1.1}$$

$$> \frac{1}{2}\mathbf{P}\Big(\sup_n \Big\| \sum_{i=1}^n r_i \boldsymbol{x}_i \Big\| < M, \varepsilon_i r_{n_i} = 1, i = 1, \ldots, k-1\Big) > \frac{1}{2} 2^{-k},$$

$$k \in \mathbf{N}.$$

Let us now define $r_i' = r_i$, for $i \in (n_i)$, and $r_i' = -r_i$, for $i \notin (n_i)$. Since (r_i) and (r_i') are identically distributed, for each $\varepsilon_i = \pm 1$, and each $k \in \mathbf{N}$,

$$\mathbf{P}\Big(\sup_n \Big\| \sum_{i=1}^n r_i' \boldsymbol{x}_i \Big\| < M, \varepsilon_i r_{n_i}' = 1, i = 1, \ldots, k\Big) > 2^{-(k+1)}. \tag{4.1.2}$$

Since $\mathbf{P}(\varepsilon_i r_{n_i} = 1, i = 1, \ldots, k) = 1/2^k$, it follows from (4.1.1), and (4.1.2), that for each $\varepsilon = \pm 1$, and each $k \in \mathbf{N}$, there exists an $\omega \in \Omega$ such that

$$\Big\| \sum_{i=1}^k \varepsilon_i \boldsymbol{x}_{n_i} \Big\| = \frac{1}{2} \Big\| \sum_{i=1}^{n_k} r_i(\omega) \boldsymbol{x}_i + \sum_{i=1}^{n_k} r_i'(\omega) \boldsymbol{x}_i \Big\| < M.$$

Therefore, the series $\sum_i \boldsymbol{x}_{n_i}$ is weakly unconditionally convergent with $\inf \|\boldsymbol{x}_{n_i}\| > 0$ and, by the Bessaga-Pelczynski Theorem,[3] there exists a subsequence $(\boldsymbol{y}_i) \subset (\boldsymbol{x}_{n_i})$ such that $(\boldsymbol{y}_i)$ is isomorphic to the canonical basis $(\boldsymbol{e}_i) \subset \boldsymbol{c}_0$. QED

The above result can be extended to the case of general series of independent random vectors with values in a Banach space $\boldsymbol{X}$.

Theorem 4.1.2.[4] *Let $1 \leq p < \infty$. The following properties of a Banach space $\boldsymbol{X}$ are equivalent:*

(i) $\boldsymbol{X}$ does not contain an isomorphic copy of $\boldsymbol{c}_0$;

(ii) $\boldsymbol{L}_p(\Omega, \Sigma, \mathbf{P}; \boldsymbol{X})$ does not contain an isomorphic copy of $\boldsymbol{c}_0$;

(iii) For each sequence $(\boldsymbol{x}_n) \subset \boldsymbol{X}$ such that the sums $\sum_{i=1}^n r_i \boldsymbol{x}_i, n \in \mathbf{N}$, are a.s. bounded, the series $\sum_i r_i \boldsymbol{x}_i$ converges a.s. ;

(iv) For each sequence (X_n) of symmetric, and independent random vectors in $\boldsymbol{X}$, the a.s. boundedness of the sums $\sum_{i=1}^n X_i$ implies the a.s. convergence of the series $\sum_i X_i$.

Proof. We will prove the following implications: $(iii) \equiv (iv)$, $(i) \implies (iii) \implies (ii) \implies (i)$.

$(iii) \equiv (iv)$. The equivalence follows immediately from the following Lemma which itself is a straightforward corollary to the Fubini Theorem, and the fact that if (X_i), and (r_i) are independent and (X_i) is sign-invariant, then (X_i), and $(r_i X_i)$ are identically distributed.

Lemma 4.1.1. *If (X_n) is a sign-invariant sequence of random vectors in a Banach space $\boldsymbol{X}$, and (r_n) is independent of (X_n), then $\sum_{i=1}^n X_i$ is a.s. bounded [convergent] if, and only if, for almost every $\omega \in \Omega$, the sums $\sum_{i=1}^n r_i X_i(\omega)$ are a.s. bounded [convergent].*

$(i) \implies (iii)$. Assume that (iii) is not satisfied, and let $(\boldsymbol{x}_i) \subset \boldsymbol{X}$ be such that $\sum_i r_i \boldsymbol{x}_i$ is a.s. bounded, but not a.s. convergent. In view of the Kahane Theorem (see, Chapter 1), $\sum_{i=1}^n r_i \boldsymbol{x}_i$ is not a Cauchy sequence in $\boldsymbol{L}_1(\boldsymbol{X})$, so that there exists

[3]See C. Bessaga and A. Pelczynski (1958).

[4]See J. Hoffmann-Jorgensen (1972/73, 1974). A variant of the proof for $(i) \implies (ii)$ can also be found in J. Hoffmann-Jorgensen (1973/74).

an $\epsilon > 0$, and a sequence $n_i \uparrow \infty$, such that, for each $i \in \mathbf{N}$, $\mathbf{E}\|X_i\| > \epsilon$, where

$$X_i := \sum_{n_i \le j < n_{i+1}} r_j \boldsymbol{x}_j, \qquad i \in \mathbf{N}.$$

By Hoffmann-Jorgensen Theorem (see, Chapter 1),

$$\mathbf{E}\Big\|\sum_{i=1}^{k} X_i\Big\| == \mathbf{E}\Big\|\sum_{i=1}^{n_{k+1}} r_i \boldsymbol{x}_i\Big\| \le \mathbf{E}\sup_n \Big\|\sum_{i=1}^{n} r_i \boldsymbol{x}_i\Big\| < \infty.$$

Therefore, in view of the above Lemma, the series $\sum_{i=1}^{n} r_i X_i(\omega)$ is a.s. bounded for almost every $\omega \in \Omega$, and because $\|X_i(\omega)\| \le 2\sup \|\sum_{i=1}^{n} r_i(\omega)\boldsymbol{x}_i\|$, we have that $\mathbf{P}(X_i \neq 0) > 0$, since $\mathbf{E}\|X_i\| \ge \epsilon$. Now, from Theorem 4.1.1, it follows that $\boldsymbol{X}$ contains an isomophic copy of $\boldsymbol{c}_0$.

$(iii) \implies (ii)$. Suppose, to the contrary, that $\boldsymbol{L}_p(\boldsymbol{X})$ contains a copy of $\boldsymbol{c}_0$. This implies the existence of constants $a, b, c > 0$, and a sequence $(X_n) \subset \boldsymbol{L}_p(\boldsymbol{X})$, such that

$$a \le \left(\mathbf{E}\|X_n\|^p\right)^{1/p} \le b, \qquad n \in \mathbf{N}, \tag{4.1.3}$$

and

$$\left(\mathbf{E}\Big\|\sum_{j=1}^{n} c_j X_j\Big\|^p\right)^{1/p} \le c \max_{1\le j\le n} |c_j|, \qquad n \in \mathbf{N},\ (c_j) \subset \mathbf{R}. \tag{4.1.4}$$

Define $X'_j = r_j X_j$, where (r_j) are independent of (X_j). Evidently, (X'_j) is sign-invariant, satisfies (4.1.3-4), and $Y_n = \sum_{j=1}^{n} X'_j$ are bounded in $\boldsymbol{L}_p(\boldsymbol{X})$. Thus, by the above Lemma, $\sum r'_j X'_j(\omega)$ are a.s. bounded for almost all $\omega \in \Omega$ (with (r'_j) being independent from (X'_j)). Therefore, (iii) cannot hold because, were $\sum r'_j X_j(\omega)$ convergent a.s., then also Y_n would converge a.s. to, say, Y (again, in view of the above Lemma), and by Fatou Lemma,

$$\mathbf{E}\|Y\| \le \liminf_n \mathbf{E}\|Y_n\|^p \le c^p.$$

Since $\mathbf{P}(\sup\|Y_n\| \ge t) \le 2\mathbf{P}(\|Y\| > t)$, we would have $\sup\|Y_n\| \in \boldsymbol{L}_p$. Therefore $Y_n \to Y$ in $\boldsymbol{L}_p(\boldsymbol{X})$ by the Lebesgue Dominated

Convergence Theorem, which would contradict $\mathbf{E}\|X_j'\|^p > a^p$ (i.e., (4.1.3)).

(ii) $\Longrightarrow$ *(i)*. This implication is obvious since $\boldsymbol{X} \subset \boldsymbol{L}_p(\Omega, \Sigma, \mathbf{P}; \boldsymbol{X})$. QED

Theorem 4.1.3.[5] *Let* $\boldsymbol{X}$ *be a Banach space which does not contain an isomorphic copy of* $\boldsymbol{c}_0$. *If* (X_i) *is a sequence of independent random vectors in* $\boldsymbol{X}$ *such that the series* $\sum_i X_i$ *converges a.s., and* $\sum_i X_i \in \boldsymbol{L}_\infty(\Omega, \Sigma, \mathbf{P}; \boldsymbol{X})$, *then the series* $\sum_i X_i$ *converges in* $\boldsymbol{L}_\infty(\Omega, \Sigma, \mathbf{P}; \boldsymbol{X})$.

Proof. Assume that $\sum_i X_i \in \boldsymbol{L}_\infty(\boldsymbol{X})$. By Hoffman-Jorgensen Theorem (Part (ii)), it follows that $\mathbf{E}\sum_i X_i = \sum_i \mathbf{E}X_i$, so that we can assume, without loss of generality, that $\mathbf{E}X_i = 0$, $i \in \mathbf{N}$.

Suppose, to the contrary, that $\sum_i X_i$ does not converge in $\boldsymbol{L}_\infty(\boldsymbol{X})$. So, in particular, there exists a constant $a > 0$, and a sequence $0 = n_0 < n_1 < \ldots$ of integers such that, for every $j \in \mathbf{N}$ we have $\operatorname{ess\,sup}\|Y_j\| > a$, where

$$Y_j = \sum_{i=n_j+1}^{n_{j+1}} X_i.$$

Evidently, $\sum_j Y_j = \sum_i X_i$ converges a.s.

Now, take a sequence (Z_i) of zero-mean random vectors in $\boldsymbol{X}$ such that, for every $j \in \mathbf{N}$,

$$Z_j = \sum_{k=1}^{\infty} \boldsymbol{x}_{jk} I_{A_{jk}}, \quad \text{and} \quad \operatorname{ess\,sup}\|Z_j - Y_j\| \le a2^{-(j+1)},$$

where $(A_{jk}), k \in \mathbf{N}$, are pairwise disjoint, $A_{jk} \in \sigma(Y_j)$, and $(x_{jk}) \subset \boldsymbol{X}$. Clearly, the series of independent random vectors $\sum_j Z_j$ converges a.s., and $\sum_j Z_j \in \boldsymbol{L}_\infty(\boldsymbol{X})$ because $\sum_j \operatorname{ess\,sup}\|Z_j - Y_j\| \le a < \infty$. Furthermore, for each $j \in \mathbf{N}$,

$$\operatorname{ess\,sup}\|Z_j\| \ge \operatorname{ess\,sup}\|Y_j\| - a2^{-(j+1)} > a/2,$$

so that there exists a subsequence $(k_j) \subset \mathbf{N}$ such that $\|\boldsymbol{x}_{jk_j}\| \ge a/2$, and $\mathbf{P}(A_{jk_j}) > 0$.

[5] See J. Hoffmann-Jorgensen (1972/73).

By the Hoffmann-Jorgensen Theorem there exists a constant $K > 0$ such that, for each $n \in \mathbf{N}$,

$$\text{ess sup}\Big\|\sum_{j=1}^{n} Z_j\Big\| \le K.$$

Then, by a symmetrization argument we get that, for every $B \subset \{0, \ldots, n\}$,

$$\text{ess sup}\Big\|\sum_{j\in B} Z_j\Big\| \le \text{ess sup}\Big\|\sum_{j=1}^{n} Z_j\Big\| \le K,$$

so that there exists $\Omega_0 \in \Sigma$, with $\mathbf{P}(\Omega_0) = 1$, such that, for all finite $B \subset \mathbf{N}$,

$$\text{ess sup}\Big\|\sum_{j\in B} Z_j\Big\| \le K.$$

Since $\mathbf{P}(A_{jk_j}) > 0$, and the events (A_{jk_j}) are independent, $\mathbf{P}\big(\bigcap_{j=1}^{n} A_{jk_j}\big) > 0$, and, for each $n \in \mathbf{N}$, there exists an $\omega_n \in \Omega_0 \cap \bigcap_{j=1}^{n} A_{jk_j}$. Because $Z_j(\omega_n) = \boldsymbol{y}_j, 0 \le j \le n$, we have that

$$\Big\|\sum_{j\in B} \boldsymbol{y}_j\Big\| \le K,$$

for each $B \subset \{1, \ldots, n\}$, and each $n \in \mathbf{N}$. Therefore, for each $\boldsymbol{x}^* \in \boldsymbol{X}^*$, we have $\sum_j |\boldsymbol{x}^*\boldsymbol{y}_j| \le 2K\|\boldsymbol{x}^*\|$, and since $\|\boldsymbol{y}_j\| \ge a/2$, the space $\boldsymbol{X}$ contains a copy of $\boldsymbol{c}_0$ in view of Bessaga-Pelczynski Theorem. A contradiction. QED

4.2 Pre-Gaussian random vectors

In this section we will consider the situation when the summands are in the domain of attraction of a Gaussian random vector with values in a Banach space $\boldsymbol{X}$.

Theorem 4.2.1.[6] *The following two properties of a Banach space $\boldsymbol{X}$ are equivalent:*

[6]See G. Pisier and J. Zinn (1978).

(i) $\boldsymbol{X}$ does not contain an isomorphic copy of $\boldsymbol{c}_0$;

(ii) For each sequence (X_n) of independent and identically distributed random vectors in $\boldsymbol{X}$ such that $\sup_n \mathbf{E}\|X_1 + \cdots + X_n\|/n^{1/2} < \infty$ there exists a Gaussian random vector $\boldsymbol{Y}$ in $\boldsymbol{X}$ such that

$$\mathbf{E}\exp[i\boldsymbol{x}^*\boldsymbol{Y}] = \exp[-\mathbf{E}(\boldsymbol{x}^*X_1)^2], \qquad \boldsymbol{x}^* \in \boldsymbol{X}^*. \tag{4.2.1}$$

Proof. *(ii)* $\Longrightarrow$ *(i)*. It is sufficient to construct an appropriate example in $\boldsymbol{c}_0$. So, let $(\boldsymbol{e}_n)$ be the canonical basis in $\boldsymbol{c}_0$, and let $X = \sum_n r_n \boldsymbol{e}_n \log^{-1/2} n$. Evidently, X is a random vector in $\boldsymbol{c}_0$ and, if $X_1, X_2, \ldots,$ are independent copies of X, then, in view of the Khinchine Inequality,

$$\mathbf{E}\|X_1+\cdots+X_n\|/n^{1/2} = n^{-1/2}\mathbf{E}\sup_k \big|r_1^{(k)}+\cdots+r_n^{(k)}\big| \log^{-1/2} k \le \text{const}.$$

On the other hand,

$$\exp[-\mathbf{E}(\boldsymbol{x}^*X)^2] = \exp[-\sum_n \alpha_n^2/\log n], \qquad \boldsymbol{x}^* = (\alpha_n) \in \boldsymbol{l}_1,$$

so that the Gaussian random vector Y would have the representation $\sum_n \gamma_n \boldsymbol{e}_n \log^{-1/2} n$, where γ_n are independent and identically distributed real symmetric Gaussian random variables. However, the latter series diverges a.s. in $\boldsymbol{c}_0$ because $\sum_n \mathbf{P}(|\gamma_n| > \log^{1/2} n) = \infty$, so that $|\gamma_n| \log^{-1/2} n \ge 1$ infinitely often with probability 1. A contradiction.

(i) $\Longrightarrow$ *(ii)*. Assume that $\boldsymbol{X}$ does not contain $\boldsymbol{c}_0$, and that (X_n) is such that $\sup_n \mathbf{E}\|X_1 + \cdots + X_n\|/n^{1/2} < \infty$. Let $(X_n^{(k)}), n, k \in \mathbf{N}$, be independent copies of a martingale of simple random vectors in $\boldsymbol{X}$, with finite range, and such that

$$\lim_{k\to\infty} \mathbf{E}\|X_1 - X_1^{(k)}\| = 0. \tag{4.2.2}$$

Then

$$\sup_N \mathbf{E}\|X_1^{(k)}+\cdots+X_n^{(K)}\|/n^{1/2} \le \sup_n \mathbf{E}\|X_1+\cdots+X_n\|/n^{1/2} < \text{const.}, \tag{4.2.3}$$

where the constant is independent of k. By the Central Limit Theorem in the finite-dimensional subspace of $\boldsymbol{X}$ spanned by $X_1^{(k)}$, there exists a Gaussian random vector $Y_1^{(k)}$ such that

$$\mathbf{E}(\boldsymbol{x}^*Y_1^{(k)})^2 = \mathbf{E}(\boldsymbol{x}^*X_1^{(k)})^2, \qquad \forall k \in \mathbf{N}, \tag{4.2.4}$$

and for which, in view of (4.2.3),

$$\sup_k \mathbf{E}\|Y_1^{(k)}\| \leq \sup_k \limsup_{n\to\infty} \mathbf{E}\|X_1^{(k)} + \ldots X_n^{(k)}\|/n^{1/2} < \text{const.} \tag{4.2.5}$$

On the other hand, from (4.2.4), and from the fact that $(X_1^{(k)})$ is a martingale, it follows that $(Y_1^{(k)} - Y_1^{(k-1)}), k \in \mathbf{N}$, form a sequence of independent Gaussian random vectors in $\boldsymbol{X}$. Now, (4.2.5) implies that the partial sums of $\sum_k (Y_1^{(k)} - Y_1^{(k-1)})$ are bounded a.s. Since, $\boldsymbol{X}$ does not contain a copy of $\boldsymbol{c}_0$, by Theorem 4.1.2, there exists $Y = \lim_{k\to\infty} Y_1^{(k)}$ (a.s., but also in $\boldsymbol{L}_2$) which is a desired Gaussian random vector in $\boldsymbol{X}$ satisfying (4.1.1), because of (4.1.2) and (4.1.4). QED

The Banach spaces $\boldsymbol{X}$ in which $\boldsymbol{c}_0$ is not finitely representable can also be characterized by the behavior of sums of $\boldsymbol{X}$-valued random vectors. These results will be discussed at length in Chapter 5 since the proofs depend on the techniques developed for spaces of cotype q.

Chapter 5

Cotypes of Banach spaces

5.1 Infracotypes of Banach spaces

Let us begin by introducing a purely geometric concept of *infracotype*[1] of a normed space and define, for $q \in [1, \infty]$, and $n \in \mathbf{N}$, the numerical constants

$$d_n^q(\boldsymbol{X}) := \inf\Big\{d \in \mathbf{R}^+ : \boldsymbol{x}_1, \ldots, \boldsymbol{x}_n \in \boldsymbol{X}, \Big(\sum_{i=1}^{n} \|\boldsymbol{x}_i\|^q\Big)^{1/q}$$
$$\leq d \sup_{\varepsilon_i = \pm 1} \Big\|\sum_{i=1}^{n} \varepsilon_i \boldsymbol{x}_i\Big\|\Big\}.$$

Definiton 5.1.1. A normed space $\boldsymbol{X}$ is said to be of *infracotype* q if there exists a constant $C > 0$, such that, for each $n \in \mathbf{N}$

$$d_n^q(\boldsymbol{X}) \leq C < \infty.$$

In other words, $\boldsymbol{X}$ is of infracotype q if, and only if, for some constant $C > 0$, and any finite sequence $(\boldsymbol{x}_i) \subset \boldsymbol{X}$,

$$\Big(\sum_{i=1}^{n} \|\boldsymbol{x}_i\|^q\Big)^{1/q} \leq d \sup_{\varepsilon_i = \pm 1} \Big\|\sum_{i=1}^{n} \varepsilon_i \boldsymbol{x}_i\Big\|.$$

[1] In the classical terminology the space is of infracotype q if, and only if, the identity operator is $(1, q)$-absolutelly summing. Results of this section are in the spirit of those obtained by G. Pisier (1973) for spaces of infratype p (see, Chapter 6.1).

It is also easy to see that $\boldsymbol{X}$ is of infracotype q if, and only if, for each unconditionally convergent series $\sum_i \boldsymbol{x}_i$ in $\boldsymbol{X}$, we have $\sum_i \|\boldsymbol{x}_i\|^q < \infty$. Also, in view of the Dvoretzky Theorem of Chapter 1, the following Proposition is evident:

Proposition 5.1.1. *(i) The space $\mathbf{R}^n$ is of infracotype 1, for each $n \in \mathbf{N}$.*

(ii) If $\boldsymbol{X}$ is infinite-dimensional and of infracotype q, then necessarily $q \geq 1$.

(iii) If $\boldsymbol{X}$ is of infracotype q, then $d_n^q(\boldsymbol{X}) \geq n^{1/q-1}$.

(iv) If $\boldsymbol{X}$ is of infracotype q, and $q_1 > q$, then $\boldsymbol{X}$ is of infracotype q_1.

The case $q = 1$ plays a special role in the investigation of the infracotypes of normed spaces and here are some properties, such as the *monotonicity and submultiplicativity*, of the sequence $d_n^1(\boldsymbol{X})$:

Proposition 5.1.2. *(i) If $\boldsymbol{X} \neq \{0\}$, then*

$$1 \leq d_n^1(\boldsymbol{X}) \leq n,$$

and if $\boldsymbol{X}$ is infinite-dimensional then

$$n^{1/2} \leq d_n^1(\boldsymbol{X}) \leq n,$$

(ii) If $n \leq m$, then $d_n^1(\boldsymbol{X}) \leq d_m^1(\boldsymbol{X})$.

(iii) For any $n, k \in \mathbf{N}$,

$$d_{nk}^1(\boldsymbol{X}) \leq d_n^1(\boldsymbol{X}) \cdot d_k^1(\boldsymbol{X}).$$

Proof. (*i*) The fact that $d_n^1(\boldsymbol{X}) \leq n$ follows from the inequality $\|\boldsymbol{x}_i\| \leq \sup\{\|\sum_i \varepsilon_i \boldsymbol{x}_i\| : \varepsilon = \pm 1\}$ which, in turn, is a consequence of the fact that, for any $\boldsymbol{x}, \boldsymbol{y} \in \boldsymbol{X}$ either $\|\boldsymbol{x}\| \leq \|\boldsymbol{x} + \boldsymbol{y}\|$, or $\|\boldsymbol{x}\| \leq \|\boldsymbol{x} - \boldsymbol{y}\|$. The inequality $d_n^1(\boldsymbol{X}) \geq n^{1/2}$ is an immediate consequence of the Dvoretzky Theorem of Chapter 1.

(*ii*) This statement follows directly from the definition of $d_n^1(\boldsymbol{X})$.

(*iii*) Let $\boldsymbol{x}_1, \ldots, \boldsymbol{x}_{nk} \in \boldsymbol{X}$. Choose $\varepsilon_j^i, (i-1)k < j \leq ik, i = 1, \ldots, n$, so that

$$\Big\| \sum_{j=ik-k+1}^{ik} \varepsilon_j^i \boldsymbol{x}_j \Big\| = \sup_{\varepsilon_j = \pm 1} \Big\| \sum_{j=ij-k+1}^{ik} \varepsilon_j \boldsymbol{x}_j \Big\|,$$

and define

$$\boldsymbol{y}_i = \sum_{j-ik-k+1}^{ik} \varepsilon_j^i \boldsymbol{x}_j, \qquad i = 1, \ldots, n.$$

By the construction itself,

$$\sum_{j=ik-k+1}^{ik} \|\boldsymbol{x}_i\| \leq d_k^1(\boldsymbol{X}) \|\boldsymbol{y}_i\|,$$

so that

$$\sum_{j=1}^{nk} \|\boldsymbol{x}_i\| \leq d_k^1(\boldsymbol{X}) \|\boldsymbol{y}_i\| \leq d_k^1(\boldsymbol{X}) d_n^1(\boldsymbol{X}) \sup_{\varepsilon_i = \pm 1} \Big\| \sum_{i=1}^{n} \varepsilon_i \boldsymbol{y}_i \Big\|$$

$$\leq d_k^1(\boldsymbol{X}) d_n^1(\boldsymbol{X}) \sup_{\varepsilon_i = \pm 1} \Big\| \sum_{i=1}^{nk} \varepsilon_i \boldsymbol{y}_i \Big\|,$$

which completes the proof of the Proposition. QED

It turns out that the sequence $d_n^1(\boldsymbol{X})$ contains complete information about the infracotype of the space $\boldsymbol{X}$. Indeed, we have the following

Proposition 5.1.3. *If $\boldsymbol{X} \neq \{0\}$ is a normed space, and $n_0 > 1$, then there exists $q_0, 1 \leq q_0 \leq \infty$ ($2 \leq q_0 \leq \infty$, if $\boldsymbol{X}$ is infinite-dimensional) such that*

$$d_{n_0}^1(\boldsymbol{X}) = n_0^{1-1/q_0},$$

and, for each $q > q_0$, $\boldsymbol{X}$ is of infracotype q.

Proof. The existence of such a q_0 follows from Proposition 5.1.2 (*i*). Using the submultiplicativity property proven in Proposition 5.1.2 (*iii*) we immediately obtain that

$$d_{n_0^k}^1(\boldsymbol{X}) \leq n_0^{k(1-1/q_0)}, \qquad k \in \mathbf{N}.$$

Put $C = n_0^{1-1/q_0}$, and consider $n \in [n_0^k, n_0^{k+1})$. Then.

$$d_n^1(\boldsymbol{X}) \le d_{n_0^{k+1}}^1(\boldsymbol{X}) \le n_0^{(k+1)(1-1/q_0)} \le Cn^{1-1/q_0}.$$

Now, take $\boldsymbol{x}_1, \ldots, \boldsymbol{x}_n \in \boldsymbol{X}$, and order them in such a way that $\|\boldsymbol{x}_{i+1}\| \le \|\boldsymbol{x}_i\|$. Then,

$$\|\boldsymbol{x}_k\| = \inf_{1\le j\le k} \|\boldsymbol{x}_j\| \le k^{-1}\sum_{j=1}^{k} \|\boldsymbol{x}_j\|$$

$$\le k^{-1} d_k^1(\boldsymbol{X}) \sup_{\varepsilon_j=\pm 1} \Big\|\sum_{j=1}^{k} \varepsilon_j \boldsymbol{x}_j\Big\| \le Ck^{-1}k^{1-1/q_0} \sup_{\varepsilon_j=\pm 1} \Big\|\sum_{j=1}^{k} \varepsilon_j \boldsymbol{x}_j\Big\|,$$

so that, if $q > q_0$, then

$$\Big(\sum_{k=1}^{n} \|\boldsymbol{x}_k\|^q\Big)^{1/q} \le C\Big(\sum_{k=1}^{\infty} k^{-q/q_0}\Big)^{1/q} \sup_{\varepsilon_j=\pm 1} \Big\|\sum_{j=1}^{n} \varepsilon_j \boldsymbol{x}_j\Big\|,$$

which shows that $\boldsymbol{X}$ is of infracotype q. QED

In the above context let us define the quantity,

$$q_{\inf}(\boldsymbol{X}) = \inf\{q : \boldsymbol{X} \text{ is of infracotype } q\}.$$

It turns out that $q_{\inf}(\boldsymbol{X})$ can be explicitely expressed in terms of the quantities $d_n^1(\boldsymbol{X})$.

Theorem 5.1.1. *Let $0 < q < \infty$, and $\boldsymbol{X} \neq \{0\}$. Then*

$$q_{\inf}(\boldsymbol{X}) = \lim_{n\to\infty} \frac{\log n}{\log\big(n/d_n^1(\boldsymbol{X})\big)}.$$

Proof. If $\boldsymbol{X}$ is of infracotype q then, by Hölder Inequality,

$$n^{1/q-1}\sum_{i=1}^{n} \|\boldsymbol{x}_i\| \le \Big(\sum_{i=1}^{n} \|\boldsymbol{x}_i\|^q\Big)^{1/q} \le C \sup_{\varepsilon_i=\pm 1} \Big\|\sum_{i=1}^{n} \varepsilon_i \boldsymbol{x}_i\Big\|,$$

so that $d_n^1(\boldsymbol{X}) \le Cn^{1-1/q}$, and

$$\limsup_{n\to\infty} \frac{\log n}{\log\big(n/d_n^1(\boldsymbol{X})\big)} \le \limsup_{n\to\infty} \frac{\log n}{(1/q)\log n - \log C} = q.$$

Therefore,

$$\limsup_{n\to\infty} \frac{\log n}{\log\bigl(n/d_n^1(\boldsymbol{X})\bigr)} \le q_{\inf}(\boldsymbol{X}).$$

If $q < q_{\inf}(\boldsymbol{X})$, then, by Proposition 5.1.3, $d_n^1(\boldsymbol{X}) > n^{1-1/q}$ for each $n \in \mathbf{N}$, and

$$\liminf_{n\to\infty} \frac{\log n}{\log\bigl(n/d_n^1(\boldsymbol{X})\bigr)} \ge \frac{\log n}{(1/q)\log n} = q \ge q_{\inf}(\boldsymbol{X}).$$

QED

5.2 Spaces of Rademacher cotype

In this section we will adapt the definition of the infracotype to the random environment by replacing the deterministic $\varepsilon_i = \pm 1$ signs used in the previous section by the random Rademacher coefficients. So, for a normed space $\boldsymbol{X}$, $q > 0$, and an arbitrary $n \in \mathbf{N}$, we define the constants,

$$c_n^q(\boldsymbol{X}) := \inf\Bigl\{c \in \mathbf{R}^+ : \boldsymbol{x}_1, \dots, \boldsymbol{x}_n \in \boldsymbol{X}, \Bigl(\sum_{i=1}^n \|\boldsymbol{x}_i\|^q\Bigr)^{1/q} \le c\Bigl(\mathbf{E}\Bigl\|\sum_{i=1}^n r_i \boldsymbol{x}_i\Bigr\|^q\Bigr)^{1/q}\Bigr\}.$$

Definition 5.2.1. The normed space $\boldsymbol{X}$ is said to be of *Rademacher cotype q* (or, simply, of *cotype q*)[2], if there exists a constant C such that, for each $n \in \mathbf{N}$,

$$c_n^q(\boldsymbol{X}) \le C < \infty.$$

[2]The notion of cotype 2 was introduced by D. Mouchtari (1973) who called it the Mazur-Orlicz Property, by E. Dubinsky, A. Pelczynski, and H.P. Rosenthal, who called it the superquadraticity of Rademacher averages, and by B. Maurey (1972/73). For general q, the notion was considered by J. Hoffmann-Jorgensen who called it weak cotype q. Results of this section are due to B. Maurey and G. Pisier (1976).

Note that, because of Kahane Theorem of Chapter 1, $\boldsymbol{X}$ is of cotype q if, and only if, there exists a constant $C < \infty$, and an $\alpha, 0 < \alpha < \infty$, such that for any finite sequence $(\boldsymbol{x}_n) \subset \boldsymbol{X}$,

$$\Big(\sum_{i=1}^{n} \|\boldsymbol{x}_i\|^q\Big)^{1/q} \leq C\Big(\mathbf{E}\Big\|\sum_{i=1}^{n} r_i \boldsymbol{x}_i\Big\|^{\alpha}\Big)^{1/\alpha}\Big\} \tag{5.2.1}$$

or, alternatively, $\boldsymbol{X}$ is of cotype q if, and only if, for each $\alpha, 0 < \alpha < \infty$ there exists a constant $C = C_\alpha$ such that, for any finite sequence $(\boldsymbol{x}_i) \subset \boldsymbol{X}$, the inequality (5.2.1) holds true.

Remark 5.2.1. If one replaces in the above definition (with apppripriate modifications) Rademacher random variables (r_i) by stable random variables (θ_i) of exponent q, then one obtains a definition of a space of stable-cotype q. However, because of the tail behavior of (θ_i), and Borel-Cantelli Lemma, each Banach space is of stable cotype q if $q < 2$. For $q = 2$, the stable cotype and Rademacher cotype coincide as we will see later on in this chapter.

Remark 5.2.2. In Banach lattices, and spaces with unconditional basis (and in spaces with local unconditional structure), if $q > 2$, then $\boldsymbol{X}$ is of infracotype q if, and only if, $\boldsymbol{X}$ is of cotype q.[3]

Proposition 5.2.1. *(i) If $\boldsymbol{X}$ is of cotype q, and $q_1 > q$, then $\boldsymbol{X}$ is of cotype q_1.*

(ii) If $\boldsymbol{X}$ is of cotype q and $\boldsymbol{X} \neq \{0\}$, then $q \geq 2$.

(iii) If $\boldsymbol{X}$ is of cotype q, then $\boldsymbol{X}$ is of infracotype q.

We omit the obvious proof of (i). The implication (ii) immediately follows from the Khinchine Inequality on the real line, and the implication (iii) follows directly from the definitions.

The properties of constants $c_n^q(\boldsymbol{X})$, are similar to the properties of constants $d_n^1(\boldsymbol{X})$ investigated in the previous section in connection of the infracotype of a Banach space.

Proposition 5.2.2. *(i) The sequence $(c_n^q(\boldsymbol{X}))$ is monotone, that is, if $n \leq m$, $n, m \in \mathbf{N}$, then $c_n^q(\boldsymbol{X}) \leq c_m^q(\boldsymbol{X})$.*

(ii) For any $q, 0 < q < \infty$, and $n, k \in \mathbf{N}$,

$$c_{nk}^q(\boldsymbol{X}) \leq c_n^q(\boldsymbol{X}) \cdot c_k^q(\boldsymbol{X}).$$

[3] See B. Maurey (1973/74).

Proof. The property (*i*) is obvious, so we will only prove (*ii*). Take $\boldsymbol{x}_1, \ldots, \boldsymbol{x}_{nk} \in \boldsymbol{X}$ and, for each $i = 1, \ldots, n$, define the random vectors

$$X_i = \sum_{j=ik-k+1}^{ik} r_j \boldsymbol{x}_j,$$

so that, if (r'_j) is a Rademacher sequence independent of (r_j) then

$$\Big(\sum_{i=1}^{n} \|X_i(\omega)\|^q\Big)^{1/q} \le c_n^q(\boldsymbol{X})\Big(\mathbf{E}'\Big\|\sum_{i=1}^{n} r'_i X_i(\omega)\Big\|^q\Big)^{1/q},$$

where $\mathbf{E}'$ denotes integration with respect to (r'_j). Furthermore, by the symmetry argument,

$$\Big(\sum_{i=1}^{nk} \|\boldsymbol{x}_i\|^q\Big)^{1/q} = \Big(\sum_{i=1}^{n} \sum_{j=ik-k+1}^{ik} \|\boldsymbol{x}_i\|^q\Big)^{1/q}$$

$$\le c_k^q(\boldsymbol{X})\Big(\sum_{i=1}^{n} \mathbf{E}\|X_i\|^q\Big)^{1/q} \le c_k^q(\boldsymbol{X}) c_n^q(\boldsymbol{X})\Big(\mathbf{E}\mathbf{E}'\Big\|\sum_{i=1}^{n} r'_i X_i(\omega)\Big\|^q\Big)^{1//q}$$

$$= c_k^q(\boldsymbol{X}) c_n^q(\boldsymbol{X})\Big(\mathbf{E}\Big\|\sum_{i=1}^{nk} r_i \boldsymbol{x}_i\Big\|^q\Big)^{1//q}. \qquad \text{QED}$$

Knowledge of the sequence $(c_n^q(\boldsymbol{X}))$ for any particular $q, 0 < q < \infty$, provides information about the cotype of $\boldsymbol{X}$.

Proposition 5.2.3. *If $\boldsymbol{X} \neq \{0\}$ is a normed space, and $n_0 > 1, 0 < q < \infty$, then there exists a $q_0 \ge q$ such that*

$$c_{n_0}^q(\boldsymbol{X}) = n_0^{1/q - 1/q_0},$$

and $\boldsymbol{X}$ is of cotype q_1, for each $q_1 > q_0$.

Proof. The existence of such a q_0 is obvious because $c_n^q \ge 1$. Using the submultiplicativity property we get the inequality

$$c_{n_0^k}^q(\boldsymbol{X}) \le n_0^{k(1/q - 1/q_0)}, \qquad k \in \mathbf{N}.$$

Now, set $C = n_0^{1/q-1/q_0}$, and select any integer $n \in [n_0^k, n_0^{k+1})$. Then,

$$c_n^q(\boldsymbol{X}) \leq c_{n_0^{k+1}}^q(\boldsymbol{X}) \leq n_0^{(k+1)(1/q-1/q_0)} \leq Cn^{1/q-1/q_0}.$$

Let $\boldsymbol{x}_1, \ldots, \boldsymbol{x}_n \in \boldsymbol{X}$, and assume that these vectors are ordered in such a way that $\|\boldsymbol{x}_{i+1}\| \leq \|\boldsymbol{x}_i\|$. In this case,

$$\|\boldsymbol{x}_k\| = \inf_{1\leq j\leq k} \|\boldsymbol{x}_j\| \leq \Big(k^{-1}\sum_{j=1}^{k} \|\boldsymbol{x}_j\|^q\Big)^{1/q}$$

$$\leq k^{-1/q} c_k^q(\boldsymbol{X}) \Big(\mathbf{E}\Big\|\sum_{j=1}^{k} r_j \boldsymbol{x}_j\Big\|^q\Big)^{1/q} \leq k^{-1/q} c_k^q(\boldsymbol{X}) \Big(\mathbf{E}\Big\|\sum_{j=1}^{n} r_j \boldsymbol{x}_j\Big\|^q\Big)^{1/q}$$

$$\leq k^{-1/q} C k^{1/q-1/q_0} \Big(\mathbf{E}\Big\|\sum_{j=1}^{n} r_j \boldsymbol{x}_j\Big\|^q\Big)^{1/q},$$

so that, if $q_1 > q_0$, then

$$\Big(\sum_{k=1}^{n} \|\boldsymbol{x}_k\|^{q_1}\Big)^{1/q_1} \leq C\Big(\sum_{k=1}^{\infty} k^{-q_1/q_0}\Big)^{1/q_1} \Big(\mathbf{E}\Big\|\sum_{j=1}^{n} r_j \boldsymbol{x}_j\Big\|^q\Big)^{1/q},$$

which, in view of (5.2.1), shows that $\boldsymbol{X}$ is of cotype q_1. QED

Now, proceeding in a way analogous to that of Section 5.1.1, let us define the quantity

$$q_{\text{rad}}(\boldsymbol{X}) = \inf\{q : \boldsymbol{X} \text{ is of cotype } q\}.$$

And again, it turns out that $q_{\text{rad}}(\boldsymbol{X})$ can be explicitly expressed in terms of the quantities $c_n^p(\boldsymbol{X})$.

Theorem 5.2.1. *Let $0 < p < \infty$, and $\boldsymbol{X} \neq \{0\}$. Then*

$$q_{\text{rad}}(\boldsymbol{X}) = \lim_{n\to\infty} \frac{\log n}{\log\big(n^{1/p}/c_n^p(\boldsymbol{X})\big)}.$$

Proof. If $\boldsymbol{X}$ is of cotype q then, in view of (5.2.1), and Hölder inequality,

$$n^{1/q-1/p}\Big(\sum_{i=1}^{n}\|\boldsymbol{x}_i\|^p\Big)^{1/p} \le \Big(\sum_{i=1}^{n}\|\boldsymbol{x}_i\|^q\Big)^{1/q} \le C\Big(\mathbf{E}\Big\|\sum_{i=1}^{n} r_i\boldsymbol{x}_i\Big\|^p\Big)^{1/p}.$$

Therefore, $c_n^p(\boldsymbol{X}) \le Cn^{1/p-1/q}$, and

$$\limsup_{n\to\infty}\frac{\log n}{\log\big(n^{1/p}/c_n^p(\boldsymbol{X})\big)} \le \limsup_{n\to\infty}\frac{\log n}{(1/q)\log n-\log C} = q$$

so that

$$\limsup_{n\to\infty}\frac{\log n}{\log\big(n^{1/p}/c_n^p(\boldsymbol{X})\big)} \le q_{\mathrm{rad}}(\boldsymbol{X}).$$

If $q < q_{\mathrm{rad}}(\boldsymbol{X})$ then, by Proposition 5.2.3, $c_n^p(\boldsymbol{X}) > n^{1/p-1/q}$, for all $n \in \mathbf{N}$, and

$$\liminf_{n\to\infty}\frac{\log n}{\log\big(n^{1/p}/c_n^p(\boldsymbol{X})\big)} \ge \frac{\log n}{(1/q)\log n} = q,$$

so that

$$\limsup_{n\to\infty}\frac{\log n}{\log\big(n^{1/p}/c_n^p(\boldsymbol{X})\big)} \ge q_{\mathrm{rad}}(\boldsymbol{X}). \qquad \text{QED}$$

As far as subspaces of spaces of cotype q are concerned, the following Proposition is evident.

Proposition 5.2.4. *If $\boldsymbol{X}$ is of cotype q, and $\boldsymbol{Y}$ is a subspace of $\boldsymbol{X}$ then $\boldsymbol{Y}$ is of cotype q, and $c_n^q(\boldsymbol{X}) \ge c_n^q(\boldsymbol{Y})$, for each $q > 0$, and $n \in \mathbf{N}$.*

Less trivial is the following result which shows circumstances under which the cotype of $\boldsymbol{X}$ is preserved by formation of spaces $\boldsymbol{L}_p(\boldsymbol{X})$.

Theorem 5.2.2. *Let (T, Σ, μ) be a σ-finite measure space. Then,*

(i) $c_n^q(\boldsymbol{X}) = c_n^q\big(\boldsymbol{L}_q(T, \Sigma, \mu; \boldsymbol{X})\big)$, and

(ii) if $p \le q$ then $\boldsymbol{X}$ is of cotype q if, and only if, $\boldsymbol{L}_p(T, \Sigma, \mu; \boldsymbol{X})$ is of cotype q.

Proof. (*i*) Evidently, $c_n^q(\boldsymbol{X}) \leq c_n^q\big(\boldsymbol{L}_q(T,\Sigma,\mu;\boldsymbol{X})\big)$. On the other hand, if $(X_i) \subset \boldsymbol{L}_q(\boldsymbol{X})$, then for each $t \in T$,

$$\sum_{i=1}^n |X_i(t)\|^q \leq (c_n^q(\boldsymbol{X}))^q \mathbf{E}\Big\|\sum_{i=1}^n r_i X_i(t)\Big\|^q.$$

Integrating both sides with respect to μ we get that $c_n^q(\boldsymbol{X}) \leq c_n^q\big(\boldsymbol{L}_q(\boldsymbol{X})\big)$, which concludes the proof of (*i*).

(*ii*) The"if" part being evident, let us assume that $\boldsymbol{X}$ is of cotype q. Then, by (5.2.1), if $(X_i) \subset \boldsymbol{L}_p(\boldsymbol{X})$ then, for each $t \in T$,

$$\Big(\sum_{i=1}^n \|X_i(t)\|^q\Big)^{1/q} \leq C\Big(\mathbf{E}\Big\|\sum_{i=1}^n r_i X_i(t)\Big\|^p\Big)^{1/p},$$

so that, because $q \geq p$,

$$\Big(\sum_{i=1}^n\Big(\int_T \|X_i(t)\|^p \mu(dt)\Big)^{q/p}\Big)^{1/q} \leq \Big(\int_T\Big(\sum_{i=1}^n \|X_i(t)\|^q\Big)^{p/q}\mu(dt)\Big)^{1/p}$$

$$\leq C\Big(\mathbf{E}\int_T\Big\|\sum_{i=1}^n r_i X_i(t)\Big\|^p \mu(dt)\Big)^{1/p}. \qquad \text{QED}$$

Corollary 5.2.1. *The following two properties of a normed space $\boldsymbol{X}$ are equivalent:*

(i) $\boldsymbol{X}$ is of cotype q;

(ii) There exists a constant $C > 0$ such that, for each finite sequence, $X_1, \ldots, X_n$, of independent, zero-mean random vectors in $\boldsymbol{L}_q(\boldsymbol{X})$,

$$\Big(\sum_{i=1}^n \mathbf{E}\|X_i\|^q\Big)^{1/q} \leq C\Big(\mathbf{E}\Big\|\sum_{i=1}^n X_i\Big\|^q\Big)^{1/q}.$$

Proof. The implication (*ii*) $\Longrightarrow$ (*i*) is obvious. The reverse implication, (*i*) $\Longrightarrow$ (*ii*) follows from the above Theorem 5.2.2 by the standard symmetrization procedure. QED

The above results permit us to give a number of concrete examples of Banach spaces which are (or, are not) of cotype q. More examples will be provided at the end of this chapter.

Example 5.2.1. (i) The space $\boldsymbol{c}_0$ is not of cotype q, for any $q < \infty$. Indeed, take $(\boldsymbol{x}_i) = (\boldsymbol{e}_i)$, where the latter is the canonical basis in $\boldsymbol{c}_0$. Then the left-hand side of (5.2.1) is equal to $n^{1/q}$, while the right-hand side (for $q = 1$), $\mathbf{E}\|\sum_i r_i \boldsymbol{e}_i\| = \mathbf{E}\max|r_i| \leq 1$. A contradiction.

(ii) As a consequence of (i), neither $\boldsymbol{l}_\infty$, nor $\boldsymbol{L}_\infty$, nor $\boldsymbol{C}[0,1]$, is of cotype q, for any $q < \infty$.

(iii) By Theorem 5.2.2, and Khinchine Inequality, for any $n \in \mathbf{N}$, the space $\mathbf{R}^n$,and any Hilbert space, are of cotype 2.

(iv) The spaces $\boldsymbol{L}_p$, and $\boldsymbol{l}_p$, are of cotype $\max(2,p)$. This fact is a direct consequence of Theorem 5.2.2 and Proposition 5.2.1.

(v) More generally, if the modulus of convexity of $\boldsymbol{X}$ (see, Chapter 3) satisfies the asymptotic condition, $\delta_{\boldsymbol{X}}(\epsilon) = O(\epsilon^q)$, as $\epsilon \to 0$, then $\boldsymbol{X}$ is of cotype q.

(vi)[4] The triple projective tensor product space $\boldsymbol{X} = \boldsymbol{l}_{p_1} \otimes_\pi \boldsymbol{l}_{p_2} \otimes_\pi \boldsymbol{l}_{p_3}$, with $1 \leq p_1 \leq p_2 \leq p_3 \leq \infty$, is not of cotype q in the following cases:

- $p_1 \leq 2$, $1/p_1 + 1/p_2 < 1/2$, and $1/p_1 + 1/p_2 + 1/p_3 \geq 1$, and for $q < 3/(1 + 1/p_2 + 1/p_3)$;
- $p_1 < 2$, $1/p_1 + 1/p_2 + 1/p_3 \leq 1$, and for $q < 3/(2 - 1/p_1)$;
- $1/p_1 + 1/p_2 + 1/p_3 < 1/2$, and for $q < 3/(1 + 1/p_1 + 1/p_2 + 1/p_3)$.

5.3 Local structure of spaces of cotype q

The following Proposition follows directly from the definitions.

Proposition 5.3.1. *Cotype q is a superproperty, i.e., if $\boldsymbol{X}$ is of cotype q and $\boldsymbol{Y}$ is finitely representable in $\boldsymbol{X}$, then $\boldsymbol{Y}$ is also of cotype q. In particular, by the local reflexivity principle (see Section 1.3) $\boldsymbol{X}$ is of cotype q if, and only if, the space $\boldsymbol{X}^{**}$ is of cotype q.*

[4]Due to O. Giladi, J. Prochno, C. Schütt, N. Tomczak-Jaegermann, and E. Werner (2017). Results on cotypes of k-fold projective tensor products of $\boldsymbol{l}_p$ spaces are also available in the paper.

The next result about the local structure is much more subtle and requires a nontrivial proof.

Theorem 5.3.1.[5] *If $\boldsymbol{X}$ is an infinite-dimensional normed space, then the canonical injection*

$$\boldsymbol{l}_{q_{\mathrm{rad}}(X)} \mapsto \boldsymbol{l}_{\infty},$$

is finitely factorable through $\boldsymbol{X}$.

Proof. If $q_{\mathrm{rad}}(\boldsymbol{X}) = 2$, then the Theorem is a corollary to the Dvoretzky-Rogers Lemma from Section 1.3 which states, in our current terminology, that the canonical injection $\boldsymbol{l}_2 \mapsto \boldsymbol{l}_\infty$ is finitely factorable through any infinite-dimensional normed space. If $q_{\mathrm{rad}}(\boldsymbol{X}) > 2$, then it is sufficient to show that, for any q such that $q_{\mathrm{rad}}(X) > q > 2$, the injection $\boldsymbol{l}_q \mapsto \boldsymbol{l}_\infty$ is finitely factorable through $\boldsymbol{X}$, because the interval of those q for which the injection $\boldsymbol{l}_q \mapsto \boldsymbol{l}_\infty$ is finitely factorable through $\boldsymbol{X}$ is closed (see, Section 1.3).

So, let $2 < q < q_{\mathrm{rad}}(\boldsymbol{X})$. Then, by Theorem 5.2.1,

$$\log c_n^q(\boldsymbol{X}) \geq \Big(\frac{1}{q} - \frac{1}{q_{\mathrm{rad}}(\boldsymbol{X})}\Big) \log n, \qquad n \in \mathbf{N}.$$

Now, because, for any sequence $(a_n) \subset \mathbf{R}^+$,

$$\limsup_n \frac{\log a_n}{\log n} \leq \limsup_n n\Big(\frac{a_n}{a_{n-1}} - 1\Big),$$

(since, $\lim_{t\to\infty}[(t^\alpha/(t-1)^\alpha) - 1] = \alpha$) there exists an $\mathbf{N}_1 \subset \mathbf{N}$, a constant $b > 0$, and a sequence $(\epsilon_n) \subset (0,1)$ such that, for each $n \in \mathbf{N}_1$,

$$\frac{1}{1+\epsilon_n} \cdot \frac{c_n^q(\boldsymbol{X})}{c_{n-1}^q(\boldsymbol{X})} \geq 1 - \frac{b}{n} \geq \Big(1 + \frac{b}{n}\Big)^{1/q}. \tag{5.3.1}$$

One can assume that $b \leq 1$, so that $n/(n+b) \geq 1/2$, for all $n \in \mathbf{N}$. By the definition of $c_n^q(\boldsymbol{X})$, for each $n \in \mathbf{N}$, there exists a finite sequence $(\boldsymbol{x}_i) \subset \boldsymbol{X}$ such that $\sum_{i=1}^n \|\boldsymbol{x}_i\|^q = n$, and

$$\Big(\mathbf{E}\Big\|\sum_{i=1}^n r_i \boldsymbol{x}_i\Big\|^q\Big)^{1/q} \leq \frac{(1+\epsilon_n)n^{1/q}}{c_n^q(\boldsymbol{X})}.$$

[5] See, B. Maurey and G. Pisier (1976).

Let us set $\alpha := \inf\{\|\boldsymbol{x}_i\| : 1 \le i \le n\}$, and let $\|\boldsymbol{x}_{i_0}\| = \alpha$. Then

$$\Big(\sum_{i \ne i_0} \|\boldsymbol{x}_i\|^q\Big)^{1/q} = (n - \alpha^q)^{1/q} \le c^q_{n-1}(\boldsymbol{X})\Big(\mathbf{E}\Big\|\sum_{i \ne i_0} r_i \boldsymbol{x}_i\Big\|^q\Big)^{1/q}$$

$$\le c^q_{n-1}(\boldsymbol{X})\Big(\mathbf{E}\Big\|\sum_{i=1}^n r_i \boldsymbol{x}_i\Big\|^q\Big)^{1/q} \le c^q_{n-1}(\boldsymbol{X})\frac{(1+\epsilon_n)n^{1/q}}{c^q_n(\boldsymbol{X})}.$$

Now, if $n \in \mathbf{N}_1$ then, by (5.3.1),

$$1 - \frac{\alpha^q}{n - \alpha^q} \ge 1 - \frac{b}{n}, \qquad \text{and} \qquad \alpha^q \ge \frac{nb}{n+b} \ge \frac{b}{2}.$$

Therefore,

$$\inf\{\|\boldsymbol{x}_i\| : 1 \le i \le n\} \ge (b/2)^{1/q}, \qquad n \in \mathbf{N}_1. \tag{5.3.2}$$

By the Contraction Principle (see, Chapter 1), for each $(a_i) \subset \mathbf{R}$,

$$\Big(\mathbf{E}\Big\|\sum_{i=1}^n a_i r_i \boldsymbol{x}_i\Big\|^q\Big)^{1/q} \le \frac{2n^{1/q}}{c^q_n(\boldsymbol{X})} \sup_{1 \le i \le n} |a_i|.$$

At this point, let us define the operators

$$U_n : \boldsymbol{l}^{(n)}_\infty \ni a = (a_i) \mapsto \sum_{i=1}^n a_i r_i \boldsymbol{x}_i \in \boldsymbol{L}_q(\Omega, \Sigma, \mathbf{P}; \boldsymbol{X}).$$

The preceding inequality implies that $\|U_n\| \le 2n^{1/q}/c^q_n(\boldsymbol{X})$. Because, in view of Theorem 5.2.2, $c^q_n(\boldsymbol{X}) = c^q_n(\boldsymbol{L}_q(\boldsymbol{X}))$, if $(a^k) \subset \boldsymbol{l}^{(n)}_\infty$, $k = 1, \ldots, n$, we have the inequalities,

$$\Big(\sum_{k=1}^n \|U_n(a^k)\|^q\Big)^{1/q} \le c^q_n(\boldsymbol{X})\Big(\mathbf{E}\Big\|\sum_{k=1}^n U_n(a^k) r_k\Big\|^q\Big)^{1/q}$$

$$\le c^q_n(\boldsymbol{X})\|U_n\|\Big(\mathbf{E}\Big\|\sum_{k=1}^n a^k r_k\Big\|^q\Big)^{1/q} \le c^q_n(\boldsymbol{X})\|U_n\| \cdot \sup_{1 \le i \le n} \sum_{k=1}^n |a^k_i|.$$

By the Nikishin Lemma (see, Chapter 1) there exists a subset $A \subset \{1, \ldots, n\}$, $|A| \ge n/2$, such that, for every (a_i) with $\operatorname{supp}(a_i) \subset A$,

$$\Big(\mathbf{E}\Big\|\sum_{i=1}^n a_i r_i \boldsymbol{x}_i\Big\|^q\Big)^{1/q} \le 2^{1/q} c^q_n(\boldsymbol{X})\|U_n\| n^{-1/q} \sigma_q((a_i)) \le 2^{1+1/q} \sigma_q((a_i)).$$

By (5.3.2), and the discussion that precedes (5.3.2) , there exist a subset $\mathbf{N}_2 \subset \mathbf{N}$, a constant $d > 0$, an integer $n \in \mathbf{N}_2$, and $\boldsymbol{x}_1^n, \ldots, \boldsymbol{x}_n^n \in \boldsymbol{X}$, such that

$$\inf_{1 \leq i \leq n} \|\boldsymbol{x}_i^n\| \geq d, \qquad \text{and} \qquad \left(\mathbf{E}\left\|\sum_{i=1}^n a_i r_i \boldsymbol{x}_i^n\right\|^q\right)^{1/q} \leq \sigma_q((a_i)). \tag{5.3.3}$$

Let k_0 be such that $dK_0^{1/2} - k_0^{1/q} > 0$, and define

$$\delta := k_0^{-1}(dK_0^{1/2} - k_0^{1/q}).$$

We shall show that if $n \in N_2, n \geq k_0$, then the sequence, $\boldsymbol{x}_1^n, \ldots, \boldsymbol{x}_n^n$, satisfies the property $P(k_0, \delta)$ of the Johnson Lemma cited in Section 1.5 of Chapter 1. Indeed, let $A \subset \{1, \ldots, n\}, |A| = k_0$. Choosing $\boldsymbol{x}_{i_0}^n$ in such a way that $\|\boldsymbol{x}_{i_0}^n\| = \sup_{i \in A} \|\boldsymbol{x}_i^n\|, i_0 \in A$, we have

$$\begin{aligned}
&\left(\mathbf{E}\left\|\sum_{i \in A} r_i \boldsymbol{x}_i^n\right\|^q\right)^{1/q} \\
&\geq \left(\mathbf{E}\left\|\boldsymbol{x}_{i_0}^n \sum_{i \in A} r_i\right\|^q\right)^{1/q} - \left(\mathbf{E}\left\|\sum_{i \in A} r_i(\boldsymbol{x}_i^n - \boldsymbol{x}_{i_0}^n)\right\|^q\right)^{1/q} \\
&\geq \sup_{i \in A} \|\boldsymbol{x}_i^n\| \left(\mathbf{E}\left|\sum_{i \in A} r_i\right|^q\right)^{1/q} - k_0 \sup_{i,j \in A} \|\boldsymbol{x}_i^n - \boldsymbol{x}_j^n\| \\
&\geq dk_0^{1/2} - k_0 \sup_{i,j \in A} \|\boldsymbol{x}_i^n - \boldsymbol{x}_j^n\|,
\end{aligned}$$

so that, in view of (5.3.3),

$$\sup_{i,j \in A} \|\boldsymbol{x}_i^n - \boldsymbol{x}_j^n\| \geq k_0^{-1}(dk_0^{1/2} - k_0^{1/q}) = \delta.$$

Now, let $m \in \mathbf{N}$, and take $n \geq N(k_0, \delta, 1, m)$ (existence thereof being guaranteed, again, by the Johnson Lemma), $n \in N_2$, and a subsequence $(\boldsymbol{x}_{i_1}^n, \ldots, \boldsymbol{x}_{i_{2m}}^n)$ such that the differences $(\boldsymbol{x}_{i_{2j}}^n - \boldsymbol{x}_{i_{2j-1}}^n), j = 1, \ldots, m$, form an unconditional sequence with constant ≤ 3, and

$$\inf_{1 \leq j \leq m} \|\boldsymbol{y}_i\| \geq \delta/2, \qquad \text{where} \qquad \boldsymbol{y}_j := \boldsymbol{x}_{i_{2j}}^n - \boldsymbol{x}_{i_{2j-1}}^n.$$

Now, it follows from (5.3.3) that, for each $(a_k) \subset \mathbf{R}$,

$$\Big\|\sum_{k=1}^{m} a_k \boldsymbol{y}_k\Big\| \le 3\Big(\mathbf{E}\Big\|\sum_{k=1}^{m} a_k r_k \boldsymbol{y}_k\Big\|^q\Big)^{1/q}$$

$$\le 3\Big(\mathbf{E}\Big\|\sum_{k=1}^{m} a_k r_k \boldsymbol{x}^n_{i_{2k}}\Big\|^q\Big)^{1/q} + 3\Big(\mathbf{E}\Big\|\sum_{k=1}^{m} a_k r_k \boldsymbol{x}^n_{i_{2k-1}}\Big\|^q\Big)^{1/q} \le 6\sigma_q((a_i)).$$

Therefore, since for every p, $1 < p < q < \infty$,

$$\sigma_q((a_i)) \le \Big(\frac{p^*}{p^* - q^*}\Big)^{1/p^*} \Big(\sum_i |a_i|^p\Big)^{1/p},$$

where $1/p + 1/p^* = 1$, we have, for each $(a_k) \subset \mathbf{R}$, that

$$\frac{\delta}{6} \sup_k |a_k| \le \Big\|\sum_{k=1}^{m} a_k \boldsymbol{y}_k\Big\| \le 6\Big(\frac{p^*}{p^* - q^*}\Big)^{1/p^*} \Big(\sum_i |a_i|^p\Big)^{1/p}.$$

Thus, for each $q < q_{\text{rad}}(\boldsymbol{X})$, the embedding $\boldsymbol{l}_q \mapsto \boldsymbol{l}_\infty$ is crudely finitely factorable through $\boldsymbol{X}$ and, in view of the James theorem (Section 1.3), it is also finitely factorable. QED

In the context of the above result it now becomes interesting to define a new parameter of of a normed space.

Definition 5.3.1. For a normed space $\boldsymbol{X}$ we define the constant $q(\boldsymbol{X})$ as the supremum of those q for which the embedding $\boldsymbol{l}_q \mapsto \boldsymbol{l}_\infty$ is finitely factorable through $\boldsymbol{X}$.

The next result shows that the constant $q(\boldsymbol{X})$ introduced above is closely tied to the other geometric parameters of $\boldsymbol{X}$ we have introduced before.

Theorem 5.3.2.[6] *For any infinite-dimensional normed space* $\boldsymbol{X}$,

$$q_{\text{rad}}(\boldsymbol{X}) = q_{\text{inf}}(\boldsymbol{X}) = q(\boldsymbol{X}).$$

[6] See B. Maurey and G. Pisier (1976).

Proof. The inequality $q_{\mathrm{rad}}(\boldsymbol{X}) \geq q_{\mathrm{inf}}(\boldsymbol{X})$ follows from Proposition 5.2.1, and the inequality $q_{\mathrm{rad}}(\boldsymbol{X}) \leq q(\boldsymbol{X})$. from Theorem 5.3.1.

The inequality $q_{\mathrm{inf}}(\boldsymbol{X}) \geq q(\boldsymbol{X})$ can be verified as follows. Since the interval of those q for which the embedding $\boldsymbol{l}_q \mapsto \boldsymbol{l}_\infty$ is finitely factorable through $\boldsymbol{X}$ is closed, the embedding $\boldsymbol{l}_{q(X)} \mapsto \boldsymbol{l}_\infty$ is finitely factorable through $\boldsymbol{X}$. Therefore, for each $n \in \mathbf{N}$ there exist $\boldsymbol{x}_1, \ldots, \boldsymbol{x}_n \in \boldsymbol{X}$ such that, for every $(a_i) \subset \mathbf{R}$

$$\frac{1}{2} \sup_i |a_i| \leq \Big\| \sum_{i=1}^n a_i \boldsymbol{x}_i \Big\| \leq \Big(\sum_{i=1}^n |a_i|^{q(\boldsymbol{X})} \Big)^{1/q(\boldsymbol{X})},$$

so that

$$\sup_{\varepsilon_i = \pm 1} \Big\| \sum_{i=1}^n \varepsilon_i \boldsymbol{x}_i \Big\| \leq n^{1/q(\boldsymbol{X})}.$$

Hence, if $q < q(\boldsymbol{X})$, then $\boldsymbol{X}$ cannot be of infratype q. Indeed, were $\boldsymbol{X}$ of infratype q, we would have

$$\frac{1}{2} n^{1/q} \leq \Big(\sum_{i=1}^n \|\boldsymbol{x}_i\|^q \Big)^{1/q} \leq \sup_{\varepsilon_i = \pm 1} \Big\| \sum_{i=1}^n \varepsilon_i \boldsymbol{x}_i \Big\| \leq n^{1/q(\boldsymbol{X})}, \qquad n \in \mathbf{N}.$$

A contradiction. Thus $q(\boldsymbol{X}) \leq q_{\mathrm{inf}}(\boldsymbol{X})$. QED

Finally, the next theorem shows the importance of parameter $q(\boldsymbol{X})$ in finite representablitiy of spaces $\boldsymbol{l}_q$ in $\boldsymbol{X}$.

Theorem 5.3.3.[7] *If $\boldsymbol{X}$ is an infinite-dimensional Banach space, then $\boldsymbol{l}_{q(X)}$ is finitely representable in $\boldsymbol{X}$.*

Proof. For the sake of simplicity, put $q = q(\boldsymbol{X})$ throughout the proof. By Theorem 5.3.1, for each $n \in \mathbf{N}$ there exist $\boldsymbol{x}_1^n, \ldots, \boldsymbol{x}_n^n \in \boldsymbol{X}$ such that , for each $(a_i) \subset \mathbf{R}$,

$$\frac{1}{2} \sup_i |a_i| \leq \Big\| \sum_i a_i \boldsymbol{x}_i^n \Big\| \leq \Big(\sum_i |a_i|^q \Big)^{1/q}. \tag{5.3.4}$$

Let $\mathcal{U}$ be a non-trivial ultrafilter on $\boldsymbol{X}$ (on can use the Banach limits instead). The formula, $\|(a_i)\| := \lim_{n \in \mathcal{U}} \| \sum_i a_i \boldsymbol{x}_i^n \|$, defines

[7]See B. Maurey and G. Pisier (1976), but the proof strongly depends on the result of J.L. Krivine (1976).

a norm on $\mathcal{S}$, say $\|.\|_1$, such that, if $(\boldsymbol{e}_i)$ is the canonical basis of $\mathcal{S}$ then, for each $(a_i) \in \mathcal{S}$,

$$\frac{1}{2}\sup_i |a_i| \leq \Big\|\sum_i a_i \boldsymbol{e}_i\Big\|_1 \leq \Big(\sum_i |a_i|^q\Big)^{1/q}.$$

Utilizing the Brunel-Sucheston procedure from Chapter 1 one can find a subsequence $(\boldsymbol{e}_n^1) \subset (\boldsymbol{e}_n)$ such that the formula,

$$\Big|\Big|\Big|\sum_i a_i \boldsymbol{e}_i\Big|\Big|\Big| := \lim_{i_1 < i_2 < \cdots : i_1 \to \infty} \Big\|\sum a_k \boldsymbol{e}_{i_k}\Big\|_1,$$

defines a new norm of $\mathcal{S}$ which satisfies (5.3.4) and which, additionally, is invariant under spreading, and such that the sequence $\boldsymbol{d}_n = \boldsymbol{e}_{2n} - \boldsymbol{e}_{2n-1}, n = 1, 2, \ldots,$ is invariant under spreading, and unconditional, with constant ≤ 2. By (5.2.4) , for each $(a_i) \subset \mathbf{R}$

$$\frac{1}{2}\sup_i |a_i| \leq \Big|\Big|\Big|\sum_i a_i \boldsymbol{d}_i\Big|\Big|\Big| \leq 2^{1/q}\Big(\sum_i |a_i|^q\Big)^{1/q}. \tag{5.3.5}$$

Denote by $\boldsymbol{Y}$ the Banach space spanned by $(\boldsymbol{d}_n) \subset \mathcal{S}$ in $\|.\|_1$ norm. The sequence $(\boldsymbol{d}_n)$ is an unconditional basis in $\boldsymbol{Y}$, and $\boldsymbol{Y}$ is finitely representable in $\boldsymbol{X}$. Consequently, for each $\epsilon > 0$, the space $\boldsymbol{Y}$ is of cotype $q + \epsilon$. Therefore, there exists a constant C_ϵ such that, for each $(a_i) \in \mathcal{S}$,

$$\|\sum_i a_i \boldsymbol{d}_i\|_1 \leq C_\epsilon^{-1}\Big(\sum_i |a_i|^{q+\epsilon}\Big)1/(q+\epsilon),$$

so that, by (5.3.5), for each $n \in \mathbf{N}$,

$$C_\epsilon^{-1} 2^{n/(q+\epsilon)} \leq \Big\|\sum_{i=1}^{2^n} a_i\Big\|_1 \leq 2^{1/q} \cdot 2^{n/q}.$$

Therefore, if we put

$$b := \inf\Big\{\lambda > 0 : \lim_n \lambda^n \Big\|\sum_{i=1}^{2^n} a_i\Big\|_1 = \infty\Big\},$$

we get that $b = 2^{-1/q}$. Now, by Krivine Theorem of Chapter 1, $\boldsymbol{l}_q$ is finitely representable in $\boldsymbol{Y}$, and thus also in $\boldsymbol{X}$. QED

5.4 Operators in spaces of cotype q

Recall (see Section 1.5) that an operator $U : \boldsymbol{X} \mapsto \boldsymbol{Y}$ is said to be q-absolutely summing if there exists a constant $C > 0$ such that, for each $(\boldsymbol{x}_i) \subset \boldsymbol{X}$,

$$\Big(\sum_i \|U\boldsymbol{x}_i\|^q\Big)^{1/q} \le C \sup_{\|\boldsymbol{x}^*\le 1} \Big(\sum_i |\boldsymbol{x}^*\boldsymbol{x}_i|^q\Big)^{1/q}$$

$$= C \sup_{\|\boldsymbol{x}^*\le 1} \Big\{\Big\|\sum_i a_i\boldsymbol{x}_i\Big\| : a_i \in \mathbf{R}, \sum_i |a_i|^p \le 1, \frac{1}{p} + \frac{1}{q} = 1\Big\}.$$

The space of q-absolutely summing operators is denoted $\Pi_q(\boldsymbol{X}, \boldsymbol{Y})$, and it is a subspace of the space $B(\boldsymbol{X}, \boldsymbol{Y})$ of all bounded operators. The minimal constant C in the above inequality will be denoted $\pi_q(U)$, and it serves as a complete norm on the space $\Pi_q(\boldsymbol{X}, \boldsymbol{Y})$.

Theorem 5.4.1.[8] *Let* $2 < q < \infty$. *Then* $\Pi_q(\boldsymbol{c}_0, \boldsymbol{X}) = B(\boldsymbol{c}_0, \boldsymbol{X})$ *if, and only if,* $q > q(\boldsymbol{X})$, *where the constant* $q(\boldsymbol{X})$ *was defined in Definition 5.3.2. In particular,*

$$q(\boldsymbol{X}) = \inf\{q : \Pi_q(\boldsymbol{c}_0, \boldsymbol{X}) = B(\boldsymbol{c}_0, \boldsymbol{X})\}.$$

Proof. We start with the "only if" implication. Assume that $\Pi_q = B$. We have to prove that the mapping $\boldsymbol{l}_q \mapsto \boldsymbol{l}_\infty$ is not finitely factorable through $\boldsymbol{X}$ (since $\boldsymbol{l}_{q(X)} \mapsto \boldsymbol{l}_\infty$ is finitely factorable through $\boldsymbol{X}$).

Suppose, to the contrary, that $\boldsymbol{l}_q \mapsto \boldsymbol{l}_\infty$ is finitely factorable through $\boldsymbol{X}$. Let (θ_i) be a sequence of i.i.d. stable random variables with the characteristic function $\mathbf{E}\exp[it\theta_i] = \exp[-|t|^p]$, and $\Omega = [0,1], 1/p + 1/q = 1$. Then, in view of the Schwartz Theorem of Chapter 1, there exists a constant C such that

$$C^{-1}(\log n)^{1/p} \le \mathbf{E}\Big(n^{-1}\sum_{i=1}^n |\theta_i|^p\Big)^{1/p}, \qquad n \in \mathbf{N}.$$

[8]See, B. Maurey and G. Pisier (1976).

For each $n \in \mathbf{N}$, define the operator,

$$Z_n : \boldsymbol{C}[0,1] \ni g \mapsto Z_n(g) = \big(\mathbf{E}(g\theta_i)\big)_{i=1}^n \in \boldsymbol{l}_q^{(n)}.$$

Denote by J_n the canonical embedding $\boldsymbol{l}_q^{(n)} \mapsto \boldsymbol{l}_\infty^{(n)}$, and set $U_n = J_n \circ Z_n$. Now, it is sufficient to prove that $\pi_q(U_n)$ tends to infinity as $n \to \infty$, whereas $\|Z_n\|$ remains bounded. Indeed, because $\boldsymbol{l}_q \mapsto \boldsymbol{l}_\infty$ is finitely factorable through $\boldsymbol{X}$, for each $n \in \mathbf{N}$ there exists a factorization $J_n = W_n \circ V_n$, with $V_n \in B(\boldsymbol{l}_q^{(n)}, \boldsymbol{X}), W_n \in B(\boldsymbol{X}, \boldsymbol{l}_\infty^{(n)})$, and $\|V_n\|, \|W_n\| \leq 2$. Then, since $\pi_q(U_n) \leq 2\pi_q(V_n \circ Z_n)$, and $\|V_n \circ Z_n\| \leq 2\|Z_n\|$, we would have

$$\lim_n \pi_q(V_n \circ Z_n) = \infty, \qquad \text{and} \qquad \sup_n \|V_n \circ Z_n\| < \infty,$$

and this would contradict the equality, $\Pi_q(\boldsymbol{C}[0,1], \boldsymbol{X}) = B(\boldsymbol{C}[0,1], \boldsymbol{X})$, in view of the Closed Graph Theorem.

Now, the boundedness of $\|Z_n\|$ follows, by transposition, from the equality $(\mathbf{E}|\sum_n a_n\theta_n|^r)^{1/r} = C^r(\sum_n |a_n|^p)^{1/p}$, and we can prove that $\pi_q(U_n) \to \infty$ as follows:

$$C^{-1}(\log n)^{1/p} \leq \mathbf{E}\Big(n^{-1}\sum_{i=1}^n |\theta_i|^p\Big)^{1/p}$$

$$= \sup\Big\{\mathbf{E}\sum_{i=1}^n n^{-1/p} g_i\theta_i : g_i \in \boldsymbol{C}[0,1], \sum_i |g_i(t)|^q \leq 1\Big\}$$

$$\leq \sup\Big\{\Big(\sum_{i=1}^n \mathbf{E} g_i\theta_i|^q\Big)^{1/q} : \sum_i |g_i(t)|^q \leq 1\Big\}$$

$$\leq \sup\Big\{\Big(\sum_{i=1}^n \|U_n g\|^q\Big)^{1/q} : \sum_i |g_i(t)|^q \leq 1\Big\} \leq \pi_q(U).$$

The proof of the "if" part is more straightforward. Indeed, if $q > p > q(\boldsymbol{X})$ then, by Theorem 5.3.2, $\boldsymbol{X}$ is of infracotype p, which means that the identity mapping, $\boldsymbol{X} \mapsto \boldsymbol{X}$, is $(p,1)$-summing. Thus, $\Pi_{p,1}(\boldsymbol{c}_0, \boldsymbol{X}) = B(\boldsymbol{c}_0, \boldsymbol{X})$ and, because for $q > p$ we have the inclusion $\Pi_q(\boldsymbol{c}_0, \boldsymbol{X}) \supset \Pi_{p,1}(\boldsymbol{c}_0, \boldsymbol{X})$ (see Chapter 1), we get the desired equality. QED

Corollary 5.4.1 *(i) Let* $2 < q < \infty$. *The canonical embedding* $\boldsymbol{l}_q \mapsto \boldsymbol{l}_\infty$ *is finitely factorable through* $\boldsymbol{X}$ *if, and only if,* $\Pi_q(\boldsymbol{c}_0, \boldsymbol{X}) \neq B(\boldsymbol{c}_0, \boldsymbol{X})$.

(ii) Let, again, $2 < q < \infty$. *If* $\Pi_q(\boldsymbol{c}_0, \boldsymbol{X}) = B(\boldsymbol{c}_0, \boldsymbol{X})$, *then there exists a* $p < q$ *such that* $\Pi_p(\boldsymbol{c}_0, \boldsymbol{X}) = B(\boldsymbol{c}_0, \boldsymbol{X})$.

Proof. The Corollary follows immediately from the above Theorem, and the fact that the set of q's for which the embedding $\boldsymbol{l}_q \mapsto \boldsymbol{l}_\infty$ is finitely factorable through $\boldsymbol{X}$ is closed (see, Chapter 1).

Remark 5.4.1. (a) By duality, $\Pi_q(\boldsymbol{c}_0, \boldsymbol{X}) = B(\boldsymbol{c}_0, \boldsymbol{X})$ if, and only if, for any Banach space $\boldsymbol{Y}$, $\Pi_p(\boldsymbol{X}, \boldsymbol{Y}) = \Pi_1(\boldsymbol{X}, \boldsymbol{Y})$, with $1/p + 1/q = 1$.

(b) In the equality $\Pi_2(\boldsymbol{c}_0, \boldsymbol{X}) = B(\boldsymbol{c}_0, \boldsymbol{X})$ on can replace $\boldsymbol{c}_0$ by $\boldsymbol{l}_\infty$, or $\boldsymbol{C}[0,1]$, and get equivalent results.

Theorem 5.4.2.[9] *The following properties of a Banach space* $\boldsymbol{X}$ *are equivalent:*

(i) There exists a $q < \infty$ *such that* $\Pi_q(\boldsymbol{c}_0, \boldsymbol{X}) = B(\boldsymbol{c}_0, \boldsymbol{X})$;

(ii) There exists a $q < \infty$ *such that* $\boldsymbol{X}$ *is of cotype* q;

(iii) The space $\boldsymbol{c}_0$ *is not finitely representable in* $\boldsymbol{X}$.

Proof. *(i)* $\Longrightarrow$ *(ii)* By Theorem 5.4.1, if $\Pi_q(\boldsymbol{c}_0, \boldsymbol{X}) = B(\boldsymbol{c}_0, \boldsymbol{X})$ then $q > q(\boldsymbol{X})$, so that $\boldsymbol{X}$ is of cotype q, by Theorem 5.3.2.

(ii) $\Longrightarrow$ *(iii)* Cotype q is a superproperty, and we know, in view of Example 5.2.1, that $\boldsymbol{c}_0$ is not of cotype q for any $q < \infty$. Therefore, a space in which $\boldsymbol{c}_0$ is finitely representable cannot be of cotype q for any $q < \infty$.

(iii) $\Longrightarrow$ *(i)* By Corollary 5.4.1(i), if $\Pi_q(\boldsymbol{c}_0, \boldsymbol{X}) \neq B(\boldsymbol{c}_0, \boldsymbol{X})$ then, for each $q < \infty$, the embedding $\boldsymbol{l}_q \mapsto \boldsymbol{l}_\infty$ is finitely factorable through $\boldsymbol{X}$. Because the interval of such q's is closed (see, Chapter 1), also the identity mapping, $\boldsymbol{l}_\infty \mapsto \boldsymbol{l}_\infty$, is finitely factorable through $\boldsymbol{X}$, that is, $\boldsymbol{c}_0$ is finitely representable in $\boldsymbol{X}$. QED

Remark 5.4.2. It is easy to see that the following statement is also equivalent to the statements *(i)* – *(iii)* of the above Theorem:

(iv) There exists a $q < \infty$ such that $\boldsymbol{X}$ is of infracotype q.

---

[9]See B. Maurey and G. Pisier (1976).

In the case of spaces of cotype 2 we also have the following result which is only a one-sided implication:

Theorem 5.4.3.[10] *If $\boldsymbol{X}$ is of cotype 2, then* $\Pi_2(\boldsymbol{c}_0, \boldsymbol{X}) = B(\boldsymbol{c}_0, \boldsymbol{X})$.

Proof. Asssume that $\Pi_2(\boldsymbol{c}_0, \boldsymbol{X}) \neq B(\boldsymbol{c}_0, \boldsymbol{X})$. Then, by a duality argument, $\Pi_1(\boldsymbol{X}^*, \boldsymbol{l}_1) \neq \Pi_2(\boldsymbol{X}^*, \boldsymbol{l}_1)$, and there exists an operator $W : \boldsymbol{X}^* \mapsto \boldsymbol{L}_1$ which is not factorable through any Hilbert space. Now, consider two cases.

Case 1. There exists a $p \in (1, 2)$ such that $W = U \circ V$, where $U : \boldsymbol{X}^* \mapsto \boldsymbol{L}_p$, and $V : \boldsymbol{L}_p \mapsto \boldsymbol{l}_1$. Because $p < 2$, and $q > 2$, the space $\boldsymbol{L}_q$ is of Rademacher type 2, and $\boldsymbol{X}^{**}$ is of cotype 2. Therefore, (see Remark 7.6.1 in Chapter 7) by duality, U admits a factorization through a Hilbert space. A contradiction.

Case 2. If $W : \boldsymbol{X}^* \mapsto \boldsymbol{l}_1$ does not factor through any $\boldsymbol{L}_p$ with $1 < p < 2$, then, by Maurey-Rosenthal Theorem (see, Chapter 1), for each $p \in (1, 2)$, and each $n \in \mathbf{N}$, there exist $\boldsymbol{x}_1^{**}, \dots, \boldsymbol{x}_n^{**} \in S_{\boldsymbol{X}^{**}}$ such that, for any $(a_i) \subset \mathbf{R}$,

$$\Big\| \sum_{j=1}^{n} a_j \boldsymbol{x}_j^{**} \Big\| \leq 2 \Big(\sum_{j=1}^{n} |a_j|^q \Big)^{1/q}, \qquad \frac{1}{p} + \frac{1}{q} = 1.$$

Therefore, were $\boldsymbol{X}$ of cotype 2, we would have the contradictory inequalities

$$cn^{1/2} \leq \mathbf{E} \Big\| \sum_{j=1}^{n} r_j \boldsymbol{x}^{**} \Big\| \leq 2n^{1/q}, \qquad n \in \mathbf{N}. \qquad \text{QED}$$

However, in spaces with unconditional basis we also have the reverse implication.

Theorem 5.4.4.[11] *Let $\boldsymbol{X}$ be a Banach space with an unconditional basis $(\boldsymbol{e}_n)$. Then $\boldsymbol{X}$ is of cotype 2 if, and only if,* $\Pi_2(\boldsymbol{c}_0, \boldsymbol{X}) = B(\boldsymbol{c}_0, \boldsymbol{X})$.

---

[10] See E. Dubinsky, A. Pelczynski and H.P. Rosenthal (1972).

[11] This result is valid even for $\boldsymbol{X}$ with local unconditional structure, see B. Maurey (1973/74), Exp. XIV-XV. Moreover, one can prove that, if $\boldsymbol{X}$ has a local unconditional structure, then $\boldsymbol{X}$ is of type 2 if, and only if, $\Pi_2(\boldsymbol{c}_0, \boldsymbol{X}^*) = B(\boldsymbol{c}_0, \boldsymbol{X}^*)$, and $\boldsymbol{l}_1$ is not finitely representable in $\boldsymbol{X}^*$.

Proof. In view of Theorem 5.4.3 it suffices to prove only the "if" part of the above statement. Denote by $(\boldsymbol{e}_n^*)$ the sequence of coefficient functionals of $(\boldsymbol{e}_n)$. It is known[12] that the equality $\Pi_2 = B$ implies that there exists a constant C such that, for all $n \in \mathbf{N}$, and all $(\boldsymbol{x}_i) \subset \boldsymbol{X}$,

$$\Big\|\sum_{n=1}^{\infty}\Big(\sum_{j=1}^{m}|\boldsymbol{e}_n^*\boldsymbol{x}_j|^2\Big)^{1/2}\boldsymbol{e}_n\Big\| \geq C\Big(\sum_{j=1}^{m}\|\boldsymbol{x}_j\|^2\Big)^{1/2}.$$

Since the basis $(\boldsymbol{e}_n)$ is unconditional, and the real line $\mathbf{R}$ is of cotype 2, there exists a constant A such that

$$\mathbf{E}\Big\|\sum_{j=1}^{m} r_j\boldsymbol{x}_j\Big\| = \mathbf{E}\Big\|\sum_{n=1}^{\infty}\sum_{j=1}^{m} r_j\boldsymbol{e}_n^*\boldsymbol{x}_j\boldsymbol{e}_n\Big\|$$

$$\geq A\mathbf{E}\Big\|\sum_{n=1}^{\infty}\Big|\sum_{j=1}^{m} r_j\boldsymbol{e}_n^*\boldsymbol{x}_j\Big|\boldsymbol{e}_n\Big\| \geq A\Big\|\sum_{n=1}^{\infty}\mathbf{E}\Big|\sum_{j=1}^{m} r_j\boldsymbol{e}_n^*\boldsymbol{x}_j\Big|\boldsymbol{e}_n\Big\|$$

$$\geq A^2\Big\|\sum_{n=1}^{\infty}\Big(\sum_{j=1}^{m}|\boldsymbol{e}_n^*\boldsymbol{x}_j|^2\Big)^{1/2}\boldsymbol{e}_n\Big\| \geq A^2C\Big(\sum_{j=1}^{m}\|\boldsymbol{x}_j\|^2\Big)^{1/2},$$

so that $\boldsymbol{X}$ is of cotype 2. QED

Remark 5.4.3. By duality, if $\boldsymbol{X}$ is of cotype 2 (and also "only if", in the case $\boldsymbol{X}$ when has an unconditional basis) then $\Pi_1(\boldsymbol{X}, \boldsymbol{Y}) = B(\boldsymbol{X}, \boldsymbol{Y})$, for any Banach space $\boldsymbol{Y}$.

The above analysis of absolutely summing operators was conducted in the context of Rademacher random multipliers. Next, we will conduct a similar analysis in the case when the multipliers (γ_i) form a sequence of independent and identically distributed Gaussian random variables with the standard $N(0,1)$ distribution .

Definition 5.4.1. The operator $U : \boldsymbol{X} \mapsto \boldsymbol{Y}$ is said to be *G-absolutely summing* if there exists a constant $C > 0$ such that, for each $n \in \mathbf{N}$, and any $(\boldsymbol{x}_i) \subset \boldsymbol{X}$,

$$\Big(\mathbf{E}\Big\|\sum_{i=1}^{n}\gamma_i U\boldsymbol{x}_i\Big\|^2\Big)^{1/2} \leq C\sup_{\|\boldsymbol{x}^*\|\leq 1}\Big(\sum_{i=1}^{n}(\boldsymbol{x}^*\boldsymbol{x}_i)^2\Big)^{1/2}.$$

[12] See E. Dubinsky, A. Pelczynski and H.P.Rosenthal (1972).

The space of G-absolutely summing operators will be denoted by $\Pi_G(\boldsymbol{X},\boldsymbol{Y})$, and the smallest C in the above inequality will be denoted by $\pi_G(U)$. Note that $\pi_G(U)$ is the complete norm on the space $\Pi_G(\boldsymbol{X},\boldsymbol{Y})$.

Theorem 5.4.5.[13] *The following properties of a Banach space* $\boldsymbol{X}$ *are equivalent:*

(i) The space $\boldsymbol{X}$ *is of cotype 2;*

(ii) For any Banach space $\boldsymbol{Y}$, $\Pi_2(\boldsymbol{Y},\boldsymbol{X}) = \Pi_G(\boldsymbol{Y},\boldsymbol{X})$*;*

(iii) If $\boldsymbol{H}$ *is a Hilbert space, then* $\Pi_2(\boldsymbol{H},\boldsymbol{X}) = \Pi_G(\boldsymbol{H},\boldsymbol{X})$.

Proof. The implication $(i) \implies (ii)$ follows from Corollary 5.2.1, and the implication $(ii) \implies (iii)$ is evident. So, let us prove the implication $(iii) \implies (i)$.

$(iii) \implies (i)$ We begin with checking that, for every finite dimensional $U : \boldsymbol{H} \mapsto \boldsymbol{X}$,

$$\pi_G(U) = \Big(\int_H \|Uh\|^2 m_H(dh)\Big)^{1/2}, \tag{5.4.1}$$

where m_H is the canonical Gaussian cylindrical measure on $\boldsymbol{H}$ with the characteristic functional

$$\int_H \exp[i(\boldsymbol{h}^*,\boldsymbol{h})]m_H(d\boldsymbol{h}) = \exp[-(1/2)\|\boldsymbol{h}^*\|^2], \qquad \boldsymbol{h},\boldsymbol{h}^* \in \boldsymbol{H}.$$

Indeed, let $U\boldsymbol{h} = \sum_{i=1}^n (\boldsymbol{e}_i,\boldsymbol{h})\boldsymbol{x}_i$, where $(\boldsymbol{e}_i)$ is an orthonormal basis in $\boldsymbol{H}$, and $\boldsymbol{x}_1,\dots,\boldsymbol{x}-n \in \boldsymbol{X}$. Denote by $P_n : \boldsymbol{H} \mapsto \mathbf{R}^n$ the operator of projection on the first n coordinates. Then $P_n^* : \mathbf{R}^n \mapsto \boldsymbol{H}$, with $\|P_n\| = \|P_n^*\| = 1$, $U = UP^*P$, and, by the definition of $\pi_G(U)$,

$$\Big(\int_H \|U\boldsymbol{h}\|^2 m_H(d\boldsymbol{h})\Big)^{1/2} = \Big(\int_{\mathbf{R}^n} \|UP^*\boldsymbol{s}\|^2 (m_H P_n^{-1})(d\boldsymbol{s})\Big)^{1/2}$$

$$= \Big(\int_{\mathbf{R}^n} \Big\|\sum_{i=1}^n \boldsymbol{s}_i U\boldsymbol{e}_i\Big\|^2 (m_H P_n^{-1})(d\boldsymbol{s})\Big)^{1/2} \le \pi_G(U),$$

where $\boldsymbol{s} = (s,\dots,s_n) \in \mathbf{R}^n$.

[13] See V. Linde and A. Pietsch (1974).

To prove the converse inequality let us define, for each $(\boldsymbol{h}_i) \subset \boldsymbol{H}$, the operator,

$$V : \mathbf{R}^k \ni \boldsymbol{t} = (t_1, \ldots, \boldsymbol{t}_k) \mapsto V\boldsymbol{t} = \sum_{i=1}^{k} t_i \boldsymbol{h}_i \in \boldsymbol{H}.$$

Since

$$\|V\| = \sup\Big\{ \Big(\sum_{i=1}^{k} (\boldsymbol{h}_i, \boldsymbol{h})^2 \Big)^{1/2} : \|\boldsymbol{h}\| \leq 1 \Big\},$$

we have

$$\Big(\int_{\mathbf{R}^k} \Big\| \sum_{i=1}^{k} t_i U \boldsymbol{h}_i \Big\|^2 (m_H P_n^{-1})(d\boldsymbol{t}) \Big)^{1/2}$$

$$= \Big(\int_{\mathbf{R}^k} \| \, U P_n^* P_n V \boldsymbol{t} \|^2 (m_H P_n^{-1})(d\boldsymbol{t}) \Big)^{1/2}$$

$$\leq \|P_n V\| \int_{\mathbf{R}^k} \|U P_n^* \boldsymbol{s}\|^2 (m_H P_n^{-1})(d\boldsymbol{t}) \Big)^{1/2}$$

$$\leq \Big(\int_H \|U\boldsymbol{h}\|^2 m_H(\boldsymbol{h}) \Big)^{1/2} \sup\Big\{ \Big(\sum_{i=1}^{n} (\boldsymbol{h}_i, \boldsymbol{h})^2 \Big)^{1/2} : \|\boldsymbol{h}\| \leq 1 \Big\},$$

so that

$$\pi_G(U) \leq \Big(\int_H \|U\boldsymbol{h}\|^2 m_H(d\boldsymbol{h}) \Big)^{1/2},$$

which proves (5.4.1).

Now, assume that $\Pi_2(\boldsymbol{H}, \boldsymbol{X}) = \Pi_G(\boldsymbol{H}, \boldsymbol{X})$. By the Closed Graph Theorem, there exists a constant $C \geq 0$, such that

$$\pi_2(U) \leq C \pi_G(U), \qquad U \in \Pi_G(\boldsymbol{H}, \boldsymbol{X}).$$

For each $(\boldsymbol{x}_i) \subset \boldsymbol{X}$, define the operator,

$$U_0 : \mathbf{R}^n \ni \boldsymbol{t} \mapsto \sum_{i=1}^{,} t_i \boldsymbol{x}_i \in \boldsymbol{X}.$$

By (5.4.1)

$$\pi_G(U_0) = \Big(\int_{R^n} \Big\| \sum_{i=1}^{n} t_i \boldsymbol{x}_i \Big\|^2 (m_H P_n^{-1})(d\boldsymbol{t}) \Big)^{1/2}.$$

Therefore,

$$\Big(\sum_{i=1}^{n}\|\boldsymbol{x}_i\|^2\Big)^{1/2}\le \pi_2(U_0P_n)\le C\pi_G(U_0P_n)$$

$$=C\Big(\int_{R^n}\Big\|\sum_{i=1}^{n}t_i\boldsymbol{x}_i\Big\|^2(m_HP_n^{-1})(d\boldsymbol{t})\Big)^{1/2}=C\Big(\mathbf{E}\Big\|\sum_{i=1}^{n}\gamma_i\boldsymbol{x}_i\Big\|^2\Big)^{1/2},$$

that is, $\boldsymbol{X}$ is of cotype 2. QED

In the next section we will address the issue of convergence of vector series in Banach spaces with Rademacher and Gaussian coefficients, and its relationship to the geometric structure of those Banach spaces.

5.5 Random series and law of large numbers

The next result provides necessary conditions for the convergence of a series of independent random vectors with values in Banach spaces of cotype q.

Theorem 5.5.1. *The following properties of a Banach space* $\boldsymbol{X}$ *are equivalent:*

(i) The space $\boldsymbol{X}$ *is of cotype* q*;*

(ii) If the series $\sum_i r_i\boldsymbol{x}_i$, $(\boldsymbol{x}_i)\subset\boldsymbol{X}$, *converges a.s. (or, in* $\boldsymbol{L}_q(\boldsymbol{X})$*), then* $\sum_i\|\boldsymbol{x}_i\|^q<\infty$*;*

(iii) For each sequence (X_i) *of independent, zero-mean random vectors in* $\boldsymbol{X}$*, if* $\sum_i X_i$ *converges in* $\boldsymbol{L}_q(\boldsymbol{X})$*, then* $\sum_i\mathbf{E}\|X_i\|^q<\infty$*.*

Proof. *(iii)* $\Longrightarrow$ *(ii)* The $\boldsymbol{L}_q$ convergence part is evident, and the almost sure convergence follows from the Ito-Nisio Theorem (see Chapter 1).

(ii) $\Longrightarrow$ *(i)* This implication follows from the Closed Graph Theorem.

(i) $\Longrightarrow$ *(iii)* This implication follows directly from Corollary 5.2.1. QED

Random series in spaces of cotype q, for some $q < \infty$, that is spaces in which $\boldsymbol{c}_0$ is not finitely representable (see Theorem 5.4.2), enjoy a number of other remarkable properties.

Theorem 5.5.2.[14] *Let (γ_i) be a sequence if i.i.d. $N(0,1)$ Gaussian random variables. The following properties of a Banach space $\boldsymbol{X}$ are equivalent:*

(i) $\boldsymbol{c}_0$ is not finitely representable in $\boldsymbol{X}$;

(ii) There exists a constant $C > 0$ such that, for each $n \in \mathbf{N}$, and each finite set $(\boldsymbol{x}_i) \subset \boldsymbol{X}$,

$$\left(\mathbf{E}\left\|\sum_i \gamma_i \boldsymbol{x}_i\right\|^2\right)^{1/2} \leq C\left(\mathbf{E}\left\|\sum_i r_i \boldsymbol{x}_i\right\|^2\right)^{1/2};$$

(iii) If the series $\sum_i r_i \boldsymbol{x}_i$, $(\boldsymbol{x}_i) \subset \boldsymbol{X}$, converges a.s. then the series $\sum_i \gamma_i \boldsymbol{x}_i$ also converges a.s.

Proof. $(i) \implies (ii)$ Because $\boldsymbol{c}_0$ is not finitely representable in $\boldsymbol{X}$, and because $q(\boldsymbol{X}) = q(\boldsymbol{L}_2(\boldsymbol{X}))$ (see, Theorem 5.2.2), it follows from Theorem 5.4.2 that there exists a $q, 2 < q < \infty$, such that $\Pi_q(\boldsymbol{c}_0, \boldsymbol{L}_2(\boldsymbol{X})) = B(\boldsymbol{c}_0, \boldsymbol{L}_2(\boldsymbol{X}))$. Therefore, by the Closed Graph Theorem, there exist a $K > 0$ such that

$$\pi_q(U) \leq K\|U\|, \qquad U \in B(\boldsymbol{c}_0, \boldsymbol{L}_2(\boldsymbol{X})).$$

Let $\boldsymbol{x}_1, \ldots, \boldsymbol{x}_n \in \boldsymbol{X}$, and define the operator,

$$U : \boldsymbol{c}_0 \ni (c_i) \mapsto U((c_i)) = \sum_{i=1}^n r_i c_i \boldsymbol{x}_i \in \boldsymbol{L}_2(\boldsymbol{X}).$$

By the Pietsch Factorization Theorem (see, Chapter 1), there exists an $(a_n) \subset \mathbf{R}^+$, with $\sum_n a_n = 1$, such that for any $(c_i) \in \boldsymbol{c}_0$

$$\|U((c_i))\| \leq \pi_q(U)\Big(\sum_i a_i|c_i|^q\Big)^{1/q} \leq K\|U\|\Big(\sum_i a_i|c_i|^q\Big)^{1/q}.$$

Therefore, after integration,

$$\left(\mathbf{E}_\gamma \mathbf{E}_r\left\|\sum_i r_i\gamma_i \boldsymbol{x}_i\right\|^2\right)^{1/2} \leq K\|U\|\left(\mathbf{E}_\gamma\Big(\sum_i a_i|\gamma_i|^q\Big)^{2/q}\right)1/2$$

[14]See, B. Maurey and G. Pisier (1976).

$$\leq K\|U\|\Big(\mathbf{E}_\gamma\sum_i a_i|\gamma_i|^q\Big)^{1/q} \leq K\|U\|\Big(\mathbf{E}|\gamma|^q\Big)^{1/q}.$$

On the other hand, (r_i), and $(r_i\gamma_i)$, are identically distributed, so that

$$\Big(\mathbf{E}\Big\|\sum_i \gamma_i \boldsymbol{x}_i\Big\|^2\Big)^{1/2} = \Big(\mathbf{E}\Big\|\sum_i r_i\gamma_i \boldsymbol{x}_i\Big\|^2\Big)^{1/2} \leq K\|U\|\Big(\mathbf{E}|\gamma|^q\Big)^{1/q}.$$

Now, by the Contraction Principle (see Chapter 1), we have

$$\|U\| = \sup\Big\{\Big(\mathbf{E}\Big\|\sum_i c_i r_i \boldsymbol{x}_i\Big\|^2\Big)^{1/2} : |c_i| \leq 1\Big\} \leq \Big(\mathbf{E}\Big\|\sum_i r_i \boldsymbol{x}_i\Big\|^2\Big)^{1/2},$$

which completes the proof of the implication $(i) \implies (ii)$.

$(ii) \implies (iii)$ This implication is evident if one takes into account the Ito-Nisio Theorem (see Chapter 1), the Closed Graph Theorem, and the fact that the a.s. convergent series, $\sum_i r_i\boldsymbol{x}_i$, and $\sum_i \gamma_i\boldsymbol{x}_i$, have all the moments finite.

$(ii) \implies (i)$ If $\boldsymbol{c}_0$ is finitely representable in $\boldsymbol{X}$ then, for each $n \in \mathbf{N}$, there exists $(\boldsymbol{x}_i) \subset \boldsymbol{X}$ such that, for every $(a_i) \subset \mathbf{R}$,

$$\frac{1}{2}\sup_i |a_i| \leq \Big\|\sum_{i=1}^n a_i\boldsymbol{x}_i\Big\| \leq \sup_i |a_i|.$$

Therefore, for each $n \in \mathbf{N}$,

$$\frac{1}{4}\mathbf{E}\sup_{1\leq i\leq n}|\gamma_i|^2 \leq \mathbf{E}\Big\|\sum_{i=1}^n \gamma_i\boldsymbol{x}_i\Big\|^2,$$

and

$$\mathbf{E}\Big\|\sum_{i=1}^n r_i\boldsymbol{x}_i\Big\|^2 \leq 1.$$

Thus (ii) is violated because, by the Borel-Cantelli Lemma, $\sup_{1\leq i\leq n}|\gamma_i|^2$ goes almost surely to $+\infty$, as $n \to \infty$, and in view of the fact that the tail of the distribution of γ_i behaves like $t\exp[-t^2]$. QED

Remark 5.5.1. (a) It follows immediately from Corollary 5.4.1 that the statements $(i)-(iii)$ of the above Theorem are also equivalent to the following statement:

(iv) For each sequence (ξ_n) of independent, zero-mean real random variables which are uniformly bounded in $\boldsymbol{L}_q(\Omega, \Sigma, \mathbf{P})$, for each $q < \infty$, and for each $(\boldsymbol{x}_n) \subset \boldsymbol{X}$ such that $\sum_n r_n \boldsymbol{x}_n$ converges a.s., also $\sum_n \xi_n \boldsymbol{x}_n$ converges a.s.

(b) The above Theorem also gives the following alternative characterization of spaces of cotype q:

$\boldsymbol{X}$ is of cotype q if, and only if, for each $p, 0 < p < \infty$, there exists a constant $C > 0$ such that, for every $n \in \mathbf{N}$, and each sequence $(\boldsymbol{x}_i) \subset \boldsymbol{X}$,

$$\Big(\sum_i \|\boldsymbol{x}_i\|^q\Big)^{1/q} \leq C\Big(\mathbf{E}\Big\|\sum_i \gamma_i \boldsymbol{x}_i\Big\|^p\Big)^{1/p}.$$

Now, we turn to the investigation of Gaussian measures on spaces that are of cotype q, for some $q < \infty$, and which, additionally, have an unconditional basis $(\boldsymbol{e}_n)$ with the coordinate functionals $(\boldsymbol{e}^*) \subset \boldsymbol{X}^*$. Recall that, for any random vector X (always zero-mean) in $\boldsymbol{X}$ such that $\mathbf{E}(\boldsymbol{x}^* X)^2 < \infty$, $\boldsymbol{x}^* \in \boldsymbol{X}^*$, there exists a symmetric (i.e., $\boldsymbol{x}^* R \boldsymbol{y}^* = \boldsymbol{y}^* R \boldsymbol{x}^*$), and positive (i.e., $\boldsymbol{x}^* R \boldsymbol{x}^* \geq 0$) *covariance operator* $R_X : \boldsymbol{X}^* \mapsto \boldsymbol{X}$.

Theorem 5.5.3.[15] *The following properties of a Banach space $\boldsymbol{X}$ with an unconditional basis $(\mathbf{e}_n)$ are equivalent:*

(i) The space $\mathbf{c}_0$ is not finitely representable in $\boldsymbol{X}$;

(ii) A positive, symmetric operator $R \in B(\boldsymbol{X}^*, \boldsymbol{X})$ is the covariance operator of a Gaussian measure on $\boldsymbol{X}$ if, and only if, the series $\sum_n (\mathbf{e}_n^* R \mathbf{e}_n^*)^{1/2} \mathbf{e}_n$ converges in $\boldsymbol{X}$.

Proof. First, observe that the convergence of $\sum_n (\boldsymbol{e}_n^* R \boldsymbol{e}_n^*)^{1/2} \boldsymbol{e}_n$ is a necessary condition in any Banach space with an unconditional basis. Indeed, we can assume (up to an equivalent renorming) that $\|.\|$ does not change under sign changes of coordinates. Therefore,

[15] See D. Mustari (1973), but the proof provided below is due to V. Mandrekar (1977).

if X is a Gaussian random vector in $\boldsymbol{X}$, then $\mathbf{E}\|\sum_n |\boldsymbol{e}_n^* X|\boldsymbol{e}_n\| = \mathbf{E}\|X\| < \infty$, so that

$$\mathbf{E}\sum_n |\boldsymbol{e}_n^* X|\boldsymbol{e}_n = \left(\frac{2}{\pi}\right)^{1/2} \sum_n (\boldsymbol{e}_n^* R\boldsymbol{e}_n^*)^{1/2}\boldsymbol{e}_n,$$

has to converge in $\boldsymbol{X}$.

Now, let $R : \boldsymbol{X}^* \overset{A}{\mapsto} \boldsymbol{H} \overset{A^*}{\mapsto} \boldsymbol{X}$, be the standard factorization of R. The convergence of $\sum_n (\boldsymbol{e}_n^* R\boldsymbol{e}_n^*)^{1/2}\boldsymbol{e}_n$ is equivalent to the convergence of $\sum_n \|A\boldsymbol{e}_n^*\|\boldsymbol{e}_n$. First, we shall prove that we have a factorization,

$$A : \boldsymbol{X}^* \overset{A_1}{\mapsto} \boldsymbol{l}_1 \overset{A_2}{\mapsto} \boldsymbol{H}, \tag{5.5.1}$$

such that $A_1^* : \boldsymbol{l}_\infty \mapsto \boldsymbol{X}$, and $\|A_2\| \leq 1$, which, in particular, would show that $A \in \Pi_1(\boldsymbol{X}^*, \boldsymbol{H})$. Indeed, because $A^*\boldsymbol{H} \subset \boldsymbol{X}$, A is a continuous operator from $(\boldsymbol{X}^*, \sigma(\boldsymbol{X}^*, \boldsymbol{X}))$ into $(\boldsymbol{H}, \sigma(\boldsymbol{H}, \boldsymbol{H}))$. Since $(\boldsymbol{e}_n)$ is unconditional, for each $\boldsymbol{x}^* \in \boldsymbol{X}^*$, the series $\boldsymbol{x}^* = \sum_n (\boldsymbol{x}^*\boldsymbol{e}_n)\boldsymbol{e}^*$ converges unconditionally in the weak topology $\sigma(\boldsymbol{X}^*, \boldsymbol{X})$. So $A\boldsymbol{x}^* = \sum_n (\boldsymbol{x}^*\boldsymbol{e}_n)A\boldsymbol{e}_n^*$, where the series converges unconditionally in the weak topology, and the metric topology on $\boldsymbol{H}$. Define,

$$A_1\boldsymbol{x}^* := \big((\boldsymbol{x}^*\boldsymbol{e}_n)\|A\boldsymbol{e}_n^*\|\big)_{n\in\mathbf{N}} \in \boldsymbol{l}_1, \qquad \boldsymbol{x}^* \in \boldsymbol{X}^*,$$

and

$$A_2(a_n) = \sum_{n-1}^{\infty} a_n\boldsymbol{h}_n, \qquad (a_n) \in \boldsymbol{l}_1,$$

where $\boldsymbol{h}_n = A\boldsymbol{e}_n^*/\|A\boldsymbol{e}_n^*\|$, if $A\boldsymbol{e}_n^* \neq 0$, and 0, otherwise. Evidently, $A_1^*(\boldsymbol{l}_\infty) \subset \boldsymbol{X}$, and $\|A_2(a_n)\| \leq \sum_n |a_n|$. This shows the validity of (5.5.1).

Since $A_1^* \in B(\boldsymbol{l}_\infty, \boldsymbol{X})$, by Theorem 5.4.2, $A_1^* \in \Pi_q(l_\infty, \boldsymbol{X})$, for some $q, 1 < q < \infty$, so that, also, $A^* \in \Pi_q(\boldsymbol{H}, \boldsymbol{X})$. Therefore $m_H(A^*)^{-1}$, where

$$\int_H \exp[i(\boldsymbol{h}^*, \boldsymbol{h})]m_H(\boldsymbol{h}) = \exp[(-1/2)\|\boldsymbol{h}^*\|^2],$$

is σ-additive, and its extension has covariance R.

(*ii*) $\implies$ (*i*) In view of Theorem 5.5.2 it suffices to show that $\sum_i \gamma_i \boldsymbol{x}_i$ converges a.s. in $\boldsymbol{X}$, whenever $\sum_i r_i \boldsymbol{x}_i$ does. By Ito-Nisio, and Kahane Theorems (see, Chapter 1), if $Y = \sum_i r_i \boldsymbol{x}_i$ converges a.s., then it also converges in $\boldsymbol{L}_2(\boldsymbol{X})$, and $\mathbf{E}\|\sum_i r_i \boldsymbol{x}_i\|^2 < \infty$. Thus, arguing as in the proof of the implication (*i*) $\implies$ (*ii*), we get that $\sum_n (\boldsymbol{e}_n^* R_Y \boldsymbol{e}_n^*)\boldsymbol{e}_n$ converges in $\boldsymbol{X}$. By assumption, there exists a Gaussian random vector with the covariance operator R_Y. Checking its finite-dimensional distributions we conclude that it has to be $\sum_i \gamma_i \boldsymbol{x}_i$, the series being convergent in distribution. However, the Ito-Nisio Theorem guarantees the a.s. convergence of $\sum_i \gamma_i \boldsymbol{x}_i$, as well. QED

Corollary 5.5.1 *The following properties of a Banach space* $\boldsymbol{X}$ *with unconditional basis* $(\mathbf{e}_n)$ *are equivalent:*

(i) The space $\mathbf{c}_0$ *is not finitely representable in* $\boldsymbol{X}$*;*

(ii) Let m_H *be the standard cylindrical measure on the Hilbert space* $\boldsymbol{H}$ *with the Fourier transform* $\exp[-\|\boldsymbol{h}\|^2]$. *The operator* $U : \boldsymbol{H} \mapsto \boldsymbol{H}$ *maps* m_H *into a* σ*-additive measure on* $\boldsymbol{X}$ *if, and only if,* $U^* \in \Pi_1(\boldsymbol{X}, \boldsymbol{H})$*;*

(iii) The functional $\exp[-\boldsymbol{x}^* R \boldsymbol{x}^*]$ *is the Fourier transform of a Gaussian measure on* $\boldsymbol{X}$ *if, and only if,* $\sum_n (\boldsymbol{e}_n^* R \boldsymbol{e}_n^*)^{1/2} \boldsymbol{e}_n$ *converges in* $\boldsymbol{X}$.

Remark 5.5.2. (a) In general, the convergence of the series $\sum_n (\boldsymbol{e}_n^* R \boldsymbol{e}_n^*)^{1/2} \boldsymbol{e}_n$ means that $R^{1//2} \in \Pi_1(\boldsymbol{X}, \boldsymbol{H})$, but if $\boldsymbol{X} = \boldsymbol{H}$, then the summability condition on R means, simply, that R is nuclear.

(b) The cylindrical measure m_H in the statement (*ii*) of the above Corollary may be replaced by any cylindrical measure invariant under unitary transformations, and such that finite-dimensional projections thereof are absolutely continuous with respect to the Lebesgue measure.

Corollary 5.5.2. *Let* $p > 0$, *and let* $\boldsymbol{X}$ *be a Banach space with unconditional basis* $(\mathbf{e}_n)$, *and such that* $\mathbf{c}_0$ *is not finitely representable in* $\boldsymbol{X}$. *Then the norms* $(\mathbf{E}\|X\|^p)^{1/p}$, *and* $\|\sum_n (\boldsymbol{e}_n^* R \boldsymbol{e}_n^*)\boldsymbol{e}_n\|$, *are equivalent on the subspace of* $\boldsymbol{L}_p(\boldsymbol{X})$ *spanned by all Gaussian random vectors* X *in* $\boldsymbol{X}$.

Proof. The Corollary follows from Theorem 5.5.3 by an obvious application of the Closed Graph Theorem and Kahane's Theorem of Chapter 1.

The above discussion directly implies the following result for Gaussian random series:

Corollary 5.5.3. *The following properties of a Banach space* $\boldsymbol{X}$ *with unconditional basis* $(\boldsymbol{e}_n)$ *are equivalent:*

(i) The space $\mathbf{c}_0$ *is not finitely representable in* $\boldsymbol{X}$*;*

(ii) Let (X_n) *be a sequence of independent zero-mean Gaussian random vectors in* $\boldsymbol{X}$ *with covariance operators* (R_n)*. Then* $\sum_i X_i$ *converges a.s. if, and only if, the series* $\sum_i(\sum_n \boldsymbol{e}_i^* R_n \boldsymbol{e}_i^*)^{1/2}\boldsymbol{e}_i$ *converges in* $\boldsymbol{X}$*;*

(iii) For each sequence $(\boldsymbol{x}_i) \subset \boldsymbol{X}$*, the series* $\sum_i \gamma_i \boldsymbol{x}_i$ *converges a.s. if, and only if,* $\sum_i(\sum_n(\boldsymbol{e}_i^*\boldsymbol{x}_n)^2)^{1/2}\boldsymbol{e}_i$ *converges in*$\boldsymbol{X}$*;*

Finally, the above Corollary and the standard Kronecker Lemma give the following *Strong Law of Large Numbers*:

Theorem 5.5.4 (SLLN).[16] *If* $\mathbf{c}_0$ *is not finitely representable in the Banach space* $\boldsymbol{X}$*, and* (X_n) *is a sequence of independent zero-mean Gaussian random vectors in* $\boldsymbol{X}$ *with covariance operators* (R_n) *such that*

$$\sum_i \Big(\sum_n n^{-2}\boldsymbol{e}_i^* R_n \boldsymbol{e}_i^*\Big)^2\Big)^{1/2}\boldsymbol{e}_i,$$

converges in $\boldsymbol{X}$*, then*

$$\lim_{n\to\infty} \frac{X_1+\cdots+X_n}{n} = 0, \qquad a.s.$$

In the remainder of this section we will concentrate on spaces of cotype 2, Gaussian measures on such spaces, and convergence of general series of independent random vectors in such spaces, as well as the Law of Large Numbers for them. Let us begin with the basic result that shows that cotype 2 of $\boldsymbol{X}$ is indispensable for the validity of the vector analogue of the classical Bochner Theorem which characterizes functions which are Fourier transforms of positive measures (on $\boldsymbol{X}$, in our case).

[16] See, S.A. Chobanyan and V.I. Tarieladze (1977).

Theorem 5.5.5. *Let $\boldsymbol{X}$ be a Banach space, and assume that there exists a topology τ on $\boldsymbol{X}^*$ such that $\phi : \boldsymbol{X}^* \mapsto \boldsymbol{C}$ is positive-definite, τ-continuous with $\phi(0) = 1$, if, and only if, there exists a probability measure m on $\boldsymbol{X}$ with*

$$\int_{\boldsymbol{X}} \exp[i\boldsymbol{x}^*\boldsymbol{x}]m(d\boldsymbol{x}) = \phi(\boldsymbol{x}^*),$$

Then $\boldsymbol{X}$ is of cotype 2.

Proof. Assume that $\boldsymbol{X}$ is of cotype 2. Then there exists a sequence $(\boldsymbol{x}_i) \subset \boldsymbol{X}$ such that the series $\sum_i r_i\boldsymbol{x}_i$ converges a.s., and $a_n = \sum_{i=1}^n \|\boldsymbol{x}_i\|^2 \uparrow \infty$. Then also $\sum_k \|\boldsymbol{x}_k\|^2/a_k = \infty$. Now, define a sequence, (Y_n), of independent random vectors in $\boldsymbol{X}$ with the distribution laws (μ_k) such that

$$\mathbf{P}(Y_k = \pm a_k\boldsymbol{x}_k/\|\boldsymbol{x}_k\|) = (1/4)\|\boldsymbol{x}_k\|^2/a_k^2,$$

$$\mathbf{P}(Y_k = 0) = 1 - \|\boldsymbol{x}_k\|^2/(2a_k^2).$$

Then, by the Borel-Cantelli Lemma, Y_k's do not converge to 0, a.s., so that the series $\sum_k Y_k$ diverges a.s.

Now, the assumption that $\sum_i r_i\boldsymbol{x}_i$ converges a.s. implies that $\phi_1(\boldsymbol{x}^*) := \prod_{k=1}^\infty \cos(\boldsymbol{x}^*\boldsymbol{x}_k)$ is the characteristic functional of a measure on $\boldsymbol{X}$, and thus is τ-continuous, by assumption. let $\delta > 0$ be such that , if $|t| \leq \pi/2, \cos t \geq 1-\delta$, then $1-t^2/2 \leq \cos t \leq 1-t^2/4$. If $\phi_1(\lambda) > 1 - \delta$, for $|\lambda| \leq 1$, then

$$\phi_1(\boldsymbol{x}^*) \leq \prod_{k=1}^\infty (1 - (1/4)(\boldsymbol{x}^*\boldsymbol{x}_k)^2)$$

$$\leq \prod_{k=1}^\infty [1 - (\|\boldsymbol{x}_k\|^2/2a^2)\cos\bigl(\boldsymbol{x}^*a_k\boldsymbol{x}_k/\|\boldsymbol{x}_k\|\bigr)] =: \phi_2(\boldsymbol{x}^*).$$

The functional ϕ_2 is also τ-continuous at 0, and positive-definite as a limit of the characteristic functions of $\mu_1 * \cdots * \mu_n$. Therefore, again by the assumption, there exists a probability measure m on $\boldsymbol{X}$ such that $\phi_2(\boldsymbol{x}^*) = \int \exp[i\boldsymbol{x}^*\boldsymbol{x}]m(d\boldsymbol{x})$. That means, however, that $\sum_k Y_k$ converges in law and, by the Ito-Nisio Theorem (see Chapter 1), also a.s. A contradiction. QED

Now, we are ready to study measures with Gaussian covariance on spaces of cotype 2.

Theorem 5.5.6. *The following properties of a Banach space* $\boldsymbol{X}$ *are equivalent:*

(i) Space $\boldsymbol{X}$ *is of cotype-2;*

(ii) Let the operator $U \in B(\boldsymbol{H}, \boldsymbol{X})$, *and let* m_H *be the standard cylindrical measure with the Fourier transform* $\exp[[-\|\boldsymbol{h}\|^2]$. *Then* $m_h U^{-1}$ *is countably additive if, and only if,* $U \in \Pi_2(\boldsymbol{H}, \boldsymbol{X})$;

(iii) For each probability measure m *on* $\boldsymbol{X}$ *such that, for each* $\boldsymbol{x}^* \in \boldsymbol{X}^*$ *the integral* $\int (\boldsymbol{x}^*\boldsymbol{x})^2 m(d\boldsymbol{x}) < \infty$, *and which has a Gaussian covariance operator, there exists* $U \in B(\boldsymbol{l}_2, \boldsymbol{X})$, *and a measure* μ *on* $\boldsymbol{l}_2$, *with* $\int \|\boldsymbol{h}\|^2 \mu(d\boldsymbol{h}) < \infty$ *such that* $m = \mu U^{-1}$.

Proof. *(i)* $\implies$ *(ii)* The "if" part is true in any Banach space $\boldsymbol{X}$ because, by Pietsch Factorization Theorem (see Chapter 1), we have the decomposition $U : \boldsymbol{H} \overset{V}{\mapsto} \boldsymbol{H} \mapsto \boldsymbol{X}$, where V is a Hilbert-Schmidt operator.

Now, assume that $U \in B(\boldsymbol{H}, \boldsymbol{X})$, and that $m_H U^{-1}$ is countably additive. Let $(\boldsymbol{g}_n) \subset \boldsymbol{H}$, and $\sum_n (\boldsymbol{g}_n, \boldsymbol{h})^2 < \infty, \boldsymbol{h} \in \boldsymbol{H}$. Denote by $(\boldsymbol{h}_n)$ an orthonormal basis in $\boldsymbol{H}$, and define the bounded operator, V, by the formula $V\boldsymbol{h} = \sum_n (\boldsymbol{h}, \boldsymbol{h}_n)\boldsymbol{g}_n$. We shall show that the series, $\sum_n \gamma_n U \boldsymbol{g}_n = \sum_n \gamma_n UV\boldsymbol{h}_n$, converges a.s. Indeed, by our assumption, UU^* is a covariance operator of a Gaussian measure. Therefore UVV^*U^* is also a Gaussian covariance, so that $\sum_n \gamma_n U\boldsymbol{g}_n$ converges in law, and thus also a.s. by the Ito-Nisio Theorem (see, Chapter 1). Because $\boldsymbol{X}$ is of cotype 2, by Corollary 5.2.1, $\sum_n \|U\boldsymbol{g}_n\|^2 < \infty$. Hence, for each $\boldsymbol{h} \in \boldsymbol{H}$, the condition $((\boldsymbol{g}_n, \boldsymbol{h})) \in \boldsymbol{l}_2$ implies that $(U\boldsymbol{g}_n) \in \boldsymbol{l}_2(\boldsymbol{X})$, so that, by the Closed Graph Theorem, $U \in \Pi_2(\boldsymbol{H}, \boldsymbol{X})$.

(ii) $\implies$ *(iii)* *Step 1:* Assume m itself is Gaussian with the covariance operator $R : \boldsymbol{X}^* \mapsto \boldsymbol{X}$, $R = A^*A$, $A \in B(\boldsymbol{X}^*, \mathbf{H})$. Evidently, m is the countably additive extension of the standard Gaussian cylindrical measure $m_H(A^*)^{-1}$. In view of *(ii)*, we have $A^* \in \Pi_2(\boldsymbol{H}, \boldsymbol{X})$. Therefore, by the Pietsch Factorization Theorem (see Chapter 1), $A^* : \boldsymbol{H} \overset{V}{\mapsto} \boldsymbol{l}_2 \overset{U}{\mapsto} \boldsymbol{X}$, where V is a Hilbert-Schmidt operator, and $U \in B(\boldsymbol{l}_2, \boldsymbol{X})$. Therefore, $R = UVV^*U^*$, where $VV^* : \boldsymbol{l}_2 \mapsto \boldsymbol{l}_2$ is nuclear. Thus m is the image, under U, of a

Gaussian measure on $\boldsymbol{l}_2$ with covariance VV^*.

Step 2: Let m be an arbitrary measure with a Gaussian covariance, and let X be a random vector in $\boldsymbol{X}$ such that its probability distributions $\mathcal{L}(X) = m$. Define the operator,

$$A : \boldsymbol{X}^* \ni \boldsymbol{x}^* \mapsto A\boldsymbol{x}^* := \boldsymbol{x}^* X \in \boldsymbol{L}_2(\mathbf{R}).$$

Then A^*A is a Gaussian covariance operator, and, by (ii), $A^* \in \Pi_2(\boldsymbol{H}, \boldsymbol{X})$. Hence, A^* admits a factorization $A^* = UV$, where $V \in \Pi_2(\boldsymbol{L}_2, \boldsymbol{l}_2)$, and $U = B(\boldsymbol{l}_2, \boldsymbol{X})$. Therefore, $V^* : \boldsymbol{l}_2 \mapsto \boldsymbol{L}_2(\mathbf{R})$ is Hilbert-Schmidt, so that there exists a random vector Y in $\boldsymbol{l}_2$ such that $\mathbf{E}\|Y\|^2 < \infty$, and $V^*\boldsymbol{h} = (\boldsymbol{h}, Y), \boldsymbol{h} \in \boldsymbol{l}_2$. Clearly, $m = \mathcal{L}(Y)U^{-1}$, which completes the proof of the implication $(ii) \implies (iii)$.

$(iii) \implies (i)$ Assume that $\sum_n \gamma_n \boldsymbol{x}_n$, $(\boldsymbol{x}_n) \subset \boldsymbol{X}$, coverges a.s. In view of (iii), every measure m of weak second order with a Gaussian covariance has second moments finite. Consider the measure m on $\boldsymbol{X}$ concentrated at the points $\pm p_n^{-1/2}\boldsymbol{x}_n, n \in \mathbf{N}$, with

$$m(\{\pm p_n^{-1/2}\boldsymbol{x}_n\}) = p_n/2, \qquad \text{where} \qquad \sum_n p_n = 1, p_n \geq 0.$$

Evidently, $R_m = R_{\sum_n \gamma_n x_n}$. Therefore, $\sum_n \|\boldsymbol{x}_n\|^2 = \int \|\boldsymbol{x}\|^2 m(\boldsymbol{x}) < \infty$, and $\boldsymbol{X}$ is of cotype 2 in view of Remark 5.5.1 (b). QED

Corollary 5.5.4. *The following properties of a Banach space* $\boldsymbol{X}$ *are equivalent:*

(i) The space $\boldsymbol{X}$ *is of cotype 2;*

(ii) For each random vector X *in* $\boldsymbol{X}$ *with Gaussian covariance, and* $\mathbf{E}(\boldsymbol{x}^*X)^2 < \infty$, *for each* $\boldsymbol{x}^* \in \boldsymbol{X}^*$, *we have* $\mathbf{E}\|X\|^2 < \infty$*;*

(iii) A function $\phi : \boldsymbol{X}^* \mapsto \boldsymbol{C}$ *is the Fourier transform of a zero-mean Gaussian measure on* $\boldsymbol{X}$ *if, and only if,*

$$\phi(\boldsymbol{x}^*) = \exp[-\boldsymbol{x}^* UVU^* \boldsymbol{x}^*), \qquad \boldsymbol{x}^* \in \boldsymbol{X}^*,$$

where $V : \boldsymbol{l}_2 \mapsto \boldsymbol{l}_2$ *is nuclear, and* $U \in B(\boldsymbol{l}_2, \boldsymbol{X})$.

(iv) There exist constants, $C_1, C_2, C_3 \geq 0$, *such that, for each random vector* X *in* $\boldsymbol{X}$, *with* $\mathbf{E}(\boldsymbol{x}^*X)^2 < \infty$, *each* $\boldsymbol{x}^* \in \boldsymbol{X}^*$,

and the covariance operator R of a Gaussian random vector Y, we have the inequalities,

$$\left(\mathbf{E}\|X\|^2\right)^{1/2} \le C_1\Pi_2((R^{1/2})^*) \le C_2\left(\mathbf{E}\|Y\|^2\right)^{1/2} \le C_3\Pi_2((R^{1/2})^*).$$

(v) Under the additional assumption that $\boldsymbol{X}$ has an unconditional basis $(\mathbf{e}_n)$, there exists a constant $C \ge 0$ such that, for each X with $\mathbf{E}(\boldsymbol{x}^*X)^2 < \infty$, each $\boldsymbol{x}^* \in \boldsymbol{X}^*$, and a Gaussian covariance operator R,

$$\left(\mathbf{E}\|X\|^2\right)^{1/2} \le C\left\|\sum_n (\mathbf{e}_n^* R\mathbf{e}_n^*)^{1/2}\mathbf{e}_n\right\|.$$

As corollaries to the above Theorem 5.5.6 we obtain the following results on the convergence of random series and the Strong Law of Large Numbers in spaces of cotype 2.

Theorem 5.5.7.[17] *Let $\boldsymbol{X}$ be of cotype 2, and let $(\mathbf{e}_n)$ be an unconditional basis in $\boldsymbol{X}$. Then, for any independent zero-mean random vectors (X_n) in $\boldsymbol{X}$ with the covariance operators (R_n) such that the series $\sum_i(\sum_n \mathbf{e}_i^* R_n\mathbf{e}_i^*)^{1/2}\mathbf{e}_i$ converges in $\boldsymbol{X}$, the series $\sum_i X_i$ converges a.s. in $\boldsymbol{X}$.*

Proof. By Theorem 5.5.3, and in view of the assumption, the operator, $R = \sum_n R_n$, is the covariance operator of a Gaussian measure. Therefore, in the factorization, $R : \boldsymbol{X}^* \overset{A}{\mapsto} \boldsymbol{H} \overset{A^*}{\mapsto} \boldsymbol{X}$, we have $A^* \in \Pi_2(\boldsymbol{H}, \boldsymbol{X})$, so that, for each $\boldsymbol{x}^* \in \boldsymbol{X}^*$, the series $\sum_n \boldsymbol{x}^*X_n$ converges a.s. By the Ito-Nisio Theorem (see Chapter 1), the series $\sum_n X_n$ converges a.s. as well. QED.

Corollary 5.5.5 (SLLN). *Let $\boldsymbol{X}$ be of cotype 2, and let $(\mathbf{e}_n)$ be an unconditional basis in $\boldsymbol{X}$. If (X_n) is a sequence of independent zero-mean random vectors in $\boldsymbol{X}$ with the covariance operators (R_n) such that the series $\sum_i(\sum_n n^{-2}\mathbf{e}_i^* R_n\mathbf{e}_i^*)^{1/2}\mathbf{e}_i$ converges in $\boldsymbol{X}$, then*

$$\lim_{n\to\infty} \frac{X_1 + \cdots + X_n}{n} = 0, \qquad \text{a.s. in } \boldsymbol{X}.$$

[17]See S.A. Chobanyan and V.I. Tarieladze (1977).

5.6 Central limit theorem, law of the iterated logarithm, and infinitely divisible distributions

In this section we will discuss how cotype 2 of a Banach space is related to the validity of the Central Limit Theorem (CLT) for random vectors of weak and strong second order, and to the Law of the Iterated Logarithm (LIL) for them. The section concludes demonstrating that the cotype q of the Banach space is equivalent to the $\min(1, x^q)$ integrability of the Levy measure for infinitely divisible probability distributions on a Banach space.

Theorem 5.6.1 (CLT).[18] *The following properties of a Banach space $\boldsymbol{X}$ are equivalent:*

(i) Space $\boldsymbol{X}$ is of cotype 2;

(ii) For any sequence (X_i) of zero-mean, independent, and identically distributed random vectors in $\boldsymbol{X}$ the probability distribution,

$$\mathcal{L}\Big(\frac{X_1 + \cdots + X_n}{n^{1/2}}\Big),$$

converges weakly (to a Gaussian measure) if, and only if, $\mathbf{E}(\boldsymbol{x}^* X_1)^2 < \infty$, for each $\boldsymbol{x}^* \in \boldsymbol{X}^*$, and X_1 has a Gaussian covariance operator.

Proof. The implication $(i) \implies (ii)$ is immediate in view of Theorem 5.5.6 since X_1 is a continuous linear image of a Y in $\boldsymbol{l}_2$ with $\mathbf{E}\|Y\|^2 < \infty$, and in the Hilbert space the Central Limit Theorem for i.i.d summands of second order holds true. (see also Chapter 7).

$(ii) \implies (i)$ Assume that if X has a Gaussian covariance operator then it satisfies the Central Limit Theorem. Then, necessarily, $\mathbf{E}\|X\| < \infty$. Now, take $\sum_n \gamma_n \boldsymbol{x}_n$ that converges a.s., and let $p_n \geq 0, \sum_n p_n = 1$. Consider a random vector X with the probability distribution concentrated at the points $\pm p_n^{-1/2} \boldsymbol{x}_n$, with $\mathbf{P}(\{\pm p_n^{-1/2} \boldsymbol{x}_n\}) = p_n/2$. Clearly, $R_X = R_{\sum \gamma_n x_n}$ and, by assumption, $\sum_n \|\boldsymbol{x}_n\| p_n^{1/2} = \mathbf{E}\|X\| < \infty$. Since this holds true

[18] See S.A. Chobanyan and V.I. Tarieladze (1977), and also N. Jain (1976).

for every sequence (p_n) such that $(p_n^{1/2}) \in \boldsymbol{l}_2$, we also have that $\sum_n \|\boldsymbol{x}_n\|^2 < \infty$, which implies that $\boldsymbol{X}$ is of cotype 2 in view of Remark 5.5.1 (b). QED

As an immediate consequence of the above result we also have the following Corollary:

Corollary 5.6.1. *A Banach space $\boldsymbol{X}$ is of cotype 2 if, and only if, for any sequence of independent, and identically distributed random vectors in $\boldsymbol{X}$, for which $\mathcal{L}((X_1+\cdots+X_n)/n^{1/2}))$ converges weakly, we have $\mathbf{E}\|X_1\|^2 < \infty$.*

Theorem 5.6.2 (LIL)[19] *The following properties of a Banach space $\boldsymbol{X}$ are equivalent:*

(i) Space $\boldsymbol{X}$ is of cotype 2;

(ii) For any sequence (X_i) of independent, identically distributed random vectors in $\boldsymbol{X}$ with $\mathbf{E}(\boldsymbol{x}^* X_1)^2 < \infty$, and Gaussian covariance operator R,

$$\mathbf{P}\Big(\lim_{n\to\infty} \text{distance}\,\Big(\frac{X_1+\cdots+X_n}{(2n\log\log n)^{1/2}}, K\Big) = 0\Big) = 1,$$

and

$$\mathbf{P}\Big(\text{cluster points of}\,\Big\{\frac{X_1+\cdots+X_n}{(2n\log\log n)^{1/2}}\Big\} = K\Big) = 1,$$

where the set $K = \{(R^{1/2})^* \boldsymbol{h} : \|\boldsymbol{h}\| < 1\}$.

Proof (*i*) $\Longrightarrow$ (*ii*) In view of Theorem 5.5.6 it is sufficient to check that the Law of the Iterated Logarithm holds true in the Hilbert space. And, indeed, it does because Hilbert space is of Rademacher type 2 (see Chapter 7).

(*ii*) $\Longrightarrow$ (*i*) The Law of the Iterated Logarithm implies that $(X_1+\cdots+X_n)/n \to 0$, a.s. This, however, even in general Banach spaces, implies that $\mathbf{E}\|X_1\| < \infty$[20].Thus we obtain that $\boldsymbol{X}$ is of

[19]See S.A. Chobanyan and V.I. Tarieladze (1977), but the result depends strongly on earlier work of J. Kuelbs (1977), and G. Pisier (1975/76) and (1976).

[20]See W.A. Woyczyński (1974).

cotype 2 exactly as in the proof of the implication $(ii) \implies (i)$ of the preceding theorem. QED

The final result of this chapter deals with the Lévy-Khinchine representation of non-Gaussian, infinitely divisible probability distributions on spaces of cotype q.

Theorem 5.6.3.[21] *The following properties of a Banach space $\boldsymbol{X}$ are equivalent:*

(i) Space $\boldsymbol{X}$ is of cotype q;

(ii) If m is a symmetric, infinitely divisible probability distribution on $\boldsymbol{X}$ with the Lévy-Khinchine representation,

$$\int_{\boldsymbol{X}} \exp[i\boldsymbol{x}^*\boldsymbol{x}]m(d\boldsymbol{x}) = \exp\Big[\int_{\boldsymbol{X}} \big(\cos(\boldsymbol{x}^*\boldsymbol{x}) - 1\big)\mu(d\boldsymbol{x})\Big],$$

then

$$\int_{\boldsymbol{X}} \min(1, \|\boldsymbol{x}\|^q)\mu(d\boldsymbol{x}) < \infty.$$

Proof. $(i) \implies (ii)$ By a standard argument[22] one can reduce the proof to showing that if μ_n are symmetric, finite, satisfy the condition

$$\mu_n(\boldsymbol{x} : \|\boldsymbol{x}\| \le 1) = k_n \uparrow \infty, \qquad \text{as} \qquad n \to \infty,$$

and if the measures m_n, determined by the formula

$$\int_{\boldsymbol{X}} \exp[i\boldsymbol{x}^*\boldsymbol{x}]m_n(d\boldsymbol{x}) = \exp\Big[\int_{\boldsymbol{X}} \big(\cos(\boldsymbol{x}^*\boldsymbol{x}) - 1\big)\mu_n(d\boldsymbol{x})\Big],$$

are unifomly tight, then

$$\sup_{n\in\mathbf{N}} \int_{\boldsymbol{X}} \|\boldsymbol{x}\|^q \mu_n(d\boldsymbol{x}) < \infty. \tag{5.6.1}$$

Define $\nu_n = \mu_n k_n^{-1}$, and $\rho_n = \nu_n^{*k_n}, n \in \mathbf{N}$ (convolution power). Then, since $\boldsymbol{X}$ is of cotype q,

$$\int_{\boldsymbol{X}} \|\boldsymbol{x}\|^q \rho_n(d\boldsymbol{x}) \ge Ck_n \int_{\boldsymbol{X}} \|\boldsymbol{x}\|^q \nu_n(d\boldsymbol{x}) = C\int_{\boldsymbol{X}} \|\boldsymbol{x}\|^q \mu_n(d\boldsymbol{x}),$$

[21]See A. Araujo and E. Gine, (1978).
[22]See K.R. Parthasarathy (1967).

so that (5.6.1) follows because

$$\sup_{n\in\mathbf{N}} \int_{\boldsymbol{X}} \|\boldsymbol{x}\|^q \rho_n(d\boldsymbol{x}) < \infty.$$

in view of the inequality

$$\int_{\boldsymbol{X}} \|\boldsymbol{x}\|^q \rho(d\boldsymbol{x}) \leq \frac{(a+1)^q + a^q(1-2^{1-q})}{1-2^q\rho_n(\|\boldsymbol{x}\| \geq a)},$$

valid for any $a > 0$, and which is a corollary to the following inequality

$$\mathbf{P}(\|X_1 + \cdots + X_n\| > a) \geq 2^{-q}\Big(1 - \frac{(a+c)^q + a^q(1-2^{1-q})}{\mathbf{E}\|X_1+\cdots+X_n\|^q}\Big),$$

valid in any Banach space for independent, zero-mean random vectors uniformly bounded by a constant c [23], and the fact that tightness of $(\rho_n) = (\nu_n^{*k_n})$ is implied by tightness of (m_n)[24].

(*ii*) $\Longrightarrow$ (*i*) We first prove that (*ii*) implies that $\boldsymbol{c}_0$ is not contained in $\boldsymbol{X}$. If $\boldsymbol{c}_0 \subset \boldsymbol{X}$ then the series $\sum_n \pi_n n^{-1/q}\boldsymbol{e}_n$, where (π_n) are i.i.d. real symmetrized Poissonian random variables, converges a.s. The sequence $(\boldsymbol{e}_n)$ here is the canonical basis in $\boldsymbol{c}_0$. Indeed, because the tail of the distribution of π_1 satisfies the inequality, $\mathbf{P}(|\pi_1| > n) \leq C/(n!n)$ we have

$$\sum_n \mathbf{P}\big(\|\pi_n n^{-1/q}\boldsymbol{e}_n\| > n^{-1-1/q}\big) \leq C\sum_n \frac{1}{n!n} < \infty,$$

so that, by Borel-Cantelli Lemma, $\sum_n \pi_n n^{-1/q}\boldsymbol{e}_n$ converges a.s. Therefore, the probability distribution $\mathcal{L}(\sum_n \pi_n n^{-1/q}\boldsymbol{e}_n)$ is infinitely divisible with the Fourier transform

$$\exp\Big[\int_{\boldsymbol{X}} (\cos(\boldsymbol{x}^*\boldsymbol{x}) - 1)\Big(\sum_{n=1}^{\infty} \delta_{(\boldsymbol{e}_n n^{-1/q})}\Big)(d\boldsymbol{x})\Big],$$

so that

$$\int_{\boldsymbol{X}} \min(1, \|\boldsymbol{x}\|^q)\Big(\sum_{n=1}^{\infty} \delta_{(\boldsymbol{e}_n n^{-1/q})}\Big)(d\boldsymbol{x}) = \sum_n \frac{1}{n} = \infty,$$

[23] See A. de Acosta and J. Samur (1979).

[24] See L. LeCam (1970), Theorem 3.

which would contradict (*ii*).

Now, because (*ii*) is a superproperty (since any measure can be weakly approximated by atomic measures), we conclude that $\boldsymbol{c}_0$ is not finitely representable in $\boldsymbol{X}$.

Therefore, in view of Theorem 5.5.2, Kahane Theorem, and Hoffmann-Jorgensen Theorem (see Chapter 1), to show (*i*) it suffices to prove that if the series $\sum_n \gamma_n \boldsymbol{x}_n$ converges a.s., then $\sum_n \|\boldsymbol{x}_n\|^q < \infty$. So, assume that $\sum_n \gamma_n \boldsymbol{x}_n$, where $(\boldsymbol{x}_n) \subset \boldsymbol{X}$, converges a.s. As we observed above, the tail of the distributions of the Poisson random variable π_1 is thinner than the tail of the Gaussian random variable γ_1, so that by the Comparison Theorem (see Chapter 1), the series $\sum_n \pi_n \boldsymbol{x}_n$ converges a.s. Evidently, $\mathcal{L}(\sum_n \pi_n \boldsymbol{x}_n)$ is an infinitely divisible probability distribution with the Fourier transform $\exp[\int_X (\cos(\boldsymbol{x}^* \boldsymbol{x}) - 1)\mu(d\boldsymbol{x})$, where $\mu = \sum_n \delta_{x_n}$. By assumption (*ii*), $\sum_n \min(1, \|\boldsymbol{x}_n\|^q) < \infty$ so that, because $(\boldsymbol{x}_n)$ is bounded (since $\sum_n \gamma_n \boldsymbol{x}_n$ converges a.s.), also $\sum_n \|\boldsymbol{x}_n\|^q < \infty$. QED

Chapter 6

Spaces of Rademacher and stable types

6.1 Infratypes of Banach spaces

For a normed space $\boldsymbol{X}$ we define numerical constants $a_n^p(\boldsymbol{X})$, $1 \leq p \leq \infty$, $n \in \mathbf{N}$, as follows:

$$a_n^p(\boldsymbol{X}) := \inf\Big\{a \in \mathbf{R}^*; \forall \boldsymbol{x}_1, \ldots, \boldsymbol{x}_n \in \boldsymbol{X}, \inf_{\varepsilon_i=\pm 1}\Big\|\sum_{i=1}^{n} \varepsilon_i \boldsymbol{x}_i\Big\| \leq a\Big(\sum_{i=1}^{n} \|\boldsymbol{x}\|^p\Big)^{1/p}\Big\}.$$

Definition 6.1.1. We shall say that the normed space $\boldsymbol{X}$ is of *infratype* p[1] if there exists a constant $C > 0$ such that, $a_n^p(\boldsymbol{X}) \leq C < \infty$, for all $n \in \mathbf{N}$.

In other words, $\boldsymbol{X}$ is of infratype p if, and only if, for some constant C, and any finite sequence $(\boldsymbol{x}_i) \subset \boldsymbol{X}$,

$$\inf_{\varepsilon_i=\pm 1}\Big\|\sum_{i=1}^{n} \varepsilon_i \boldsymbol{x}_i\Big\| \leq C\Big(\sum_{i=1}^{n} \|\boldsymbol{x}\|^p\Big)^{1/p}.$$

[1]The notion of infratype is due to G. Pisier (1973), and it was also investigated in B. Maurey and G. Pisier (1976).

Remark 6.1.1. It is easy to see that $\boldsymbol{X}$ is of infratype p if, and only if, for each sequence $(\boldsymbol{x}_i) \subset \boldsymbol{X}$ with $\sum_i \|\boldsymbol{x}_i\|^p < \infty$, there exists a sequence $\varepsilon = \pm 1$ such that the series $\sum_i \varepsilon_i \boldsymbol{x}_i$ converges.

The following Proposition is evident.

Proposition 6.1.1. *(i) Any normed space is of infratype 1.*

(ii) If $\boldsymbol{X}$ is of infratype p, and $1 \leq p_1 < p$, then $\boldsymbol{X}$ is of infratype p_1.

(iii) If $\boldsymbol{X}$ is of infratype p, then $a_n^\infty(\boldsymbol{X}) \leq Cn^{1/p}$, for some $C > 0$.

The properties of the sequence $(a_n^\infty(\boldsymbol{X}))$ stated below will be useful in developing the theory of spaces of type p in the remainder of the chapter.

Proposition 6.1.2. *(i) If $\boldsymbol{X} = \{0\}$, then $a_n^\infty(\boldsymbol{X}) = 0$, for each $n \in \mathbf{N}$, but if $\boldsymbol{X} \neq \{0\}$ then*

$$1 \leq a_n^\infty(\boldsymbol{X}) \leq n, \qquad n \in \mathbf{N},$$

and if, additionally, $\boldsymbol{X}$ is infinite-dimensional, then

$$n^{1/2} \leq a_n^\infty(\boldsymbol{X}) \leq n, \qquad n \in \mathbf{N};$$

(ii) **Monotonicity:** *If $n \leq m$, $n, m \in \mathbf{N}$, then*

$$a_n^\infty(\boldsymbol{X}) \leq a_m^\infty(\boldsymbol{X});$$

(iii) **Subadditivity:** *For any $n, k \in \mathbf{N}$,*

$$a_{n+k}^\infty(\boldsymbol{X}) \leq a_n^\infty(\boldsymbol{X}) + a_k^\infty(\boldsymbol{X});$$

(iv) **Submultiplicativity:** *For any $n, k \in \mathbf{N}$,*

$$a_{nk}^\infty(\boldsymbol{X}) \leq a_n^\infty(\boldsymbol{X}) \cdot a_k^\infty(\boldsymbol{X}).$$

Proof. (*i*) Only the fact that in infinite-dimensional $\boldsymbol{X}$ we have the inequality $a_n^\infty(\boldsymbol{X}) \geq n^{1/2}$ is non-trivial, but it follows immediately from the Dvoretzky Theorem of Chapter 1. Indeed,

Dvoretzky Theorem states that for each $n \in \mathbf{N}$, and any $\epsilon > 0$, there exist $\boldsymbol{x}_1, \ldots, \boldsymbol{x}_n \in \boldsymbol{X}$ such that, for any $a_1, \ldots, a_n \in \mathbf{R}$,

$$\Big(\sum_{i=1}^{n} |a_i|^2\Big)^{1/2} \leq \Big\|\sum_i a_i \boldsymbol{x}_i\Big\| \leq (1+\epsilon)\sum_{i=1}^{n} |a_i|,$$

which gives the necessary estimate.

(*ii*), and (*iii*), have straightforward proofs which we omit.

(iv) To prove the submultiplicativity of $a_n^\infty(\boldsymbol{X})$ take $\boldsymbol{x}_1, \ldots, \boldsymbol{x}_{nk} \in \boldsymbol{X}$. Then, for each $i = 1, \ldots, n$, select ε_j^i, $(i-1)k < j \leq ik$, such that

$$\Big\| \sum_{j=ik-k+1}^{ik} \varepsilon_j^i \boldsymbol{x}_j \Big\| = \inf_{\varepsilon_i = \pm 1} \Big\| \sum_{j=ik-k+1}^{ik} \varepsilon_j \boldsymbol{x}_j \Big\|,$$

and define $\boldsymbol{y}_i = \sum_{j=ik-k+1}^{ik} \varepsilon_j^i \boldsymbol{x}_j, i = 1, \ldots, n$. By the very construction,

$$\|\boldsymbol{y}_i\| \leq a_k^\infty(\boldsymbol{X}) \sup_{(i-1)k<j\leq ik} \|\boldsymbol{x}_j\|,$$

so that

$$\inf_{\varepsilon_i=\pm 1}\Big\|\sum_{i=1}^{n} \varepsilon_i \boldsymbol{y}_i\Big\| \leq a_n^\infty(\boldsymbol{X}) \sup_{1\leq i\leq n} \|\boldsymbol{y}_i\| \leq a_n^\infty(\boldsymbol{X}) a_k^\infty(\boldsymbol{X}) \sup_{1\leq j\leq nk} \|\boldsymbol{x}_j\|.$$

Therefore,

$$\inf_{\varepsilon_j=\pm 1}\Big\|\sum_{i=1}^{nk} \varepsilon_j \boldsymbol{x}_j\Big\| \leq \inf_{\varepsilon_i=\pm 1}\Big\|\sum_{i=1}^{n} \varepsilon_i \boldsymbol{y}_i\Big\| \leq a_n^\infty(\boldsymbol{X}) a_k^\infty(\boldsymbol{X}) \sup_{1\leq j\leq nk} \|\boldsymbol{x}_j\|,$$

which proves (*iv*). QED

Information contained in the sequence $(a_n^\infty(\boldsymbol{X}))$ completely determines the infratype of the space $\boldsymbol{X}$. Indeed, we have the following result:

Proposition 6.1.3. *If $\boldsymbol{X} \neq \{0\}$ is a normed linear space, and $n_0 > 1$, then there exists $p_0, 1 \leq p_0 \leq \infty$ ($1 \leq p_0 \leq 2$ if the space $\boldsymbol{X}$ is infinite-dimensional), such that*

$$a_{n_0}^\infty(\boldsymbol{X}) = n_0^{1/p_0},$$

and $\boldsymbol{X}$ is of infratype p, for each $p < p_0$.

Proof. The existence of such a p_0 is implied by Proposition 6.1.2 (i). Now, let $p < p_0$, and $(\boldsymbol{x}_n)$ be a finite sequence in $\boldsymbol{X}$. For each $k \in \mathbf{N}$, we define the set of integers, A_k, as follows:

$$A_k := \Big\{ n : \frac{(\sum_n \|\boldsymbol{x}_n\|^p)^{1/p}}{n_0^{(k+1)/p}} < \|\boldsymbol{x}_n\| \leq \frac{(\sum_n \|\boldsymbol{x}_n\|^p)^{1/p}}{n_0^{k/p}} \Big\}.$$

Evidently, by the triangle inequality,

$$\inf_{\varepsilon_n = \pm 1} \Big\| \sum_n \varepsilon_n \boldsymbol{x}_n \Big\| \leq \sum_{k=0}^{\infty} \inf_{\varepsilon_n = \pm 1} \Big\| \sum_{n \in A_k} \varepsilon_n \boldsymbol{x}_n \Big\|,$$

from which it follows that

$$\inf_{\varepsilon_n = \pm 1} \Big\| \sum_n \varepsilon_n \boldsymbol{x}_n \Big\| \leq \sum_{k=0}^{\infty} a^{\infty}_{|A_k|}(\boldsymbol{X}) \frac{(\sum_n \|\boldsymbol{x}_n\|^p)^{1/p}}{n_0^{k/p}}. \tag{6.1.1}$$

However, on the other hand,

$$\Big(\sum_n \|\boldsymbol{x}_n\|^p \Big)^{1/p} \geq \Big(\sum_{n \in A_k} \|\boldsymbol{x}_n\|^p \Big)^{1/p} \geq |A_k|^{1/p} \frac{(\sum_n \|\boldsymbol{x}_n\|^p)^{1/p}}{n_0^{(k+1)/p}},$$

so that $|A_k| \leq n_0^{k+1}$ (we assume here that $\sum_n \|\boldsymbol{x}_n\|^p \neq 0$). Now, by Proposition 6.1.2 (ii) and (iv),

$$a^{\infty}_{|A_k|}(\boldsymbol{X}) \leq a^{\infty}_{n_0^{k+1}}(\boldsymbol{X}) \leq (a^{\infty}_{n_0}(\boldsymbol{X}))^{k+1} = n_0^{(k+1)/p_0},$$

and, by (6.1.1),

$$\begin{aligned} \inf_{\varepsilon_n = \pm 1} \Big\| \sum_n \varepsilon_n \boldsymbol{x}_n \Big\| &\leq \sum_{k=0}^{\infty} n_0^{(k+1)/p_0} \frac{(\sum_n \|\boldsymbol{x}_n\|^p)^{1/p}}{n_0^{k/p}} \\ &= n_0^{1/p_0} \sum_{k=0}^{\infty} \big[n_0^{(1/p_0 - 1/p)} \big]^k \cdot \Big(\sum_n \|\boldsymbol{x}_n\|^p \Big)^{1/p}. \end{aligned}$$

This proves that $\boldsymbol{X}$ is of infratype p, because the series on the right converges in view of the assumption, $p < p_0$. QED

Finally, in view of the above Proposition, it is of interest to calculate the largest value of p such that $\boldsymbol{X}$ is of infratype p.

Definition 6.1.2.

$$p_{\text{itype}}(\boldsymbol{X}) := \sup\{p : \boldsymbol{X} \text{ is of infratype } p\}.$$

It turns out that the above parameter can be explicitly expressed in terms of the constants $a_n^\infty(\boldsymbol{X})$.

Theorem 6.1.1.

$$p_{\text{itype}}(\boldsymbol{X}) = \lim_{n\to\infty} \frac{\log n}{\log[a_n^\infty(\boldsymbol{X})]}.$$

Proof. If $\boldsymbol{X}$ is of infratype p then, by Proposition 6.1.1(*iii*),

$$\liminf_{n\to\infty} \frac{\log n}{\log[a_n^\infty(\boldsymbol{X})]} \geq \liminf_{n\to\infty} \frac{\log n}{\log C + (1/p)\log n} = p,$$

so that

$$\liminf_{n\to\infty} \frac{\log n}{\log[a_n^\infty(\boldsymbol{X})]} \geq p_{\text{itype}}(\boldsymbol{X}).$$

If $q > p_{\text{itype}}(\boldsymbol{X})$, in view of Proposition 6.1.3, for each $n \in \mathbf{N}$, $a_n^\infty(\boldsymbol{X}) \geq n^{1/q}$, so that

$$\limsup_{n\to\infty} \frac{\log n}{\log[a_n^\infty(\boldsymbol{X})]} \leq \limsup_{n\to\infty} \frac{\log n}{(1/q)\log n} = q,$$

and

$$\limsup_{n\to\infty} \frac{\log n}{\log[a_n^\infty(\boldsymbol{X})]} \leq p_{\text{itype}}(\boldsymbol{X}).$$ QED

6.2 Banach spaces of Rademacher-type p

For a normed space $\boldsymbol{X}$ we define numerical constants $b_n^p(\boldsymbol{X})$, $p \geq 1$, $n \in \mathbf{N}$, as follows:

$$b_n^p(\boldsymbol{X}) :=$$

$$\inf\Big\{b\in\mathbf{R}^+ : \forall \boldsymbol{x}_1,\ldots,\boldsymbol{x}_n\in \boldsymbol{X}, \Big(\mathbf{E}\Big\|\sum_{i=1}^n r_i\boldsymbol{x}_i\Big\|^p\Big)^{1/p} \le b\Big(\sum_{i=1}^n \|\boldsymbol{x}_i\|^p\Big)^{1/p}\Big\}.$$

Definition 6.2.1. We shall say that $\boldsymbol{X}$ is of *Rademacher-type p* [2] if there exists a constant $C > 0$ such that, $b_n^p(\boldsymbol{X}) \le C < \infty$, for all $n \in \mathbf{N}$.

Remark 6.2.1. The Kahane Theorem (see, Chapter 1) implies that a Banach space $\boldsymbol{X}$ is of Rademacher-type p if, and only if, there exists a constant C and an $\alpha, 0 < \alpha < \infty$, such that, for any finite sequence $(\boldsymbol{x}_i) \subset \boldsymbol{X}$

$$\Big(\mathbf{E}\Big\|\sum_{i=1}^n r_i\boldsymbol{x}_i\Big\|^\alpha\Big)^{1/\alpha} \le C\Big(\sum_{i=1}^n \|\boldsymbol{x}_i\|^p\Big)^{1/p}, \tag{6.2.1}$$

or, alternatively, $\boldsymbol{X}$ is of Rademacher-type p if, and only if, for each $\alpha, 0 < \alpha < \infty$, there exists a constant $C = C_\alpha$ such that, for any finite sequence $(\boldsymbol{x}_i) \subset \boldsymbol{X}$, the inequality (6.2.1) holds true.

Remark 6.2.2. One can also define the concept of a Banach space of Rademacher-type Φ, where Φ is an Orlicz-type function. Namely, a Banach space $\boldsymbol{X}$ is said to be of Rademacher-type Φ if, for every $p \ge 1$, there exists a constant C such that, for any sequence $\boldsymbol{x}_1, \ldots, \boldsymbol{x}_n \in \boldsymbol{X}$,

$$\Big(\mathbf{E}\Big\|\sum_i r_i\boldsymbol{x}_i\Big\|^p\Big)^{1/p} \le C\inf\Big\{c > 0 : \sum_i (\Phi(\|\boldsymbol{x}_i\|/c)) \le 1\Big\}.$$

Many results valid in spaces of Rademacher-type p may be generalized to spaces of Rademacher-type Φ.

The following proposition partially explains the relationship between the infratype and the Rademacher-type of a normed space.

Proposition 6.2.1. *(i) Any normed space $\boldsymbol{X}$ is of Rademacher-type 1, and if the dimension of $\boldsymbol{X}$ is greater than 1, and $\boldsymbol{X}$ is of Rademacher-type p, then $p \le 2$;*

[2] The concept of Rademacher-type 2 is due to E. Dubinsky, A. Pelczynski and H.P. Rosenthal (1972). For general p the notion was introduced by J. Hoffmann-Jorgensen (1972/73).

(ii) If $\boldsymbol{X}$ is of Rademacher-type p, and $1 \leq p_1 < p$, then $\boldsymbol{X}$ is of Rademacher-type p_1;

(iii) If $\boldsymbol{X}$ is of Rademacher-type p, then it is also of infratype p.

We omit the obvious proofs of the first part of (i), and of (ii) and (iii). The second part of (i) is an immediate corollary to the Khinchine Inequality (see Introduction).

Now, we turn to the investigation of properties of the sequences of constants $b_n^p(\boldsymbol{X})$, $1 \leq p \leq \infty$.

Proposition 6.2.2.[3] *(i) If $\boldsymbol{X} = \{0\}$, then $b_n^p(\boldsymbol{X}) = 0, n \in \mathbf{N}$. If $\boldsymbol{X} \neq \{0\}$ then*

$$1 \leq a_n^p(\boldsymbol{X}) \leq b_n^p(\boldsymbol{X}) \leq n^{1-1/p}, \qquad n \in \mathbf{N},$$

and, if $\boldsymbol{X}$ is infinite-dimensional, then

$$n^{1/2} \leq a_n^\infty(\boldsymbol{X}) \leq n^{1/p} b_n^p(\boldsymbol{X}) \leq n, \qquad n \in \mathbf{N}.$$

(ii) **Monotonicity:** *If $n \leq m$, $n, m \in \mathbf{N}$, then*

$$b_n^p(\boldsymbol{X}) \leq b_m^p(\boldsymbol{X});$$

(iii) **Subadditivity:** *For any $n, k \in \mathbf{N}$,*

$$b_{n+k}^p(\boldsymbol{X}) \leq b_n^p(\boldsymbol{X}) + b_k^p(\boldsymbol{X});$$

(iv) **Submultiplicativity:** *For any $n, k \in \mathbf{N}$, and $1 \leq p < \infty$,*

$$b_{nk}^p(\boldsymbol{X}) \leq b_n^p(\boldsymbol{X}) \cdot b_k^p(\boldsymbol{X});$$

Proof. Statements $(i), (ii)$, and (iii), have straightforward proofs based on Proposition 6.1.2, and we omit them.

(iv) Take $\boldsymbol{x}_1, \ldots, \boldsymbol{x}_n \in \boldsymbol{X}$, and for each $i = 1, \ldots, n$, define random vectors

$$X_i = \sum_{j=ik-k+1}^{ik} r_j \boldsymbol{x}_j,$$

[3]This result is due to G. Pisier (1973).

so that, if (r'_j) is a Rademacher sequence independent of (r_j), we have the inequality,

$$\Big(\mathbf{E}'_r\Big\|\sum_{i=1}^{n} r'_i X_i(\omega)\Big\|^p\Big)^{1/p} \le b_n^p(\boldsymbol{X})\Big(\sum_{i=1}^{n}\|X_i(\omega)\|^p\Big)^{1/p}, \qquad \forall\omega\in\Omega,$$

(here, $\mathbf{E}'_r$ denotes the integration with respect to (r'_j)), from which, by symmetry, we obtain the estimate,

$$\begin{aligned}\Big(\mathbf{E}\Big\|\sum_{j=1}^{nk} r_j\boldsymbol{x}_j\Big\|^p\Big)^{1/p} &= \Big(\mathbf{E}\mathbf{E}'\Big\|\sum_{i=1}^{n} r'_i X_i\Big\|^p\Big)^{1/p} \le b_n^p(\boldsymbol{X})\Big(\sum_{i=1}^{n}\mathbf{E}\|X_i\|^p\Big)^{1/p}\\ &\le b_n^p(\boldsymbol{X})\Big(\sum_{i=1}^{n}(b_k^p(\boldsymbol{X}))^p\sum_{j=ik-k+1}^{ik}\|\boldsymbol{x}_j\|^p\Big)^{1/p}\\ &= b_n^p(\boldsymbol{X})\cdot b_k^p(\boldsymbol{X})\cdot\Big(\sum_{i=1}^{nk}\|\boldsymbol{x}_i\|^p\Big)^{1/p}.\end{aligned}$$

QED

As we saw in the case of infratype, the knowledge of the sequence $(b_n^q(\boldsymbol{X}))$, for any particular $q, 1\le q<\infty$, provides complete information about the Rademacher-type of $\boldsymbol{X}$.

Proposition 6.2.3. *If $\boldsymbol{X}\ne\{0\}$ is a normed space, and $n_0 > 1, 1\le q<\infty$, then there exists a p_0, $1\le p_0\le q$ ($1\le p_0\le 2$, if $\boldsymbol{X}$ is infinite-dimensional) such that*

$$b_{n_0}^q(\boldsymbol{X}) = n_0^{1/p_0-1/p},$$

and $\boldsymbol{X}$ is of Rademacher-type p, for each $p<p_0$.

Proof. The existence of such p_0 is an immediate consequence of Proposition 6.2.2 (i). By submultiplicativity,

$$b_{n_0^k}^q(\boldsymbol{X}) \le n_0^{k(1/p_0-1/q)}, \qquad k\in\mathbf{N}.$$

Now, set $C=n_0^{1/p_0-1/q}$, and take any $n\in[n_0^k, n_0^{k+1})$. Then,

$$b_{n_0^k}^q(\boldsymbol{X}) \le b_{n_0^{k+1}}^q(\boldsymbol{X}) \le n_0^{(k+1)(1/p_0-1/q)} \le Cn^{1/p_0-1/q}.$$

so that, for any finite subset $A \subset \mathbf{N}$,

$$\Big(\mathbf{E}\Big\|\sum_{n\in A} r_n \boldsymbol{x}_n\Big\|^q\Big)^{1/q} \le C|A|^{1/p_0}\Big(\sum_{n\in A}\|\boldsymbol{x}_n\|^q/|A|\Big)^{1/q}$$

$$\le C|A|^{1/p_0}\sup_{n\in A}\|\boldsymbol{x}_n\|. \tag{6.2.2}$$

At this point, take a finite sequence $(\boldsymbol{x}_i) \subset \boldsymbol{X}$, and $p < p_0$. For each k, define the set of integers, A_k, as follows:

$$A_k := \Big\{n \subset \mathbf{N};\Big(\sum_i \|\boldsymbol{x}_i\|^p\Big)^{1/p}2^{-(k+1)/p} < \|\boldsymbol{x}_n\| \le \Big(\sum_i\|\boldsymbol{x}_i\|^p\Big)^{1/p}2^{-k/p}\Big\}.$$

Then, $|A_k| \le 2^{k+1}$, and in view of (6.2.2),

$$\Big(\mathbf{E}\Big\|\sum_i r_i\boldsymbol{x}_i\Big\|^q\Big)^{1/q} \le \sum_{k=0}^{\infty}\Big(\mathbf{E}\Big\|\sum_{i\in A_k} r_i\boldsymbol{x}_i\Big\|^q\Big)^{1/q}$$

$$\le \sum_{k=0}^{\infty} C|A_k|^{1/p_0}\sup_{n\in A_k}\|\boldsymbol{x}_n\|$$

$$\le \sum_{k=0}^{\infty} C2^{(k+1)/p_0}\Big(\sum_i\|\boldsymbol{x}_i\|^p\Big)^{1/p}2^{-k/p}$$

$$\le C2^{1/p_0}\sum_{k=0}^{\infty}2^{k(1/p_0-1/p)}\Big(\sum_i\|\boldsymbol{x}_i\|^p\Big)^{1/p},$$

so that $\boldsymbol{X}$ is of Rademacher-type p in view of the Remark 6.2.1. QED

Like in the case of infratype we can now calculate the supremum of p's for which the space $\boldsymbol{X}$ is of Rademacher-type p in terms of the constants $b_n^q(\boldsymbol{X})$.

Definition 6.2.2.

$$p_{\text{Rtype}}(\boldsymbol{X}) := \sup\{p : \boldsymbol{X} \text{ is of Rademacher type } p\}.$$

Theorem 6.2.1.[4] *Let $q < \infty$, and let $\boldsymbol{X} \ne \{0\}$. Then*

$$p_{\text{Rtype}}(\boldsymbol{X}) = \lim_{n\to\infty}\frac{\log n}{\log(n^{1/q}b_n^q(\boldsymbol{X}))}.$$

[4]This result is due to G. Pisier (1973).

Proof. If $\boldsymbol{X}$ is of Rademacher-type p then, by Remark 6.2.1, and Hölder's Inequality, there exists a constant $C > 0$ such that, for any finite $(\boldsymbol{x}_i) \subset \boldsymbol{X}$,

$$\left(\mathbf{E}\Big\|\sum_{i=1}^{n} r_i\boldsymbol{x}_i\Big\|^q\right)^{1/q} \le C\Big(\sum_{i=1}^{n}\|\boldsymbol{x}_i\|^p\Big)^{1/p} \le Cn^{1/p-1/q}\Big(\sum_{i=1}^{n}\|\boldsymbol{x}_i\|^q\Big)^{1/q},$$

so that $b_n^q(\boldsymbol{X}) \le Cn^{1/p-1//q}$, and

$$\liminf_{n\to\infty}\frac{\log n}{\log(n^{1/q}b_n^q(\boldsymbol{X}))} \ge \liminf_{n\to\infty}\frac{\log n}{\log C + (1/p)\log n} = p.$$

Therefore,

$$\liminf_{n\to\infty}\frac{\log n}{\log(n^{1/q}b_n^q(\boldsymbol{X}))} \ge p_{\rm Rtype}(\boldsymbol{X}).$$

If $p > p_{\rm Rtype}(\boldsymbol{X})$, then, by Proposition 6.2.3, $b_n^q(\boldsymbol{X}) \ge n^{1/p-1//q}$, $n \in \mathbf{N}$, and

$$\limsup_{n\to\infty}\frac{\log n}{\log(n^{1/q}b_n^q(\boldsymbol{X}))} \le p.$$

Hence,

$$\liminf_{n\to\infty}\frac{\log n}{\log(n^{1/q}b_n^q(\boldsymbol{X}))} \le p_{\rm Rtype}(\boldsymbol{X}). \qquad \text{QED}$$

In the next step we will verify how the Rademacher-type is preserved under standard operations on normed spaces. These results will enable us to give a number of concrete examples of spaces of Rademacher-type p at the end of this section.

Proposition 6.2.4. *(i) If $\boldsymbol{X}$ is of Rademacher-type p, and $\boldsymbol{Y}$ is a subspace of $\boldsymbol{X}$, then $\boldsymbol{Y}$ is of Rademacher-type p as well.*

(ii) If $\boldsymbol{X}$ is of Rademacher-type p, and $\boldsymbol{Y}$ is a closed subspace of $\boldsymbol{X}$, then the quotient space $\boldsymbol{X}/\boldsymbol{Y}$ is also of Rademacher-type p.

Proof. The statement (i) is obvious, so let us prove (ii). Let $\boldsymbol{X}$ be of Rademacher-type p, $\bar{\boldsymbol{x}}_i, \ldots, \bar{\boldsymbol{x}}_n \in \boldsymbol{X}/\boldsymbol{Y}$, and let $\pi : \boldsymbol{X} \mapsto \boldsymbol{X}/\boldsymbol{Y}$ be the standard surjection. For any $\epsilon > 0$, one can find

$\boldsymbol{x}_1, \ldots, \boldsymbol{x}_n \in \boldsymbol{X}$ such that $\pi(\boldsymbol{x}_i) = \bar{\boldsymbol{x}}_i$, and $\|\boldsymbol{x}_i\| \leq (1+\epsilon)\|\bar{\boldsymbol{x}}_i\|$, so that

$$\mathbf{E}\Big\|\sum_i r_i \bar{\boldsymbol{x}}_i\Big\|^p = \mathbf{E}\Big\|\pi\Big(\sum_i r_i \bar{\boldsymbol{x}}_i\Big)\Big\|^p \leq \mathbf{E}\Big\|\sum_i r_i \boldsymbol{x}_i\Big\|^p$$

$$\leq C\sum_i \|\boldsymbol{x}_i\|^p \leq C(1+\epsilon)^p \sum_i \|\bar{\boldsymbol{x}}_i\|^p. \qquad \text{QED}$$

Theorem 6.2.2.[5] *Let (T, Σ, μ) be a σ-finite measure space. Then,*

(i) The constants $b_n^p(\boldsymbol{X}) = b_n^p(\boldsymbol{L}^p(T, \Sigma, \mu; \boldsymbol{X}))$, $1 \leq p < \infty$;

(ii) If $p \leq q < \infty$, then $\boldsymbol{X}$ is of Rademacher-type p if, and only if, $\boldsymbol{L}^q(T, \Sigma, \mu; \boldsymbol{X})$ is of Rademacher-type p.

Proof. (*i*) Since $\boldsymbol{X}$ can be identified with a subspace of $\boldsymbol{L}_p(\boldsymbol{X})$, we have $b_n^p(\boldsymbol{X}) \leq b_n^p(\boldsymbol{L}^p(\boldsymbol{X}))$. To prove the converse inequality let $X_1, \ldots, X_n \in \boldsymbol{L}^p(\boldsymbol{X})$. Then, for each $t \in T$,

$$\mathbf{E}\Big\|\sum_i r_i X_i(t)\Big\|^p \leq (b_n^p(\boldsymbol{X}))^p \sum_i \|X_i(t)\|^p.$$

Integrating this inequality with respect to measure μ we obtain the inequality

$$\mathbf{E}\Big\|\sum_i r_i X_i\Big\|^p \leq (b_n^p(\boldsymbol{X}))^p \sum_i \|X_i\|^p,$$

which shows that $b_n^p(\boldsymbol{X}) \geq b_n^p(\boldsymbol{L}^p(\boldsymbol{X}))$.

(*ii*) The "if" part is obvious because $\boldsymbol{X}$ is a subspace of $\boldsymbol{L}^q(\boldsymbol{X})$. The "only if' part follows from (*i*) in the case $p = q$, and in the case $p < q < \infty$ it follows from the fact that if $X_1, \ldots, X_n \in \boldsymbol{L}^q(\boldsymbol{X})$, then

$$\Big(\mathbf{E}\Big\|\sum_i r_i X_i\Big\|_{L^q}^q\Big)^{1/q} = \Big(\int \mathbf{E}\Big\|\sum_i r_i X_i(t)\Big\|^q d\mu\Big)^{1/q}$$

$$\leq C\Big[\int \Big(\sum_i \|X_i(t)\|^p\Big)^{q/p} d\mu\Big]^{p/q \cdot 1/p}$$

[5]See B. Maurey and G. Pisier (1976).

$$\leq C\Big[\sum_i\Big(\int \|X_i(t)\|^{p\cdot q/p}\Big)^{p/q}\Big]^{1/p} \leq C\Big[\sum_i \|X_i\|^p_{L^q}\Big]^{1/p}.$$

Above, we utilized Remark 6.2.1, and the triangle inequality for the norm in $\boldsymbol{l}_{q/p}$. QED

The next result deals with the so-called "three-space problem" which provides an additional exploration of the quotient spaces of spaces of Rademacher-type which have been already investigated in Proposition 6.2.4.

Theorem 6.2.3.[6] *Let $1 < p < q \leq 2$, and let $\boldsymbol{Y}$ be a closed subspace of a Banach space $\boldsymbol{X}$. If $\boldsymbol{Y}$, and $\boldsymbol{X}/\boldsymbol{Y}$, are of Rademacher-type q, then $\boldsymbol{X}$ is of Rademacher-type p.*

The proof of the above theorem will rely on the folllowing Lemma which can be viewed as a generalization of the submultiplicativity property from Proposition 6.2.2 (iv).

Lemma 6.2.1. *Let $\boldsymbol{X}$ be a Banach space, and $\boldsymbol{Y}$ be a closed subspace of $\boldsymbol{X}$.*

(i) If $1 < p \leq 2$, then

$$b^p_{nk}(\boldsymbol{X}) \leq b^P_n(\boldsymbol{Y})b^P_k(\boldsymbol{X}) + b^P_n(\boldsymbol{Y})b^P_k(\boldsymbol{X}/\boldsymbol{Y}) + b^P_n(\boldsymbol{X})b^P_k(\boldsymbol{X}/\boldsymbol{Y}),$$

for all $n, k \in \mathbf{N}$;

(ii) If $\boldsymbol{Y}$ and $\boldsymbol{X}/\boldsymbol{Y}$ are of Rademamcher-type q, then there exist constants C, and γ, such that

$$b^q_n(\boldsymbol{X}) \leq (\log n)^\gamma, \qquad n \geq 2.$$

Proof. As above, π stands for the canonical projection of $\boldsymbol{X}$ onto $\boldsymbol{X}/\boldsymbol{Y}$, with the norm

$$\|\pi(\boldsymbol{x})\|_{X/Y} = \inf\{\|\boldsymbol{x}+\boldsymbol{y}\| : \boldsymbol{y} \in \boldsymbol{Y}\}.$$

Now, let $\boldsymbol{x}_1, \ldots, \boldsymbol{x}_{nk} \in \boldsymbol{X}$, and define

$$X_i = \sum_{j=ik-k+1}^{ik} r_j \boldsymbol{x}_j.$$

[6]See G. Pisier (1975), and P. Enflo, J. Lindenstrauss, and G. Pisier (1975).

Then, for each $\omega \in \Omega$, $i = 1, 2, \ldots, n$,, and $\gamma > 0$, there exists a $Y_i(\omega) \in \boldsymbol{Y}$ such that

$$\|X_i(\omega) + Y_i(\omega)\|_X \leq \|\pi(X_i(\omega))\|_{X/Y} + \gamma.$$

Then, choosing (r_i') to be independent of (r_i), by the convexity of the norm, we obtain the inequality

$$\Big(\mathbf{E}'\Big\|\sum_{i=1}^{n} r_i' X_i(\omega)\Big\|^p\Big)^{1/p}$$

$$\leq \Big(\mathbf{E}'\Big\|\sum_{i=1}^{n} r_i' Y_i(\omega)\Big\|^p\Big)^{1/p} + \Big(\mathbf{E}'\Big\|\sum_{i=1}^{n} r_i'(X_i(\omega) + Y_i(\omega))\Big\|^p\Big)^{1/p},$$

so that, by definition of $b_n^p(\boldsymbol{X})$, and $b_n^p(\boldsymbol{Y})$,

$$\Big(\mathbf{E}'\Big\|\sum_{i=1}^{n} r_i' X_i(\omega)\Big\|^p\Big)^{1/p}$$

$$\leq b_n^p(\boldsymbol{Y})\Big(\sum_{i=1}^{n} \|Y_i(\omega)\|^p\Big)^{1/p} + b_n^p(\boldsymbol{X})\Big(\sum_{i=1}^{n} \|X_i(\omega) + Y_i(\omega)\|^p\Big)^{1/p},$$

Furthermore, $\|Y_i(\omega)\| \leq \|X_i(\omega)\| + \|Y_i(\omega) + X_i(\omega)\|$ and, on the other hand, $\|X_i(\omega) + Y_i(\omega)\| \leq \|\pi(X_i(\omega))\| + \gamma$, so that we have

$$\Big(\mathbf{E}'\Big\|\sum_{i=1}^{n} r_i' X_i(\omega)\Big\|^p\Big)^{1/p} \leq b_n^p(\boldsymbol{Y})\Big(\sum_{i=1}^{n} \|X_i(\omega)\|^p\Big)^{1/p}$$

$$+\big(b_n^p(\boldsymbol{Y}) + b_n^p(\boldsymbol{X})\big)\Big[\Big(\sum_{i=1}^{n} \|\pi(X_i(\omega))\|^p\Big)^{1/p} + \gamma n^{1/p}\Big].$$

After integration, and employing an obvious symmetry argument, we obtain that

$$\Big(\mathbf{E}\Big\|\sum_{i=1}^{nk} r_i \boldsymbol{x}_i\Big\|^p\Big)^{1/p} = \Big(\mathbf{E}\mathbf{E}'\Big\|\sum_{i=1}^{nk} r_i' X_i\Big\|^p\Big)^{1/p}$$

$$\leq b_n^p(\boldsymbol{Y})\Big(\sum_{i=1}^{n} \mathbf{E}\|X_i\|^p\Big)^{1/p}$$

$$+\big(b_n^p(\boldsymbol{Y})+b_n^p(\boldsymbol{X})\big)\Big[\Big(\sum_{i=1}^{n}\mathbf{E}\|\pi(X_i)\|^p\Big)^{1/p}+\gamma n^{1/p}\Big]$$

$$\leq\Big[b_n^p(\boldsymbol{Y})b_k^p(\boldsymbol{X})+\big(b_n^p(\boldsymbol{Y})+b_n^p(\boldsymbol{X})\big)b_k^p(\boldsymbol{X}/\boldsymbol{Y})\Big]\Big(\sum_{i=1}^{nk}\|\boldsymbol{x}_j\|^p\Big)^{1/p}$$

$$+\big(b_n^p(\boldsymbol{Y})+b_n^p(\boldsymbol{X})\big)\gamma n^{1/p},$$

which proves (i) in view of the arbitrariness of γ.

(ii) Let $C_1=\sup_n b_n^q(\boldsymbol{X})<\infty$, and $C_2=\sup_n b_n^q(\boldsymbol{X}/\boldsymbol{Y})<\infty$. In view of (i),

$$b_{nk}^q\boldsymbol{X})\leq C_1b_k^q(\boldsymbol{X})+C_1C_2+b_n^q(\boldsymbol{X})C_2,\qquad n\in\mathbf{N}.$$

Since $1\leq b_n^q(\boldsymbol{X})$, $n\in\mathbf{N}$, (see Proposition 6.2.1),

$$b_{n^2}^q(\boldsymbol{X})\leq(C_1+C_1C_2+C_2)b_n^q(\boldsymbol{X}).$$

Chose γ so that $2^\gamma=C_1+C_1C_2+C_2$, and set $c_n=b_n^q(\boldsymbol{X})log^{-\gamma}n$, $n=2,3,\ldots,$ Then the above inequality reads

$$C_{n^2}\leq C_n,\qquad n\in\mathbf{N}.\tag{6.2.3}$$

Now, let n be an integer ≥ 2. Choose $k\in\mathbf{N}$ such that

$$N_k:=2^{2^k}\leq n<2^{2^{k+1}}=N_k^2.$$

Since b_n^p are increasing (see Proposition 6.2.2),

$$c_n=b_n^p(\boldsymbol{X})\log^{-\gamma}n\leq b_{N_k^2}^p(\boldsymbol{X})\log^{-\gamma}n\leq 2^\gamma b_{N_k^2}^p(\boldsymbol{X})\log^{-\gamma}N_k^2=2^\gamma c_{N_k^2}.$$

However, (6.2.3) implies that, for each $k\geq 0$, we have $c_{N_k}\leq c_{N_0}=c_2$, so that $c_n\leq 2^\gamma c_2$, which proves (ii). QED

Proof of Theorem 6.2.3. By Theorem 6.2.1, and Lemma 6.2.1 (ii),

$$p_{\text{Rtype}}(\boldsymbol{X})=\lim_{n\to\infty}\frac{\log n}{\log\big(n^{1/q}b_n^q(\boldsymbol{X})\big)}$$

$$>\lim_{n\to\infty}\frac{\log n}{q^{-1}\log n+\log C+\log[\gamma^{1/2}\log n]}=q,$$

so that, by Proposition 6.2.1(*ii*), the space $\boldsymbol{X}$ is of Rademacher-type p, for each $p, 1 \leq p < q$. QED

To conclude this section we will provide a number of examples of spaces which are of Rademacher-type p, as well as the spaces that are not of Rademacher-type p. More examples can be found in the later parts of this chapter.

Example 6.2.1. (*i*) The real line $\mathbf{R}$ is of Rademacher-type p for any $p \in [1, 2]$. In fact, it is sufficient to show that $\mathbf{R}$ is of Rademacher-type 2, and this fact is a direct consequence of the Khinchine Inequality (see Introduction).

(*ii*)[7] If $\boldsymbol{X}$ is p-smoothable, i.e., it has an equivalent norm with the modulus of smoothness, $p(\tau)$, satisfying the condition $p(\tau) = O(\tau^p), \tau \to 0$, then $\boldsymbol{X}$ is of Rademacher-type p (see Chapter 3).

(*iii*) In particular, $\boldsymbol{L}_p$, and $\boldsymbol{l}_p$, are of Rademacher-type $\min(p, 2)$, for $1 < p < \infty$, and if $1 \leq q < p \leq 2$, then $\boldsymbol{L}_q$, and $\boldsymbol{l}_q$, are not of Rademacher-type p. In other words, $p_{\text{Rtype}}(\boldsymbol{L}_p) = \min(p, 2)$. This fact follows from Theorem 6.2.2 (*i*).

(*iv*)[8] Let $1 \leq p < \infty$. By $\boldsymbol{C}^p(\boldsymbol{H})$ denote the Banach space of compact operators, A, on the Hilbert space $\boldsymbol{H}$ such that

$$\|A\|_p := \left(\text{tr}(A^*A)^{p/2}\right)^{1/p} < \infty.$$

Note that, if $p \neq 2$, then $\boldsymbol{C}^p(\boldsymbol{H})$ is not isomorphic with any subspace of $\boldsymbol{L}_p$. However, $p_{\text{Rtype}}(\boldsymbol{C}^p(\boldsymbol{H})) = \min(p, 2)$, too. This fact follows directly from (*ii*), and the evaluation of the modulus of smoothness for $\boldsymbol{C}^p(\boldsymbol{H})$.

(*v*) The spaces $\boldsymbol{c}_0, \boldsymbol{l}_1$, and $\boldsymbol{l}_\infty$, are not of Rademacher-type p, for any $p \in (1, 2]$.

Finally, we have a general result relating the Rademacher-type of the Banach space $\boldsymbol{X}$ to the cotype of its dual space, $\boldsymbol{X}^*$.

Theorem 6.2.4.[9] *If Banach space* $\boldsymbol{X}$ *is of Rademacher-type* p, *then its dual,* $\boldsymbol{X}^*$ *is of cotype* q, *with* $1/p + 1/q = 1$.

[7]See W.A. Woyczynski (1973).
[8]See N. Tomczak-Jaegerman (1974).
[9]See J. Hoffmann-Jorgensen (1972/73), and G. Pisier (1973/74).

Proof. If $\boldsymbol{X}$ is of Rademacher-type p then, by Remark 6.2.1, there exists a constant C such that, for any $n \in N$, and any $(\boldsymbol{x}_i) \subset \boldsymbol{X}$,

$$\Big(\mathbf{E}\Big\|\sum_{i=1}^{n} r_i \boldsymbol{x}_i\Big\|^2\Big)^{1/2} \leq C\Big(\sum_{i=1}^{n} \|\boldsymbol{x}_i\|^p\Big)^{1/p}.$$

Let $\boldsymbol{x}_1^*, \ldots, \boldsymbol{x}_n^* \in \boldsymbol{X}^*$. Then, there exists an $\epsilon > 0$ such that, for any $\boldsymbol{x}_1, \ldots, \boldsymbol{x}_n \in \boldsymbol{X}$,

$$\sum_{i=1}^{n} \boldsymbol{x}_i^* \boldsymbol{x}_i \geq \Big(\sum_{i} \|\boldsymbol{x}_i^*\|^q\Big)^{1/q} - \epsilon, \qquad \text{and} \qquad \sum_{i=1}^{n} \|\boldsymbol{x}_i\|^p \leq 1.$$

Therefore,

$$\begin{aligned}\Big(\sum_{i} \|\boldsymbol{x}_i^*\|^q\Big)^{1/q} &\leq \sum_{i=1}^{n} \boldsymbol{x}_i^* \boldsymbol{x}_i + \epsilon = \mathbf{E}\Big(\sum_{i=1}^{n} r_i \boldsymbol{x}_i^*\Big)\Big(\sum_{i=1}^{n} r_i \boldsymbol{x}_i\Big) + \epsilon \\ &\leq \Big(\mathbf{E}\Big\|\sum_{i=1}^{n} r_i \boldsymbol{x}_i^*\Big\|^2\Big)^{1/2}\Big(\mathbf{E}\Big\|\sum_{i=1}^{n} r_i \boldsymbol{x}_i\Big\|^2\Big)^{1/2} \\ &+\epsilon \leq C\Big(\mathbf{E}\Big\|\sum_{i=1}^{n} r_i \boldsymbol{x}_i^*\Big\|^2\Big)^{1/2} + \epsilon.\end{aligned}$$

Hence, since $\epsilon > 0$ was arbitrary, $\boldsymbol{X}^*$ is of cotype q. QED

Remark 6.2.3. (*a*) Note that the dual space of the space of cotype q does not have to be of Rademacher-type p, $1/p + 1/q = 1$. For example $\boldsymbol{l}_1$ is of cotype 2, but $\boldsymbol{l}_\infty$ is not of Rademacher-type 2. However, if $\boldsymbol{X}$ is a Banach lattice with $p(\boldsymbol{X}) > 1$, then $\boldsymbol{X}$ is of Rademacher-type p if, and only if, $\boldsymbol{X}^*$ is of cotype q.[10]

(*b*) If $\boldsymbol{X}_1$, and $\boldsymbol{X}_2$, are two Banach spaces contained in a locally convex topological vector space, and $0 \leq \theta \leq 1, 1 \leq p \leq \infty$, then, if $\boldsymbol{X}_1$ is of Rademacher-type p_1, $\boldsymbol{X}_2$ is of Rademacher-type p_2, and $1/p = (1-\theta)/p_1 + \theta/p_2$, then the Lions-Petree interpolation space $\mathbf{Int}(\boldsymbol{X}_1, \boldsymbol{X}_2)_{\theta,\mathbf{p}}$ is of Rademacher-type p.[11]

[10]See B. Maurey (1973/74).

[11]See B. Beauzamy (1974/75).

6.3 Local structures of spaces of Rademacher-type p

We begin with a statement that follows directly from the definition.

Proposition 6.3.1. *Rademacher-type p is a super-property, i.e., if the Banach space $\boldsymbol{X}$ is of Rademacher-type p, and space $\boldsymbol{Y}$ is finitely representable in $\boldsymbol{X}$, then $\boldsymbol{Y}$ is of Rademacher-type p as well.*

A factor critical for the investigation of the local structure of spaces of Rademacher-type p is the following result which deals with the concept of finite factorability (see Chapter 1).

Theorem 6.3.1.[12] *If $\boldsymbol{X}$ is an infinite-dimensional normed space, then the canonical injection $\boldsymbol{l}_1 \mapsto \boldsymbol{l}_{\mathrm{Rtype}}(X)$ is finitely factorable through $\boldsymbol{X}$.*

Proof. If $p_{\mathrm{Rtype}}(\boldsymbol{X}) = 2$, then the theorem is a corollary to the Dvoretzky-Rogers Lemma (see, Chapter 1) which states, in our current terminology, that $\boldsymbol{l}_1 \mapsto \boldsymbol{l}_2$ is finitely factorable through any infinite-dimensional normed space.

If $p_{\mathrm{Rtype}}(\boldsymbol{X}) < 2$, then it is sufficient to show that, for any q satisfying the inequality, $p_{\mathrm{Rtype}}(\boldsymbol{X}) < q < 2$, the mapping $\boldsymbol{l}_1 \mapsto \boldsymbol{l}_q$ is finitely factorable through $\boldsymbol{X}$ because the interval of those q's for which $\boldsymbol{l}_1 \mapsto \boldsymbol{l}_q$ is finitely factorable through $\boldsymbol{X}$ is closed (see Chapter 1). So, let $p_{\mathrm{Rtype}}(\boldsymbol{X}) < q < 2$. The assertion of the theorem is then an immediate corollary to the following Lemma (Parts (i), and (iii)).[13]

Lemma 6.3.1. *(i) If $1 \le q < 2$, and the space $\boldsymbol{Y}$ is the completion of the space $\boldsymbol{S}$[14] under the norm $|.|$ which is invariant under spreading, and such that*

$$n^{-1/q}\Big(\mathbf{E}\Big|\sum_{i=1}^{n} \boldsymbol{e}_i r_i\Big|^q\Big)^{1/q} \ge \delta > 0, \qquad n \in \mathbf{N}, \tag{6.3.1}$$

[12]See B. Maurey and G. Pisier (1976).

[13]The technical Part (ii) is needed in the proof of Part (iii).

[14]The space of real-valued sequences with a finite number of terms different from zero.

then the embedding $\boldsymbol{l}_1 \mapsto \boldsymbol{l}_q$ is finitely factorable through $\boldsymbol{Y}$.

(ii) Let $p_{\mathrm{Rtype}}(\boldsymbol{X}) < q < 2$. Then, there exist an infinite set $\mathbf{N}_1 \subset \mathbf{N}$, and a sequence $\epsilon_n \in (0,1)$, $n = 1,2,\dots$, such that, for each $n \in \mathbf{N}_1$ there exists $\boldsymbol{x}_1,\dots \boldsymbol{x}_n \in \boldsymbol{X}$ such that, for each $(\alpha_i) \in \mathbf{R}^n$, with $\sum_i |\alpha_i| = 1$,

$$
\begin{aligned}
Cb_n^q(\boldsymbol{X})n^{1/q-1} &\leq \big[(1-\epsilon_n)b_n^q(\boldsymbol{X}) - b_{n-1}^q(\boldsymbol{X})\big]n^{1/q} \\
&\leq \Big(\mathbf{E}\Big\|\sum_i \alpha_i r_i \boldsymbol{x}_i\Big\|^q\Big)^{1/q} \qquad (6.3.2)
\end{aligned}
$$

where

$$
C = (q - p_{\mathrm{Rtype}}(\boldsymbol{X})\big[2q\, p_{\mathrm{Rtype}}(\boldsymbol{X}) \cdot b_2^q(\boldsymbol{X})\big]^{-1}.
$$

(iii) There exists an invariant under spreading norm $|.|$ on $\boldsymbol{S}$, satisfying (6.3.1), and such that $\boldsymbol{Y}$ is finitely representable in $\boldsymbol{X}$.

Proof. (*i*) Let (γ_n) be a sequence of independent $N(0,1)$ Gaussian random variables which are also independent of the sequence (r_n). In view of (6.3.1),

$$
0<\delta\leq n^{-1/q}\Big(\mathbf{E}\Big|\sum_{i=1}^n \boldsymbol{e}_i r_i\Big|^q\Big)^{1/q} = n^{-1/q}(\mathbf{E}|\gamma_1|)^{-1}\Big(\mathbf{E}\Big|\sum_{i=1}^n \boldsymbol{e}_i r_i \mathbf{E}|\gamma_1|\Big|^q\Big)^{1/q}
$$

The above inequality, and the assumption that $q < 2$ implies that

$$
\limsup_{n\to\infty} n^{-1/q}\Big(\mathbf{E}\Big|\sum_{k=1}^n (\boldsymbol{e}_{2k} - \boldsymbol{e}_{2k-1})\gamma_{2k-1}\Big|^q\Big)^{1/q} \qquad (6.3.3)
$$

$$
\geq \limsup_{n\to\infty} 2^{1/q}(2n)^{-1/q}\Big(\mathbf{E}\Big|\sum_{k=1}^{2n} \boldsymbol{e}_k\gamma_k\Big|^q\Big)^{1/q}
$$

$$
= n^{-1/q}\Big(\mathbf{E}\Big|\sum_{k=1}^n \boldsymbol{e}_{2k}(\gamma_{2k} - \gamma_{2k-1})\Big|^q\Big)^{1/q}
$$

$$
= \limsup_{n\to\infty}\Big[2^{1/q}(2n)^{-1/q}\Big(\mathbf{E}\Big|\sum_{k=1}^{2n} \boldsymbol{e}_k\gamma_k\Big|^q\Big)^{1/q}
$$

$$
-2^{1/2}n^{-1/q}\Big(\mathbf{E}\Big|\sum_{k=1}^{n} \boldsymbol{e}_k\gamma_k\Big|^q\Big)^{1/q}\Big]
$$

$$\geq \mathbf{E}|\gamma_1|(2^{1/q} - 2^{1/2})\delta > 0.$$

The last equality above took advantage of the invariance under spreading of the norm $|.|$.

Now, we shall show that the canonical embedding $\boldsymbol{l}_{q'} \mapsto \boldsymbol{l}_\infty$, $1/q + 1/q' = 1$, is finitely factorable through the dual space $\boldsymbol{G}^*$, where

$$G := \overline{\text{span}}\{\boldsymbol{u}_k = \boldsymbol{e}_{2k} - e_{2k-1} : k \in N\} \subset \boldsymbol{Y}.$$

Indeed, were it not the case then, by Theorem 5.3.2, there would exist an $r' > q$ such that $\boldsymbol{G}^*$ is of cotype r', i.e., there exists a constant $K > 0$ such that, for each $n \in \mathbf{N}$, and each $\alpha_1, \ldots, \alpha_n \in \mathbf{R}$,

$$\Big(\sum_i |\alpha_i|^{r'}\Big)^{1/r'} \leq K\mathbf{E}\Big\|\sum_i \alpha_i \boldsymbol{u}_i^* r_i\Big\| \leq 2K\Big\|\sum_i \alpha_i \boldsymbol{u}_i^*\Big\|,$$

where $\boldsymbol{u}_k^*(\sum_i \alpha_i \boldsymbol{u}_i) := \alpha_k$, and, by transposition, with $1/r + 1/r' = 1$,

$$\Big|\sum_i \alpha_i \boldsymbol{u}_i\Big| \leq 2K\Big(\sum_i |\alpha_i|^r\Big)^{1/r}.$$

Above, we used the fact that $(\boldsymbol{u}_k)$ is a (monotone) unconditional basic sequence with constant 2. Then, however,

$$\Big(\mathbf{E}\Big|\sum_{k=1}^n \boldsymbol{u}_k\gamma_k\Big|^q\Big)^{1//q} \leq 2K\Big[\mathbf{E}\Big(\sum_k |\gamma_k|^r\Big)^{q/r}\Big]^{1/q}$$

$$\leq 2K\mathbf{E}\Big(\sum_{k=1}^n |\gamma_k|^r\Big)^{1/r} \leq 2K(\mathbf{E}|\gamma_1|^r)^{1/r} \cdot n^{1/r},$$

which contradicts (6.3.3).

Hence, by the definition of finite factorability, for each $n \in \mathbf{N}$, and each $\epsilon > 0$, there exists a sequence $\boldsymbol{x}_1^*, \ldots, \boldsymbol{x}_n^* \in \boldsymbol{G}^*$ such that, for each $\alpha_1, \ldots, \alpha_n \in \mathbf{R}$,

$$(1-\epsilon)\sup_i |\alpha_i| \leq \Big\|\sum_i \alpha_i \boldsymbol{x}_i^*\Big\| \leq \Big(\sum_{i=1}^n |\alpha_i|^{q'}\Big)^{1/q'}.$$

For a certain $\boldsymbol{H} \subset \boldsymbol{G}$, $\text{span}[\boldsymbol{x}_1^*, \ldots, \boldsymbol{x}_n^*] = (\boldsymbol{G}/\boldsymbol{H})^*$. By transposition, there exist $\bar{\boldsymbol{x}}_i, \ldots, \bar{\boldsymbol{x}}_n \in \boldsymbol{G}/\boldsymbol{H}$ such that, for all $\alpha_1, \ldots, \alpha_n \in \mathbf{R}$,

$$\Big(\sum_{i=1}^{n} |\alpha_i|^q\Big)^{1/q} \leq \Big\|\sum_i \alpha_i \bar{\boldsymbol{x}}_i\Big\|_{\boldsymbol{G}/\boldsymbol{H}} \leq (1-\epsilon)^{-1} \sum_{i=1}^{n} |\alpha_i|,$$

which concludes the proof of part (i) because, for each $\bar{\boldsymbol{X}}_i$, we can pick a representative $\boldsymbol{X}_i$ such that $|\boldsymbol{x}_i| \leq (1+\epsilon)\|\bar{\boldsymbol{x}}_i\|_{\boldsymbol{G}/\boldsymbol{H}}$.

(ii) By Lemma 6.2.3, $b_n^q(\boldsymbol{X}) \geq n^{1/p_{\text{Rtype}}(X)-1/q}, n \in \mathbf{N}$, Since, for any $(a_n) \subset \mathbf{R}^+$,

$$\limsup_{n\to\infty} \frac{\log a_n}{\log n} \leq \limsup_{n\to\infty} n\Big(\frac{a_n}{a_{n-1}} - 1\Big),$$

(because $\lim_{t\to\infty} t(t^\alpha/(t-1)^\alpha - 1) = \alpha$), setting $a_n = b_n^q(\boldsymbol{X})$ we get that

$$\frac{1}{p_{\text{Rtype}}(\boldsymbol{X})} - \frac{1}{q} \leq \limsup_{n\to\infty} n\Big(\frac{b_n^q(\boldsymbol{X})}{b_{n-1}^q(\boldsymbol{X})} - 1\Big),$$

so that there exists $\mathbf{N}_1 \subset \mathbf{N}$, and a sequence $(\epsilon_n) \subset (0,1)$, such that, for all $n \in \mathbf{N}_1$,

$$\frac{1}{2}\Big(\frac{1}{p_{\text{Rtype}}(\boldsymbol{X})} - \frac{1}{q}\Big) \leq n\Big[(1-\epsilon_n)\frac{b_n^q(\boldsymbol{X})}{b_{n-1}^q(\boldsymbol{X})} - 1\Big],$$

which implies that

$$\frac{2qp_{\text{Rtype}}(\boldsymbol{X})}{q - p_{\text{Rtype}}(\boldsymbol{X})} \geq \frac{1}{n} b_{n-1}^q(\boldsymbol{X})\big[(1-\epsilon_n)b_n^q(\boldsymbol{X}) - b_{n-1}^q(\boldsymbol{X})\big]^{-1}.$$

Finally, in view of the submultiplicativity property, $b_n^q(\boldsymbol{X}) \leq b_2^q(\boldsymbol{X}) \cdot b_{n-1}^q(\boldsymbol{X})$, $n \geq 1$, and this yields the inequality

$$\frac{1}{n} b_n^q(\boldsymbol{X})\big[(1-\epsilon_n)b_n^q(\boldsymbol{X}) - b_{n-1}^q(\boldsymbol{X})\big]^{-1} \leq \frac{2qp_{\text{Rtype}}(\boldsymbol{X}) \cdot b_2^q(\boldsymbol{X})}{q - p_{\text{Rtype}}(\boldsymbol{X})} = C,$$

which gives the first inequality in (6.3.2).

Now, let $n \in \mathbf{N}_1$. By definition of $b_n^q(\boldsymbol{X})$, there exist $\boldsymbol{x}_1, \ldots, \boldsymbol{x}_n \in \boldsymbol{X}$ such that $\sum_i \|\boldsymbol{x}_i\|^q = n$, and

$$\left(\mathbf{E}\Big\|\sum_{i=1}^n r_i \boldsymbol{x}_i\Big\|^q\right)^{1/q} \geq (1-\epsilon_n) b_n^q(\boldsymbol{X}) n^{1/q}.$$

If $(\alpha_i) \in \mathbf{R}^n, \sum_i |\alpha_i| = 1$, then

$$(1-\epsilon_n) b_n^q(\boldsymbol{X}) n^{1/q} \leq \left(\mathbf{E}\Big\|\sum_{i=1}^n r_i |\alpha_i| \boldsymbol{x}_i\Big\|^q\right)^{1/q}$$

$$+\left(\mathbf{E}\Big\|\sum_{i=1}^n r_i (1-|\alpha_i|) \boldsymbol{x}_i\Big\|^q\right)^{1/q}$$

$$\leq \left(\mathbf{E}\Big\|\sum_{i=1}^n r_i \alpha_i \boldsymbol{x}_i\Big\|^q\right)^{1/q} + \sup_{(t_i)\in R_+^n, \sum t_i = 1} \left(\mathbf{E}\Big\|\sum_{i=1}^n r_i (1-t_i) \boldsymbol{x}_i\Big\|^q\right)^{1/q}.$$

Furthermore, the function $(t_i) \mapsto (\mathbf{E}\|\sum_i r_i(1-t_i)\boldsymbol{x}_i\|^q)^{1/q}$ is a convex, continuous function on the convex set $\{(t_i) \in \mathbf{R}_+^n : \sum t_i = 1\}$, so that its maximum is attained at an extreme point of this set, i.e., at a point which has all components except one equal to zero. Therefore,

$$(1-\epsilon_n) b_n^q(\boldsymbol{X}) n^{1/q} \leq \left(\mathbf{E}\Big\|\sum_{i=1}^n r_i \alpha_i \boldsymbol{x}_i\Big\|^q\right)^{1/q} + \sup_{1\leq j\leq n} \left(\mathbf{E}\Big\|\sum_{i=1, i\neq j}^n r_i \boldsymbol{x}_i\Big\|^q\right)^{1/q}$$

$$\leq \left(\mathbf{E}\Big\|\sum_{i=1}^n r_i \alpha_i \boldsymbol{x}_i\Big\|^q\right)^{1/q} + b_{n-1}^q(\boldsymbol{X}) n^{1/q},$$

which proves the second inequality in (6.3.2).

(*iii*) Take $n \in \mathbf{N}_1$, and $\boldsymbol{x}_1, \ldots, \boldsymbol{x}_n$, as in part (*ii*). Denote by $\tilde{\mathbf{R}}_n$ the subspace of $\boldsymbol{L}_q(\boldsymbol{X})$ spanned by $r_i \boldsymbol{x}_i, i = 1, \ldots, n$, and define the operator

$$U_n : \tilde{\mathbf{R}}_n \ni \sum_{i=1}^n \alpha_i r_i \boldsymbol{x}_i \mapsto (\alpha_i) \in \boldsymbol{l}_1^{(n)}.$$

If $\boldsymbol{y}_1, \ldots, \boldsymbol{y}_n \in \tilde{\mathbf{R}}_n$, then

$$\sum_{i=1}^{n} \sup_{1\le k\le n} |U_n(\boldsymbol{y}_k)(i)| \le \sum_{i=1}^{n} \mathbf{E}\Big|\sum_{k=1}^{n} r_k U_n(\boldsymbol{y}_k)(i)\Big|$$

$$\le \Big(\mathbf{E}\Big\|\sum_{k=1}^{n} r_k U_n(\boldsymbol{y}_k)\Big\|^q_{\boldsymbol{l}_1^{(n)}}\Big)^{1/q} \le \|U_n\|\Big(\mathbf{E}\Big\|\sum_{k=1}^{n} \boldsymbol{y}_k r_k\Big\|^q_{\tilde{R}_n}\Big)^{1/q},$$

wherefrom it follows that

$$\sum_{i=1}^{n} \sup_{1\le k\le n} |U_n(\boldsymbol{y}_k)(i)| \le b_n^q(\boldsymbol{X})\|U_N\|\Big(\sum_{k=1}^{n} \|\boldsymbol{y}_k\|^q\Big)^{1/q},$$

because $b_n^q(\boldsymbol{X}) = b_n^q(\boldsymbol{L}(\boldsymbol{X}))$, according to Theorem 6.2.5 (i). Therefore, by Nikishin's Lemma (see Chapter 1), there exists a set $A_n \subset \{1.\ldots, n\}$, such that $|A_n| \ge n/2$, and such that, for each (α_i) with $\alpha_i = 0$ for $i \notin A_n$,

$$\lambda_q((\alpha_i)) \le 2^{1/q} n^{-1/q'} b_n^q(\boldsymbol{X})\|U_n\|\Big(\mathbf{E}\Big\|\sum_i \alpha_i r_i \boldsymbol{x}_i\Big\|^q\Big)^{1/q},$$

where (see Chapter 1)

$$\lambda_p((a_i)) = \big(\sup_{c>0} c^p|\{I : |a_i| > c\}|\big)^{1/p},$$

and $|A|$ denotes the cardinality of the set A.

If we define $B_n = \{i : \|\boldsymbol{x}_i\| \le 4^{1/q}\}$, then we have that $|B_n| \ge 3n/4$, and $|A_n \cap B_n| \ge n/4$, because $\sum_i \|\boldsymbol{x}_i\|^q = n$. For $n \in \mathbf{N}_1$, the second inequality in Part (ii) can be written as follows:

$$\|U_n\| \le n^{-1/q}\big[(1-\epsilon_n)b_n^q(\boldsymbol{X}) - b_{n-1}^q(\boldsymbol{X})\big]^{-1},$$

which, in conjunction with the first inequality in Part (ii) yields that, for each $n \in \mathbf{N}_1$, and $\alpha_1, \ldots, \alpha_n$ with the support in $A_n \cap B_n$,

$$\lambda_q((\alpha_i)) \le 2^{1/q} C^{-1}\Big(\mathbf{E}\Big\|\sum_{i\in A_n\cap B_n} \alpha_i r_i \boldsymbol{x}_i\Big\|^q\Big)^{1/q},$$

and

$$|A_n \cap B_n| \geq \frac{n}{4}, \qquad i \in A_n \cap B_n \implies \|\boldsymbol{x}_i\| \leq 4^{1/q}.$$

Hence, one can find an infinite subset $\mathbf{N}_2 \subset \mathbf{N}$, and a constant $\delta > 0$ such that, for each $n \in \mathbf{N}_2$, there exist $\boldsymbol{x}_1^n, \ldots, \boldsymbol{x}_n^n \in \boldsymbol{X}$ such that, for all $(\alpha_i) \in \mathbf{R}^n$,

$$\delta \cdot \lambda_q((\alpha_i)) < \left(\mathbf{E}\Big\|\sum_{i=1}^{n} \alpha_i \boldsymbol{x}_i^n r_i\Big\|^q\right)^{1/q} \leq \sum_{i=1}^{n} |\alpha_i|.$$

(This means, essentially, that the canonical embedding $\lambda_q^{(n)} \mapsto \boldsymbol{l}_1^{(n)}$ is factorable through subspaces of $\boldsymbol{L}_q(\boldsymbol{X})$, for infinitely many n's.)

Now, according to a classical theorem of Banach, there exists a continuous linear functional μ on $\boldsymbol{l}_\infty(\mathbf{N}_2)$ such that $\|\mu\| = 1$ and, for every $(\beta_n) \in \boldsymbol{l}_\infty(\mathbf{N}_2)$,

$$\liminf_{n \in \mathbf{N}_2} \beta_n \leq \mu(\beta_n) \leq \limsup_{n \in \mathbf{N}_2} \beta_n.$$

Define on $\boldsymbol{S}$ the seminorm,

$$|||\sum_i \alpha_i \boldsymbol{e}_i||| = \mu\Big[\Big(\Big\|\sum_{i=1}^{n} \alpha_i \boldsymbol{x}_i^n\Big\|\Big)_{n \in \mathbf{N}_2}\Big].$$

Evidently, for each $(\alpha_i) \in \boldsymbol{S}$,

$$\delta\lambda_q((\alpha_i)) \leq \left(\mathbf{E}\Big|\Big|\Big|\sum_i \alpha_i \boldsymbol{e}_i r_i\Big|\Big|\Big|^q\right)^{1/q} \leq \sum_i |\alpha_i|. \tag{6.3.4}$$

Let $\bar{\boldsymbol{e}}_i$ be the image of $\boldsymbol{e}_i$ in $\boldsymbol{S}/|||.|||$. Clearly, (6.3.4) holds in $\boldsymbol{S}/|||.|||$ too. Using the Brunel-Sucheston construction (see Chapter 1) we can find a subsequence $(\bar{\boldsymbol{e}}'_n) \subset (\bar{\boldsymbol{e}}_n)$ such that,

$$\Big|\sum_{i=1}^{n} \alpha_i \boldsymbol{e}_i\Big| := \lim_{i_1 < i_2 < \cdots < i_n, i_1 \to \infty} \Big|\Big|\Big|\sum_{k=1}^{n} \alpha_k \boldsymbol{e}'_{i_k}\Big|\Big|\Big|$$

exists and defines a seminorm on $\boldsymbol{S}$. This seminorm is a norm if, and only if, $(\bar{\boldsymbol{e}}'_n)$ does not contain a Cauchy sequence (see Chapter

1), and that is exactly what is happening in our case. Indeed, if A is a subset of $\mathbf{N}$ containing k_0 elements, with $k_o > 1, \delta k_0^{1/q} - k_0^{1/2} \geq 1$, then

$$\delta k_0^{1/q} \leq \Big(\mathbf{E}\Big|\Big|\Big|\sum_{n\in A} \bar{\boldsymbol{e}}_n r_n\Big|\Big|\Big|^q\Big)^{1/q}$$

$$\leq \Big(\mathbf{E}\Big|\Big|\Big|\bar{e}_{n_0}\sum_{n\in A} r_n\Big|\Big|\Big|^q\Big)^{1/q} + \Big(\mathbf{E}\Big|\Big|\Big|\sum_{n\in A}(\bar{\boldsymbol{e}}_n - \bar{\boldsymbol{e}}_{n_0} r_n\Big|\Big|\Big|^q\Big)^{1/q}$$

$$\leq k_0^{1/2} + k_0 \sup_{i,j\in\mathcal{A}} |||\bar{\boldsymbol{e}}_i - \bar{\boldsymbol{e}}_j|||,$$

which shows that $\sup_{i,j\in\mathcal{A}} |||\bar{\boldsymbol{e}}_i - \bar{\boldsymbol{e}}_j||| \geq 1/k_0$.

By Brunel-Sucheston theory (see Chapter 1), the normed space $(\boldsymbol{S}, |.|)$ is finitely representable in $S/|||.|||$ and, because the latter is finitely representable in $\boldsymbol{X}$, also $(\boldsymbol{S}, |.|)$ is finitely representable in $\boldsymbol{X}$.

The norm $|.|$ is invariant under spreading (see Chapter 1), and the inequality (6.3.4) gives, for each $(\alpha_n) \in \boldsymbol{S}$, the inequality,

$$\delta \cdot \lambda_q((\alpha_n)) \leq \Big(\mathbf{E}\Big|\sum_n \alpha_n \boldsymbol{e}_n r_n\Big|^q\Big)^{1/q} \leq \sum_n |\alpha_n|,$$

so that

$$n^{-1/q}\Big(\mathbf{E}\Big|\sum_{i=1}^n \boldsymbol{e}_i r_i\Big|\Big)^{1/q} \geq \delta,$$

which proves Part (iii), and concludes the proof of Theorem 6.3.1, as well. QED

At this point we will explore the relationship between the parameter $p_{\text{Rtype}}(\boldsymbol{X})$ of a Banach space $\boldsymbol{X}$, finite factorability through $\boldsymbol{X}$, and finite representability of spaces $\boldsymbol{l}_p$ in $\boldsymbol{X}$. For this purpose let us define the new parameter,

$$p(\boldsymbol{X}) := \inf\{p : \text{embedding } \boldsymbol{l}_1 \mapsto \boldsymbol{l}_p \text{ is finitely factorable through } \boldsymbol{X}\}. \tag{6.3.5}$$

Corollary 6.3.1. *For any infinite-dimensional normed space* $\boldsymbol{X}$,

$$p_{\text{Rtype}}(\boldsymbol{X}) = p_{\inf}(\boldsymbol{X}) = p(\boldsymbol{X}).$$

Proof. Evidently, $p(\boldsymbol{X}) \leq p_{\text{Rtype}}(\boldsymbol{X})$ by Theorem 6.3.1, and the inequalities $p_{\text{Rtype}}(\boldsymbol{X}) \leq p_{\inf}(\boldsymbol{X}) \leq p(\boldsymbol{X})$, follow directly from the definitions. QED

Theorem 6.3.2. *If* $\boldsymbol{X}$ *is an infinite-dimensional Banach space, then* $\boldsymbol{l}_{p(\boldsymbol{X})}$ *is finitely representable in* $\boldsymbol{X}$.

Proof. By Theorem 6.3.1, for each $n \in \mathbf{N}$ there exist $\boldsymbol{x}_1^n, \ldots, \boldsymbol{x}_n^n \in \boldsymbol{X}$ such that, for each $(\alpha_i) \in \mathbf{R}^n$,

$$\Big(\sum_i |\alpha_i|^p\Big)^{1/p} \leq \Big\|\sum_i \alpha_i \boldsymbol{x}_i^n\Big\| \leq 2\sum_i |\alpha_i|.$$

Let U be a non-trivial ultrafilter on $\mathbf{N}$ (one can use here a Banach limit instead). The formula, $\lim_{n\in U} \|\sum_i \alpha_i \boldsymbol{x}_i^n\|$, defines the norm on the space $\boldsymbol{S}$, say $\|.\|_1$, such that if $(\boldsymbol{e}_i)$ is a canonical basis of $\boldsymbol{S}$, then, for each $(\alpha_i) \in \boldsymbol{S}$,

$$\Big(\sum_i |\alpha_i|^p\Big)^{1/p} \leq \Big\|\sum_i \alpha_i \boldsymbol{e}_i\Big\|_1 \leq 2\sum_i |\alpha_i|. \tag{6.3.6}$$

Using the Brunel-Sucheston procedure (see Chapter 1) one can find a subsequence $(\boldsymbol{e}'_n) \subset (\boldsymbol{e}_n)$ such that the formula

$$\Big|\sum_i \alpha_i \boldsymbol{e}_i\Big| := \lim_{i_1<i_2<\ldots;i_1\to\infty} \Big\|\sum_k \alpha_k \boldsymbol{e}'_{i_k}\Big\|,$$

defines the new norm on $\boldsymbol{S}$ which satisfies (6.3.6), and which, additionally, is invariant under spreading, and such that the sequence $\boldsymbol{u}_n = e_{2n} - \boldsymbol{e}_{2n-1}$, $n = 1, 2, \ldots$, is also invariant under spreading, and unconditional with constant ≤ 2. Because of (6.3.6), for each (α_i),

$$2^{1/p}\Big(\sum_i |\alpha_i|^p\Big)^{1/p} \leq \Big|\sum_i \alpha_i \boldsymbol{u}_i\Big| \leq 4\sum_i |\alpha_i|. \tag{6.3.7}$$

Denote by $\boldsymbol{Y}$ the Banach space spanned by $(\boldsymbol{u}_n) \subset \boldsymbol{S}$ (in the $\|.\|_1$ norm). The sequence forms an unconditional basis in $\boldsymbol{Y}$, and $\boldsymbol{Y}$ is finitely representable in $\boldsymbol{X}$. Consequently, for each $\epsilon > 0$ ($\epsilon < p$), the space $\boldsymbol{Y}$ is of Rademacher-type $p - \epsilon$. Therefore, there exists a constant C_ϵ such that, for every $(\alpha_i) \in \boldsymbol{S}$,

$$\Big\|\sum_i \alpha_i \boldsymbol{u}_i\Big\|_1 \le C_\epsilon \Big(\sum_i |\alpha_i|^{p-\epsilon}\Big)^{1/(p-\epsilon)}.$$

From (6.3.7), and the above property, one can deduce that, for each $n \in \mathbf{N}$,

$$2^{1/p} \cdot 2^{n/p} \le \Big\|\sum_{i=1}^{2^n} \boldsymbol{u}_i\Big\|_1 \le C_\epsilon 2^{n/(p-\epsilon)},$$

so that, if we define

$$b := \inf\Big\{\lambda > 0 : \lim_{n\to\infty} \lambda^n \Big\|\sum_{i=1}^{2^n} \boldsymbol{u}_i\Big\|_1 = +\infty\Big\},$$

we get that $b = 2^{-1/p}$. By the Krivine Theorem (see Chapter 1), $\boldsymbol{l}_p$ is finitely representable in $\boldsymbol{Y}$, and thus also in $\boldsymbol{X}$. QED

6.4 Operators on Banach spaces of Rademacher-type p

In this section we discuss properties of operators acting from a Banach space $\boldsymbol{X}$ of Rademacher-type p into the space $\boldsymbol{L}_0(T, \Sigma, \mu; \boldsymbol{Y})$ of measurable functions on (T, Σ, μ) with values in a Banach space $\boldsymbol{Y}$.

Recall that the Lorentz space Λ_p of $\boldsymbol{X}$-valued functions on T is defined by the condition

$$\Lambda_p(T, \Sigma, \mu; \boldsymbol{X}) := \{f \in \boldsymbol{L}_0(T, \Sigma, \mu; \boldsymbol{X}) : \Lambda_p(X, \mu) < \infty\},$$

where

$$\Lambda_p(f, \mu) := \sup_{c>0} \mu(\|f\| > c)^{1/p},$$

and the gauge

$$J_\epsilon(f,\mu) := \inf\{c>0; \mu(\|f\|>c)<\epsilon\}, \qquad 0<\epsilon<1.$$

Definition 6.4.1. (*a*) We say that a set $A \subset \boldsymbol{L}_0(T,\Sigma,\mu)$ is *almost bounded in* Λ_p if, for every $\epsilon>0$, there exists a measurable set $T_\epsilon \subset T$ such that $\mu(T-T_\epsilon)\le\epsilon$, and such that the set $\{fI_{T_\epsilon} : X\in A\}$ is bounded in $\Lambda_p(T,\Sigma,\mu;\boldsymbol{X})$.

(*b*) We say that the linear operator $U : \boldsymbol{X} \mapsto \boldsymbol{L}_0(T,\Sigma,\mu;\boldsymbol{Y})$ is *almost continuous into* $\Lambda_q(T,\Sigma,\mu;\boldsymbol{Y})$ if the image by U of the unit ball in $\boldsymbol{X}$ is almost bounded in $\Lambda_q(T,\Sigma,\mu;\boldsymbol{Y})$.

Theorem 6.4.1.[15] *If* $\boldsymbol{X}$ *is a Banach space of Rademacher-type* p*,* $1\le p\le 2$*, and* $\boldsymbol{Y}$ *is a Banach space, then each linear continuous operator from* $\boldsymbol{X}$ *into* $\boldsymbol{L}_0(T,\Sigma,\mu;\boldsymbol{Y})$*, where* μ *is a finite measure, is almost continuous from* $\boldsymbol{X}$ *into* $\Lambda_p(T,\Sigma,\mu;\boldsymbol{Y})$*.*

The proof of Theorem 6.4.1 will rely on the following two Lemmas which provide alternative characterizations of the concepts of almost boundedness and almost continuity defined above.

Lemma 6.4.1. *A set* $A\subset \boldsymbol{L}_0(T,\Sigma,\mu;\mathbf{R})$ *is almost bounded in* $\Lambda_1(\mathbf{R})$ *if, and only if, for each* $\epsilon\in(0,1)$*, there exists a constant* C_ϵ *such that, for all* $(c_n)\in \boldsymbol{l}_1$*, and all* $(f_n)\subset A$*,*

$$J_\epsilon(\sup_n |c_n f_n|,\mu) \le C_\epsilon \sum_n |c_n|. \tag{6.4.1}$$

Proof. "If". By assumption, in view of the inequality (6.4.1), for each $\epsilon\in(0,1)$, there exists a constant $C>0$ such that, for all $(f_n)\subset A$, and $(c_n)\in \boldsymbol{l}_1$,

$$\sum_n |c_n|\le 1 \implies \mu\{\sup|c_n f_n|>C\}\le\epsilon. \tag{6.4.2}$$

For the sake of brevity let's introduce the concept of the N-set: we shall say that a measurable set $B\subset T$ is an N-set if $\mu(b)>0$, and if there exists a function $f\in A$ such that, for μ-almost all $t\in B$,

[15]This result is due to B. Maurey (1973/74), Exp. IV and V.

$\mu(B)|f(t)| > C$. Denote by F the set of families (B_i) of pairwise a.s. disjoint N-sets. Assume that F is non-empty, and order F by the relation : $(B_i) < (C_i)$ if (B_i) is a subfamily of (C_i). Note that F is inductive in this order.

Let (B_i) be a maximal element of F. Because μ is finite, and B_i's are disjoint, we can assume that $i \in \mathbf{N}$.

By the definition of an N-set, for each $i \in \mathbf{N}$, there exist $f_i \in A$ such that $\mu(B_i)|f_i| > C$ on B_i, so that $\sup_i \mu(B_i)|f_i| > C$ on $B = \bigcup_i B_i$. Put $c_i = \mu(B_i)$. We have $\sum_i |c_i| \leq 1$, and by (6.4.2), $\mu(B) \leq \epsilon$. Define $T_\epsilon := T \setminus B$, and let $c > 0$, and $f \in A$.

If we define $D := \{t \in T_\epsilon : |f(t)| > Cc\}$, then $\mu(D) \leq 1/c$. Otherwise D would be an N-set (on D, one would have $|f| > Cc \geq C/\mu(D)$) disjoint with B, which would contradict the maximality of (B_i). Hence, for each $c > 0$, and each $f \in A$,

$$\mu\{|f\chi_{T_\epsilon}| > Cc\} \leq 1/c, \tag{6.4.3}$$

which proves that A is almost bounded in Λ_1 in the case of nonempty F. If F is empty, then there is no N-set. Let $c > 0, f \in A$, and put $D = \{t \in T : |f(t)| > Cc\}$. Then $\mu(D) \leq 1/c$ since, in the opposite case D would be an N-set. Therefore, one can take $T_\epsilon = T$, and the proof of the implication "If" is over.

"Only if". In view of the assumption, for each $\epsilon > 0$, there exists a measurable $T_\epsilon \subset T$ with $\mu(T \setminus T_\epsilon) \leq \epsilon$, such that for each $c > 0$, and each $f \in A$ the inequality (6..4.3) holds true. let $(c_n) \subset \mathbf{R}$, $\sum_n |c_n| \leq 1$, and $(f_n) \subset A$. Put

$$B_n = \{t \in T_\epsilon : |c_n f_n(t)| > C/c\}.$$

In view of (6.4.3), $\mu(B_n) \leq \epsilon|c_n|$, and $\mu(\bigcup_n B_n) \leq \epsilon$. But,

$$\{t : \sup_n |c_n f_n(t)| > C/c\} \subset (T \setminus T_\epsilon) \cup (\bigcup_n B_n),$$

so that

$$\mu\{t : \sup_n |c_n f_n(t)| > C/c\} \leq 2\epsilon,$$

which implies (6.4.1). QED

Lemma 6.4.2. *If $\boldsymbol{X}$, and $\boldsymbol{Y}$, are Banach spaces, (T, Σ, μ) is a probability space, and $0 < q < \infty$, then the linear continuous operator U from $\boldsymbol{X}$ into $\mathbf{L}_0(T, \Sigma, \mu; \boldsymbol{Y})$ is almost continuous from $\boldsymbol{X}$ into $\Lambda_q(T, \Sigma, \mu; \boldsymbol{Y})$ if, and only if, for each $\epsilon > 0$, there exists a constant C_ϵ such that, for each $(\boldsymbol{x}_n) \subset \boldsymbol{X}$,*

$$J_\epsilon(\sup_n \|U(\boldsymbol{x}_n)\|, \mu) \leq C_\epsilon\Big(\sum_n \|\boldsymbol{x}_n\|^q\Big)^{1/q}. \tag{6.4.4}$$

Proof. First of all, notice that a set A is almost bounded in $\Lambda_q(\boldsymbol{Y})$ if, and only if, the set $\{\|f\|^q : f \in A\}$ is almost bounded in $\Lambda_1(\mathbf{R})$. To complete the proof with help of Lemma 6.4.1 it is sufficient to observe that the condition (6.4.4) is equivalent to the condition (6.4.1) for $f_n = \|U(\boldsymbol{x}_n)\|^q$, with $\|\boldsymbol{x}_n\| \leq 1$, because

$$J_\epsilon(\sup_n |c_n f_n|, \mu) = J_\epsilon(\sup_n |c_n| \|U(\boldsymbol{x}_n)\|^q, \mu)$$

$$= \Big(J_\epsilon\big(\sup_n \|U(|c_n|^{1/q}\boldsymbol{x}_n)\|, \mu\big)\Big)^q \leq C^q \sum_n |c_n|. \qquad \text{QED}$$

Now, we are ready to turn to the proof of Theorem 6.4.1.

Proof of Theorem 6.4.1. Let us begin by restating the Fubini-type inequality: If $\gamma + \delta \leq \alpha\beta$, then

$$J_\alpha\big(J_\beta(f(t,s), \mu(dt)), \nu(ds)\big) \leq J_\gamma\big(J_\delta(f(t,s), \nu(ds)), \mu(dt)\big).$$

Now, let $\epsilon > 0$ be given and choose γ, and δ, so that $\gamma + \delta \leq \epsilon/3$. By continuity of the operator U,

$$J_\delta(U(\boldsymbol{x}), \mu) \leq K_\delta \|\boldsymbol{x}\|, \qquad \boldsymbol{x} \in \boldsymbol{X},$$

and, because the space $\boldsymbol{X}$ is of Rademacher-type p,

$$J_\gamma\Big(\sum_i \boldsymbol{x}_i r_i, \mathbf{P}\Big) \leq K_\gamma\Big(\sum_i \boldsymbol{x}_i\|^p\Big)^{1/p},$$

for each finite $(\boldsymbol{x}_i) \subset \boldsymbol{X}$.

Also, observe that, for any Banach space $\boldsymbol{X}$, and $\boldsymbol{x}_1, \ldots, \boldsymbol{x}_n \in \boldsymbol{X}$,

$$\sup_i \|\boldsymbol{x}_i\| \le J_\alpha\Big(\sum_i \boldsymbol{x}_i r_i, \mathbf{P}\Big), \tag{6.4.5}$$

whenever $\alpha < 1/2$. Indeed, let $\beta = 2\alpha < 1$. Then

$$2\|\boldsymbol{x}_i\| = J_\beta(2\boldsymbol{x}_i r_i, \mathbf{P}) = J_\beta\Big[\big(\boldsymbol{x}_i r_i + \sum_{j\neq i} \boldsymbol{x}_j r_j\big) + \big(\boldsymbol{x}_i r_i - \sum_{j\neq i} \boldsymbol{x}_j r_j\big), \mathbf{P}\Big]$$

$$\le J_\beta\big(\boldsymbol{x}_i r_i + \sum_{j\neq i} \boldsymbol{x}_j r_j, \mathbf{P}\big) + J_\alpha\big(\boldsymbol{x}_i r_i - \sum_{j\neq i} \boldsymbol{x}_j r_j, \mathbf{P}\big) = 2J_\alpha\big(\sum_j \boldsymbol{x}_j r_j, \mathbf{P}\big).$$

Using (6.4.5), and Fubini's Inequality, we get that

$$J_\epsilon(\sup_i \|U(\boldsymbol{x}_i)\|, \mu) \le J_\epsilon\Big(J_{1/3}\big(\sum_i U(\boldsymbol{x}_i) r_i, \mathbf{P}\big), \mu\Big)$$

$$\le J_\gamma\Big(J_\delta\big(U\big(\sum \boldsymbol{x}_i r_i\big), \mu\big), \mathbf{P}\Big) \le K_\delta J_\gamma\Big(\Big\|\sum_i \boldsymbol{x}_i r_i\Big\|, \mathbf{P}\Big)$$

$$\le K_\delta K_\gamma \Big(\sum_i \|\boldsymbol{x}_i\|^p\Big)^{1/p},$$

which, in view of Lemma 6.4.2, concludes the proof of Theorem 6.4.1. QED

6.5 Banach spaces of stable-type p and their local structures

In analogy to the Rademacher-type spaces we will now introduce the concept of a Banach space of stable-type p in which the sequence (r_i) Rademacher random variables is replaced by the sequence (ξ_i) of independent, identically distributed, p-stable random variables with the characteristic function

$$\mathbf{E} e^{it\xi_i} = e^{-|t|^p}, \qquad t \in \mathbf{R}, \qquad 0 \le p \le 2. \tag{6.5.1}$$

Again, we will take advantage of the numerical constants, called $s_n^p(\boldsymbol{X})$ in this case, related to the concept of stable-type p,

$$s_n^p := \inf\Big\{s \in \mathbf{R}^+ : \forall \boldsymbol{x}_1, \ldots, \boldsymbol{x}_n \in \boldsymbol{X}, \big(\mathbf{E}\Big\|\sum_{i=1}^n \xi_i \boldsymbol{x}_i\Big\|^{p/2}\big)^{2/p} \le s\big(\sum_{i=1}^n \|\boldsymbol{x}_i\|^p\big)^{1/p}\Big\}.$$

Definition 6.5.1.[16] We shall say that a normed space $\boldsymbol{X}$ is of *stable-type* p if there exists a constant $C > 0$ such that, for all $n \in \mathbf{N}$, $s_n^p(\boldsymbol{X}) \le C < \infty$.

Remark 6.5.1. In view of Hoffmann-Jorgensen Theorem (see Chapter 1) $\boldsymbol{X}$ is of stable-type p if, and only if, there exists a constant $C > 0$ such that, for each $\alpha, 0 \le \alpha < p^*$ ($p^* = p$, if $p < 2$, and $p^* = \infty$, if $p = 2$), and for each finite $(\boldsymbol{x}_i) \subset \boldsymbol{X}$,

$$\Big(\mathbf{E}\Big\|\sum_i \xi_i \boldsymbol{x}_i\Big\|^\alpha\Big)^{1/\alpha} \le C\Big(\sum_i \|\boldsymbol{x}_i\|^p\Big)^{1/p}, \tag{6.5.2}$$

or, alternatively, $\boldsymbol{X}$ is of stable-type p if, and only if, for each $\alpha, 0 \le \alpha < p^*$, there exists a constant $C > 0$ such that, for each finite $(\boldsymbol{x}_i) \subset \boldsymbol{X}$, the inequality (6.5.2) holds true. In the case $\alpha = 0$, the inequality (6.5.2) has the usual interpretation: $\boldsymbol{X}$ is of stable-type p if, and only if, for each $\epsilon \in (0,1)$ there exists a constant $C > 0$ such that, for every finite $(\boldsymbol{x}_i) \subset \boldsymbol{X}$,

$$J_\epsilon\Big(\Big\|\sum_i \xi_i \boldsymbol{x}_i\Big\|, P\Big) \le C\Big(\sum_i \|\boldsymbol{x}_i\|^p\Big)^{1/p}. \tag{6.5.3}$$

The relationship between Rademacher and stable types is explained in the following Proposition. In the case $p = 2$ both notions coincide and this case will be discussed separately in detail in Chapter 7.

---

[16]Investigation of spaces of stable-type was initiated by J. Hoffman-Jorgensen (1972/73), B. Maurey (1972/73), Exp, VII, X, XI, and B. Maurey and G, Pisier (1973).

Proposition 6.5.1. *(i) If $\boldsymbol{X}$ is of stable-type $p, 1 \leq p \leq 2$, then it is also of Rademacher-type p.*

(ii) If $\boldsymbol{X}$ is of Rademacher-type $p, 1 \leq p \leq 2$, and $q < p$, then $\boldsymbol{X}$ is of stable-type q.

Proof. (i) Because each normed space is of Rademacher-type 1, we can assume that $p > 1$. Let $\xi_1, \ldots, \xi_n$ be independent and identically distributed p-stable random variables with the characteristic function (6.5.1), and the Rademacher sequence (r_i) be independent of (ξ_i). If $\boldsymbol{X}$ is of stable-type p, then

$$\mathbf{E}\Big\|\sum_i r_i \boldsymbol{x}_i\Big\| = (\mathbf{E}_\xi|\xi_i|)^{-1}\mathbf{E}\Big\|\sum_i r_i \mathbf{E}_\xi|\xi_i|\boldsymbol{x}_i\Big\|$$

$$\leq (\mathbf{E}_\xi|\xi_i|)^{-1}\mathbf{E}\Big\|\sum_i r_i|\xi_i|\boldsymbol{x}_i\Big\|$$

$$\leq (\mathbf{E}_\xi|\xi_i|)^{-1}\mathbf{E}\Big\|\sum_i \xi_i\boldsymbol{x}_i\Big\| \leq (\mathbf{E}_\xi|\xi_i|)^{-1}C\Big(\sum_i \|\boldsymbol{x}_i\|^p\Big)^{1/p},$$

because (ξ_i), and $(r_i|\xi_i|)$ have identical distributions. This proves that $\boldsymbol{X}$ is of Rademacher-type p in view of Remark 6.2.1.

(ii) Recall that if (ξ_i) are independent α-stable random variables and $\beta < \alpha$, then

$$\Big(\sum_i |r_i|^\alpha\Big)^{1/\alpha}(\mathbf{E}|\xi_i|^\beta)^{1/\beta} = \Big(\mathbf{E}\Big|\sum_i r_i\xi_i\Big|^\beta\Big)^{1/\beta}. \tag{6.5.4}$$

Now, let $0 < r < q < p$, (ξ_i) be a sequence of independent q-stable random variables, and (r_i) be independent of (ξ_i). Since $\boldsymbol{X}$ is of Rademacher-type p, by Remark 6.2.1,

$$\Big(\mathbf{E}\Big\|\sum_i r_i\boldsymbol{x}_i\Big\|^r\Big)^{1/r} \leq C\Big(\sum_i \|\boldsymbol{x}_i\|^p\Big)^{1/p},$$

and, for a fixed $\omega \in \Omega$,

$$\mathbf{E}_r\Big\|\sum_i r_i\xi_i(\omega)\boldsymbol{x}_i\Big\|^r \leq C^r\Big(\sum_i \|\boldsymbol{x}_i\|^p|\xi_i(\omega)|^p\Big)^{r/p},$$

from which, by integration, and (6.5.4) (used twice),

$$\Big(\mathbf{E}\Big\|\sum_i r_i\xi_i\boldsymbol{x}_i\Big\|^r\Big)^{1/r} \le C\Big[\mathbf{E}\Big(\sum_i \|\boldsymbol{x}_i\|^p|\xi_i|^p\Big)^{r/p}\Big]^{1/r}$$

$$= C(\mathbf{E}\|\eta_1\|^r)^{-1/r}\Big(\mathbf{E}\Big|\sum_i \|\boldsymbol{x}_i\|\xi_i\eta_i\Big|^r\Big)^{1/r}$$

$$= C(\mathbf{E}\|\eta_i\|^r)^{-1/r}(\mathbf{E}\|\xi_1\|^r)^{1/r}\Big(\mathbf{E}\Big(\sum_i \|\boldsymbol{x}_i\|^q|\eta_i|^q\Big)^{r/q}\Big)^{1/r},$$

where (η_i) are independent, identically distributed p-stable random variables independent of (ξ_i). Because (ξ_i), and $(r_i\xi_i)$, are identically distributed, and in view of the Jensen's Inequality $(r/q < 1\,!)$,

$$\Big(\mathbf{E}\Big\|\sum_i \xi_i\boldsymbol{x}_i\Big\|^r\Big)^{1/r} \le C_1\Big(\mathbf{E}\Big(\sum_i \|\boldsymbol{x}_i\|^q|\eta_i|^q\Big)^{r/q}\Big)^{1/r}$$

$$\le C_1\Big(\mathbf{E}\sum_i \|\boldsymbol{x}_i\|^q|\eta_i|^q\Big)^{1/q} = C_1(\mathbf{E}\|\eta_i\|^q)^{1/q}\Big(\sum_i \|\boldsymbol{x}_i\|^q\Big)^{1/q},$$

again by (6.5.4), so that $\boldsymbol{X}$ is of stable-type q. QED

The following Corollary is a straightforward consequence of the Proposition 6.5.1, and the results contained in Sections 6.2 and 6.3.

Corollary 6.5.1. *(i) Each normed space is of stable-type* p*, whenever* $0 < p < 1$.

(ii) If $\boldsymbol{X}$ *is of stable-type* p*, and* $p_1 < p$*, then* $\boldsymbol{X}$ *is also of stable-type* p_1.

(iii) The spaces $\boldsymbol{L}_p$*, and* $\boldsymbol{l}_p$*, are of stable-type* q*, for each* q *such that* $0 < q < p \le 2$.

On the other hand we have the following examples of spaces that are not of stable-type p even though they are of Rademacher-type p.

Example 6.5.1. If $0 < p < 2$, then the spaces $\boldsymbol{L}_p$, and $\boldsymbol{l}_p$ are not of stable-type p. Indeed, were (say) $\boldsymbol{l}_p$ of stable-type p, for each each $\alpha \in (0,1)$ there would exist a constant $K > 0$ such that

$$J_\alpha\Big(\Big\|\sum_i \alpha_i\boldsymbol{e}_i\xi_i\Big\|, \mathbf{P}\Big) \le K\Big(\sum_i \|\alpha_i\boldsymbol{e}_i\|^p\Big)^{1/p},$$

for each $(\alpha_i) \subset \mathbf{R}$, with $(\boldsymbol{e}_i)$ being, as usual, the canonical basis iin $\boldsymbol{l}_p$. Thus,

$$J_\alpha\Big(\Big(\sum_i |\alpha_i\xi_i|^p\Big)^{1/p}, \mathbf{P}\Big) \le K\Big(\sum_i |\alpha_i|^p\Big)^{1/p},$$

which would imply that the series $\sum_i |\alpha_i \xi_I|^p$ converges a.s. whenever $\sum_i |\alpha_i|^p < \infty$. But this implication is not true (see, e.g., Schwartz Theorem, Chapter 1).

For $p \ge 2$ the situation is different.

Example 6.5.2. If $2 \le p < \infty$, then the spaces $\boldsymbol{L}_p$ and $\boldsymbol{l}_p$, are of stable-type 2. To check this fact, let (γ_i) be a sequence of independent $N(0,1)$ Gaussian (i.e., 2-stable) random variables, and $(\alpha_i) \subset \mathbf{R}$. In view of the elementary probability result, there exists a constant $C > 0$, such that

$$\Big(\mathbf{E}\Big|\sum_i \alpha_i\gamma_i\Big|^p\Big)^{1/p} \le C\Big(\sum_i |\alpha_i|^2\Big)^{1/2} \tag{6.5.5}$$

If $(\boldsymbol{x}_i) \subset \boldsymbol{L}_p(T, \Sigma, \mu), p \ge 2$, then

$$\Big(\mathbf{E}\Big\|\sum_i \gamma_i \boldsymbol{x}_i\Big\|^p\Big)^{1/p} = \Big(\mathbf{E}\int_T\Big|\sum_i \gamma_i \boldsymbol{x}_i(t)\Big|^p \mu(dt)\Big)^{1/p}$$

$$= \Big(\int_T\Big(\mathbf{E}\Big|\sum_i \gamma_i x_i(t)\Big|^p\Big)^{(1/p)p} \mu(dt)\Big)^{1/p}$$

$$\le C\Big(\int_T\Big(\sum_i |\boldsymbol{x}_i(t)|^2\Big)^{p/2} \mu(dt)\Big)^{1/p}$$

$$\le C\Big[\sum_i\Big(\int_T |\boldsymbol{x}_i(t)|^p \mu(dt)\Big)^{2/p}\Big]^{1/2} = C\Big(\sum_i \|\boldsymbol{x}_i\|^2\Big)^{1/2},$$

with the next to the last inequality being implied by the fact that $p/2 \ge 1$. QED

Actually, Example 6.5.1 can be generalized to produce the following important geometric characterization of spaces of stable-type p, for $1 \le p < 2$.

Theorem 6.5.1.[17] *A normed space* $\boldsymbol{X}$ *is of stable-type* $p, 1 \leq p < 2$, *if, and only if,* $p < p(\boldsymbol{X})$, *that is if, and only if, the canonical embedding* $\boldsymbol{l}_1 \mapsto \boldsymbol{l}_p$ *is not finitely factorable through* $\boldsymbol{X}$.

Proof. If $p < p(\boldsymbol{X})$ then Proposition 6.5.1 (*ii*), and Corollary 6.3.1, immediately imply that $\boldsymbol{X}$ is of stable-type p. Conversely, if $p \geq p(\boldsymbol{X})$ then, by Theorem 6.3.1, and Corollary 6.3.1, the embedding $\boldsymbol{l}_1 \mapsto \boldsymbol{l}_p$ is finitely factorable through $\boldsymbol{X}$, so that for each $n \in \mathbf{N}$ there exist $\boldsymbol{x}_1, \ldots, \boldsymbol{x}_n \in \boldsymbol{X}$ such that, for all $(\alpha_i) \in \mathbf{R}^n$,

$$\Big(\sum_i |\alpha_i|^p\Big)^{1/p} \leq \Big\|\sum_i \alpha_i \boldsymbol{x}_i\Big\| \leq 2\sum_i |\alpha_i|.$$

In particular, $1 \leq \|\boldsymbol{x}_i\| \leq 2$. Now, let (ξ_I) be standard independent p-stable random variables, and $r < p$. Then, were $\boldsymbol{X}$ of stable-type p we would have that

$$\Big(\mathbf{E}\Big(\sum_{i=1}^n |\alpha_i|^p|\xi_i|^p\Big)^{r/p}\Big)^{1/r} \leq \Big(\mathbf{E}\Big\|\sum_{i=1}^n \alpha_i\xi_i\boldsymbol{x}_i\Big\|^r\Big)^{1/r}$$

$$\leq C\Big(\sum_{i=1}^n |\alpha_i|^p\|\boldsymbol{x}_i\|^p\Big)^{1/p} \leq 2C\Big(\sum_i |\alpha_i|^p\Big)^{1/p},$$

so that, again, $\sigma_i|\alpha_i\xi_i|^p$ would converge a.s. whenever $\sum_i |\alpha_i|^p < \infty$, the implication that is not true in view of the Schwartz Theorem (see Chapter 1). QED

Corollary 6.5.2. *(i) A normed space* $\boldsymbol{X}$ *is of stable-type* $p, 1 \leq p < 2$, *if, and only if,* $\boldsymbol{l}_p$ *is not finitely representable in* $\boldsymbol{X}$.

(ii) The interval of those p*'s for which* $\boldsymbol{X}$ *is of stable-type* p *is open whenever* $p(\boldsymbol{X}) < 2$, *that is, if* $\boldsymbol{X}$ *is of stable-type* $p, p < 2$, *then* $\boldsymbol{X}$ *is of stable-type* q, *for some* $q > p$.

Proof. Part (*i*) follows directly from the above Theorem 6.5.1, and Theorem 6.3.2. Part (*ii*) is also a direct consequence of the above Theorem, and the fact that the set of those p's for which the embedding $\boldsymbol{l}_1 \mapsto \boldsymbol{l}_p$ is finitely factorable through $\boldsymbol{X}$ is closed (see Chapter 1). QED

[17]This result is due to B. Maurey and G. Pisier (1976).

In the next Theorem we provide a characterization of subspaces of the space $\boldsymbol{L}_q$ which are of stable-type p.

Theorem 6.5.2.[18] *Let $1 \leq p < 2$, and let $\boldsymbol{X}$ be a closed subspace of the space $\boldsymbol{L}_q(T, \Sigma, \mu)$, where μ is a finite measure. Then, the following four conditions are equivalent:*

(i) $\boldsymbol{X}$ is of stable-type p;

(ii) $\boldsymbol{X}$ does not contain a subspace isomorphic to $\boldsymbol{l}_p$;

(iii) $\boldsymbol{X}$ does not contain a complemented subspace isomorphic to $\boldsymbol{l}_p$;

(iv) For some $r, 0 \leq r < p$, the topologies of $\boldsymbol{L}_p$, and $\boldsymbol{L}_r$, coincide on $\boldsymbol{X}$.

Proof. $(i) \implies (ii)$ follows directly from Example 6.5.1.

$(ii) \implies (iii)$ is obvious.

$(iii) \implies (iv)$ This implication is an immediate corollary to the well known Kadec-Pelczynski Theorem which states that if the sequence $(\boldsymbol{x}_n) \subset \boldsymbol{L}_p, p \geq 1$, is such that, for each $\epsilon > 0$,

$$(\boldsymbol{x}_n) \subset \big\{\boldsymbol{x} \in \boldsymbol{L}_p : \mu\{\boldsymbol{x}(t) \geq \epsilon\|\boldsymbol{x}\|\} \geq \epsilon\big\},$$

then, for each $\delta > 0$, there exists a subsequence $(\boldsymbol{y}_k) = (\boldsymbol{x}_{n_k})$ such that $(\boldsymbol{y}_k/\|\boldsymbol{y}_k\|)$ is a basic sequence, equivalent to the canonical basis in $\boldsymbol{l}_p$ with constant $(1 + \delta)$, and such that the span of $(\boldsymbol{x}_n)$ is $(1 + \delta)$-complemented in $\boldsymbol{L}_p$.

$(iv) \implies (i)$ Assume, first, that $0 < r < p$, and that the topologies of $\boldsymbol{L}_r$, and $\boldsymbol{L}_p$, coincide on $\boldsymbol{X}$, that is, there exists a constant $C > 0$ such that, for all $\boldsymbol{x} \in \boldsymbol{X}$,

$$\|\boldsymbol{x}\| \leq C\Big(\int_T |\boldsymbol{x}(t)|^r \mu(dt)\Big)^{1/r}.$$

If (ξ_i) is a sequence of independent, identically distributed p-stable random variables, and $(\boldsymbol{x}_i) \subset \boldsymbol{X}$, then

$$\Big(\mathbf{E}\Big\|\sum_i \xi_i \boldsymbol{x}_i\Big\|^r\Big)^{1/r} \leq C\Big(\mathbf{E}\int_T \Big|\sum_i \xi_i \boldsymbol{x}_i(t)\Big|^r \mu(dt)\Big)^{1/r}$$

[18]This result is due to B. Maurey (1972/73), Exp. X, and XI.

$$= C\Big(\int_T \mathbf{E}\Big|\sum_i \xi_i \boldsymbol{x}_i(t)\Big|^r \mu(dt)\Big)^{1/r}$$

$$= C\Big(\int_T \mathbf{E}|\xi_i|^r \Big(\sum_i |\boldsymbol{x}_i(t)|^p\Big)^{r/p} \mu(dt)\Big)^{1//r}$$

$$\leq C\mathbf{E}|\xi_i|^r \Big(\int_T \sum_i |\boldsymbol{x}_i(t)|^p \mu(dt)\Big)^{1/p} = C\mathbf{E}|\xi_i|^r \Big(\sum_i \|\boldsymbol{x}_i\|^p\Big)^{1/p},$$

in view of (6.5.4) , and since $r/p < 1$, so that $\boldsymbol{X}$ is of stable-type p.

In the case $r = 0$ we proceed in a similar fashion. By assumption, for each $\beta \in (0,1)$, there exists a constant $C > 0$ such that, for each $\boldsymbol{x} \in \boldsymbol{X}$,

$$\|\boldsymbol{x}\| \leq C J_\beta(|\boldsymbol{x}(t)|, \mu).$$

Utilizing the Fubini Inequality (see Chapter 1), we get that for each $\alpha \in (0, 1/4)$,

$$J_{4\alpha}\Big(\Big\|\sum_i \xi_i \boldsymbol{x}_i\Big\|, \mathbf{P}\Big) \leq C J_{4\alpha}\Big[\Big(J_{1/2}\Big(\Big|\sum_i \xi_i \boldsymbol{x}_i(t)\Big|, \mu\Big), \mathbf{P}\Big]$$

$$\leq C J_\alpha\Big[\Big(J_\alpha\Big(\Big|\sum_i \xi_i \boldsymbol{x}_i(t)\Big|, \mu\Big), \mathbf{P}\Big]$$

$$= C J_\alpha\Big[J_\alpha(\xi_1)\Big(\sum_i |\boldsymbol{x}_i(t)|^p\Big)^{1/p}, \mu\Big] \leq C_1\Big(\sum_i \|\boldsymbol{x}_i\|^p\Big)^{1/p},$$

so that, again, $\boldsymbol{X}$ is of stable-type p. QED

Remark 6.5.2. Note that the above result evidently fails for $p = 2$.

To conclude this section let us verify how the stable-type property behaves under standard operations on normed linear spaces.

Proposition 6.5.2. *(i) If $\boldsymbol{X}$ is of stable-type p and $\boldsymbol{Y}$ is a subspace of $\boldsymbol{X}$, then $\boldsymbol{Y}$ is of stable-type p, as well.*

(ii) If $\boldsymbol{X}$ is of stable-type p and $\boldsymbol{Y}$ is a closed subspace of $\boldsymbol{X}$, then $\boldsymbol{X}/\boldsymbol{Y}$ is of stable-type p.

Proof. Part ((*i*) is evident. The staighforward proof of (*ii*) is analogous to the proof of Proposition 6.2.4(ii) and will be omitted. QED

Theorem 6.5.3. *If* $1 \le p < 2$, *and* $\boldsymbol{X}$ *is of stable-type* p, *the space* $\boldsymbol{L}_q(\boldsymbol{X})$ *is of stable-type* p *whenever* $q > p$.

Proof. The fact that $\boldsymbol{X}$ is of stable-type p, $p < 2$, implies the existence of an r, $p < r < q$, such that $\boldsymbol{X}$ is of stable-type r (see Corollary 6.5.2 (*ii*)). Now, Proposition 6.5.1(*i*)implies that $\boldsymbol{X}$ is of Rademacher-type r, so that, by Theorem 6.2.5 (*ii*), $\boldsymbol{L}_q(\boldsymbol{X})$ is of Rademacher-type r as well, and therefore, by Proposition 6.5.1 (*ii*), $\boldsymbol{L}_q(\boldsymbol{X})$ is of stable-type p. QED

Remark 6.5.3. If $\boldsymbol{X}$ is of stable-type 2, then $\boldsymbol{L}_q(\boldsymbol{X})$ is of stable-type 2, whenever $q \ge 2$. We shall prove this fact in Chapter 7.

Finally, here is yet another procedure which produces new spaces of stable-type p

Proposition 6.5.3. *Denote by* $[r_n] \otimes \boldsymbol{X}$ *the subspace of* $\boldsymbol{L}_2(\Omega, \mathbf{P}; \boldsymbol{X})$ *spanned by the sums* $\sum_n r_n \boldsymbol{x}_n$, $(\boldsymbol{x}_n) \subset \boldsymbol{X}$. *Then* $\boldsymbol{X}$ *is of stable-type* p *if, and only if,* $[r_n] \otimes \boldsymbol{X}$ *is of stable-type* p.

Proof. The "if" part is evident. So, let's prove the "only if" part. Consider the sequence (ξ_i) of independent and identically distributed p-stable random variables independent of the sequence $(X_i) \subset [r_n] \otimes \boldsymbol{X}$. Then, if $r < p$, by Kahane Theorem (see Chapter 1),

$$\Big(\mathbf{E}_\xi \Big\| \sum_n X_n \xi_n \Big\|^r_{L_2(X)}\Big)^{1/r} \le C \Big(\mathbf{E}_\xi \Big\| \sum_n X_n \xi_n \Big\|^r_{L_r(X)}\Big)^{1/r}$$

$$\le C \Big(\mathbf{E} \Big\| \sum_n X_n \xi_n \Big\|^r\Big)^{1/r} \le C_1 \Big(\mathbf{E}_r \Big(\sum_n \|X_n\|^p\Big)^{r/p}\Big)^{1/r}$$

$$\le C_1 \Big(\sum_n \|X_n\|^p_{L_p(X)}\Big)^{1/p} \le C_1 \Big(\sum_n \|X_n\|^p_{L_2(X)}\Big)^{1/p}. \qquad \text{QED}$$

The "three space problem" for spaces of stable-type has a solution even neater than the analogous problem for spaces of Rademacher-type.

Theorem 6.5.4. *Let $1 \le p < 2$, and $\boldsymbol{Y}$ be a closed subspace of $\boldsymbol{X}$. If both spaces, $\boldsymbol{Y}$, and $\boldsymbol{X}/\boldsymbol{Y}$, are of stable-type p, then $\boldsymbol{X}$ is of stable-type p as well.*

Proof. By Corollary 6..5.2 (*ii*), there exists a q, $p < q < 2$, such that $\boldsymbol{Y}$, and $\boldsymbol{X}/\boldsymbol{Y}$, are of stable-type q, and thus also of Rademacher-type q. By Theorem 6.2.3, $\boldsymbol{X}$ is of Rademacher-type q_1, for each $q_1 < q$. In particular, if $p < q$, this implies that $\boldsymbol{X}$ is of stable-type p in view of Proposition 6.5.1(*ii*). QED

6.6 Operators on spaces of stable-type p

We begin with the result on factorization of operators from a space of stable-type p into the Lebesgue space $\boldsymbol{L}_q$.

Theorem 6.6.1.[19] *Let $\boldsymbol{X}$ be a normed space of stable-type p, $p \ge 1$, and let $0 < q \le p \le 2$. Then each linear continuous operator $U : \boldsymbol{X} \mapsto \boldsymbol{L}_q(T,\mu)$, where μ is a finite measure, can be factored as follows:*

$$\boldsymbol{X} \overset{V}{\mapsto} \boldsymbol{L}_p(T,\mu) \overset{T_g}{\mapsto} \boldsymbol{L}_q(T,\mu),$$

where V is linear and continuous, and T_g is the operator of multiplication by a function $g \in \boldsymbol{L}_r(T,\mu)$, with $1/q = 1/p + 1/r$.

Proof. We can assume that $q < p$. In view of Maurey Theorem (see Chapter 1) it is sufficient to prove that

$$\sum_n \|\boldsymbol{x}_n\|^p < \infty \quad \Longrightarrow \quad \int_T \Big(\sum_n |U(\boldsymbol{x}_n)(t)|^p\Big)^{q/p} \mu(dt) < \infty.$$

Indeed, as in the proof of Proposition 6.5.1(*ii*), if (ξ_i) is a sequence of independent and identically distributed p-stable random variables, and $\boldsymbol{X}$ is of stable-type p, then

$$\Big(\int_T \Big(\sum_n |U(\boldsymbol{x}_i)(t)|^p\Big)^{q/p} \mu(dt)\Big)^{1/q}$$

[19]Most of the results in this Section are due to B. Maurey (1974).

$$
\begin{aligned}
&= \left(\mathbf{E}|\xi_1|^q\right)^{-1/q}\left(\int_T \mathbf{E}\Big|\sum_i U(\boldsymbol{x}_i)(t)\xi_i\Big|^q \mu(dt)\right)^{1/q} \\
&= \left(\mathbf{E}|\xi_1|^q\right)^{-1/q}\left(\mathbf{E}\Big\|\sum_i U(\boldsymbol{x}_i)\xi_i\Big\|^q\right)^{1/q} \\
&\le \left(\mathbf{E}|\xi_1|^q\right)^{-1/q}\|U\|\left(\mathbf{E}\Big\|\sum_i \boldsymbol{x}_i\xi_i\Big\|^q\right)^{1/q} \\
&\le C\left(\mathbf{E}|\xi_1|^q\right)^{-1/q}\|U\|\left(\sum_i \|\boldsymbol{x}_i\|^q\right)^{1/q},
\end{aligned}
$$

which concludes the proof of the theorem. QED

In the next step we will discuss the relationship between spaces of stable-type and absolutely summing operators which were introduced in Section 1.5. The next result is a corollary to Theorem 6.6.1.

Proposition 6.6.1 *Let the dual space $\boldsymbol{X}^*$ be of stable-type p, $p \ge 1$, and let $0 < q \le p \le 2$. Then there exists a constant $C > 0$ such that, for each operators from $\boldsymbol{X}$ into a Banach space $\boldsymbol{Y}$,*

$$\pi_q(U) \le C\pi_p(U),$$

where π_q, and π_p, denote the norms in the spaces of q-, and p-, absolutely summing operators, respectively, defined in Section 1.5.

Proof. By Theorem 6.6.1, for any $U : \boldsymbol{X}^* \mapsto \boldsymbol{L}_q$, we have the factorization $U = t_g \circ V$, where $V : \boldsymbol{X}^* \mapsto \boldsymbol{L}_p$ is bounded, T_g is a multiplication by a function $g \in \boldsymbol{L}_r$, $1/q + 1/p + 1/r$, and $||g||_{L_r} \le 1$, and $\|V\| \le C\|U\|$. Therefore, if $\boldsymbol{x}_1, \ldots, \boldsymbol{x}_n \in \boldsymbol{X}$, one can find $\alpha_1, \ldots, \alpha_n \in \mathbf{R}$, and $\boldsymbol{y}_1, \ldots, \boldsymbol{y}_n \in \boldsymbol{X}$, such that $\boldsymbol{x}_i = \alpha_i \boldsymbol{y}_i$, $\sum_i |\alpha_i|^r \le 1$, and

$$\sup_{\|\boldsymbol{x}^*\|\le 1}\left(\sum_i |\boldsymbol{x}^*\boldsymbol{y}_i|^p\right)^{1/p} \le C \sum_{\|\boldsymbol{x}^*\|\le 1}\left(\sum_i |\boldsymbol{x}^*\boldsymbol{x}_i|^q\right)^{1/q}. \tag{6.6.1}$$

Indeed, assuming that not all of $\boldsymbol{x}_i$'s are equal to 0, in view of Theorem 6.6.1, one can factor the operator

$$\boldsymbol{X}^* \ni \boldsymbol{x}^* \mapsto (\boldsymbol{x}^*\boldsymbol{x}_i) \in \boldsymbol{l}_q^{(n)},$$

as follows: $U = \alpha \circ V$, where α is the diagonal operator of multiplication, with $\sum_i |\alpha_i|^r \leq 1$, and $V : \boldsymbol{X}^* \mapsto \boldsymbol{l}_p^{(n)}$, with $\|V\| \leq C\|U\|$. Taking $\boldsymbol{y}_i = \boldsymbol{x}_i/\alpha_i$, we get

$$\sup_{\|\boldsymbol{x}^*\|\leq 1} \Big(\sum_i |\boldsymbol{x}^*\boldsymbol{y}_i|^p\Big)^{1/p} \leq \|V\| \leq C\|U\| = C \sup_{\|\boldsymbol{x}^*\|\leq 1} \Big(\sum_i |\boldsymbol{x}^*\boldsymbol{x}_i|^q\Big)^{1/q},$$

which establishes (6.6.1). Now, if $W : \boldsymbol{X} \mapsto \boldsymbol{Y}$, and $\boldsymbol{x}_1, \ldots \boldsymbol{x}_n \in \boldsymbol{X}$, using (6.6.1), and the Hölder Inequality, we get that

$$\Big(\sum_i \|W(\boldsymbol{x}_i)\|^q\Big)^{1/q} = \Big(\sum_i \|\alpha_i W(\boldsymbol{y}_i)\|^q\Big)^{1/q}$$

$$\leq \Big(\sum_i |\alpha_i|^r\Big)^{1/r} \Big(\sum_i \|W(\boldsymbol{y}_i)\|^p\Big)^{1/p} \leq \pi_p(W) \sup_{\|\boldsymbol{x}^*\|\leq 1} \Big(\sum_i |\boldsymbol{x}^*\boldsymbol{y}_i|^p\Big)^{1/p}$$

$$\leq C\pi_p(W) \sup_{\|\boldsymbol{x}^*\|\leq 1} \Big(\sum_i |\boldsymbol{x}^*\boldsymbol{x}_i|^q\Big)^{1/q}. \qquad \text{QED}$$

Theorem 6.6.2. *Let $1 \leq p < 2$. The space $\boldsymbol{X}$ is of stable type p if, and only if, there exists a constant $C > 0$ such that, for every quotient space $\boldsymbol{X}^*/\boldsymbol{Z}$, and each linear operator U acting from $\boldsymbol{X}^*/\boldsymbol{Z}$ into a Banach space $\boldsymbol{Y}$,*

$$\pi_{1/2}(U) \leq C\pi_p(U),$$

i.e., in particular, for any Banach space $\boldsymbol{Y}$, the operator space $\Pi_p(\boldsymbol{X}^*, \boldsymbol{Y}) = \Pi_0(\boldsymbol{X}^*, \boldsymbol{Y})$.

In the proof of the theorem we shall have need of the following lemma (also, note that the choice of $1/2$ above is arbitrary because, for $0 < \alpha < 1$, all the norms π_α are equivalent).

Lemma 6.6.1. *Let $1 \leq p < 2$, and $r < p$. Denote by W_n the operator from $\boldsymbol{l}_q^{(n)}$ into $\boldsymbol{l}_p^{(n)}$ $(1/p + 1/q = 1)$ defined by the formula,*

$$W_n((\alpha_i)) = (n^{-1/p}\alpha_i).$$

Then $\pi_r(W_n) \geq K(\log n)^{1/p}$, where the constant K does not depend on n.

Proof. By Pietsch Theorem (see Chapter 1), there exists a probability measure μ on the unit ball of $\boldsymbol{l}_p^{(n)}$ such that, for all $(\alpha_i) \in \boldsymbol{l}_q^{(n)}$,

$$\Big(n^{-1}\sum_i |\alpha_i|^{[}\Big)^{1/p} \leq \pi_r(W_n)\Big[\int |(\alpha_i)(\beta_i)|^r \mu(d(\beta_i))\Big]^{1/r}.$$

If (ξ_i) are independent and identically distributed p-stable random variables then, utilizing the above inequality with $\alpha_i = \xi_i(\omega)$, we get that

$$\Big(\mathbf{E}\Big(n^{-1}\sum_{i=1}^n |\xi_i|^p\Big)^{r/p}\Big)^{1/r} \leq \pi_r(W_n)\Big[\int \mathbf{E}\Big|\sum_i \beta_i\xi_i\Big|^r \mu(d(\beta_i))\Big]^{1/r},$$

so that, in view of the properties of p-stable laws,

$$c_1^{-1}(\log n)^{1/p} \leq \pi_r(W_n)\Big[\int \mathbf{E}\Big|\sum_i \beta_i\xi_i\Big|^r \mu(d(\beta_i))\Big]^{1/r}$$

$$= C\pi_r(W_n)\Big[\Big(\sum_i |\beta_i|^p\Big)^{r/p}\mu(d(\beta_i))\Big]^{1/r} \leq C\pi_r(W_n). \qquad \text{QED}$$

Proof of Theorem 6.6.2. If $\boldsymbol{X}$ is of stable-type p then, by Theorem 6.5.2, $p > p(\boldsymbol{X})$, and the canonical embedding $\boldsymbol{l}_1 \mapsto \boldsymbol{l}_p$ is finitely factorable through $\boldsymbol{X}$, i.e., for each $n \in \mathbf{N}$ there exist $\boldsymbol{x}_1, \ldots, \boldsymbol{x}_n \in \boldsymbol{X}$ such that, for all $\alpha_a, \ldots, \alpha_n \in \mathbf{R}$,

$$\Big(\sum_i |\alpha_i|^p\Big)^{1/p} \leq \Big\|\sum_{i=1}^n \alpha_i \boldsymbol{x}_i\Big\| \leq 2\sum_i |\alpha_i|.$$

Put $\boldsymbol{Z} = \{\boldsymbol{x}^* \in \boldsymbol{X}^* : \boldsymbol{x}^*\boldsymbol{x}_i = 0, i = 1, \ldots, n\}$. In the space $\boldsymbol{X}^*/\boldsymbol{Z}$ (which is the dual of the span of $\{\boldsymbol{x}_1, \ldots, \boldsymbol{x}_n\}$), one can find $\boldsymbol{y}_1^*, \ldots, \boldsymbol{y}_n^*$ such that, for every $\alpha_1, \ldots, \alpha_n \in \mathbf{R}$,

$$2^{-1}\sup|\alpha_i| \leq \Big\|\sum_i \alpha_i \boldsymbol{y}_i^*\Big\| \leq \Big(\sum_i |\alpha_i|^q\Big)^{1/q}, \qquad 1/p + 1/q = 1.$$

This means that there exist operators U_n, and V_n,

$$\boldsymbol{l}_q^{(n)} \overset{U_n}{\mapsto} \boldsymbol{X}^*/\boldsymbol{Z} \overset{V_n}{\mapsto} \boldsymbol{l}_\infty^{(n)},$$

such that $\|U_n\| \leq 1, \|V_n\| \leq 2$, and which factor the embedding $\boldsymbol{l}_q^{(n)} \mapsto \boldsymbol{l}_\infty^{(n)}$.

Denote by $\tilde{W}_n : \boldsymbol{l}_\infty^{(n)} \mapsto \boldsymbol{l}_p^{(n)}$, defined by the equality,

$$\tilde{W}_n((\alpha_i)) = (n^{-1/p}\alpha_i).$$

In the notation of the preceding Lemma,

$$W_n = \tilde{W}_n \circ V_n \circ U_n, \qquad \pi_p(\tilde{W}_n) \leq 1,$$

and, hence, for the operator $V = \tilde{W}_n \circ V_n : \boldsymbol{X}^*/\boldsymbol{Z} \mapsto \boldsymbol{l}_p^{(n)}$, one has the inequalities,

$$K^{-1}(\log n)^{1/p} \leq \pi_{1/2}(W_n) \leq \pi_{1/2}(V)\|U_n\| \leq \pi_{1/2}(V),$$

where $\pi_p(V) \leq \|V_n\| \leq 2$. Therefore, the inequality $\pi_{1/2}(U) \leq X\pi_p(U)$ would be impossible for the operator U from the quotient of the space $\boldsymbol{X}^*$.

To prove the converse we use the Proposition 6.6.1 which states that, if $\boldsymbol{X}^*$ is of stable-type p, then $\pi_{1/2}(U) \leq C\pi_p(U)$, for each U acting on $\boldsymbol{X}$. If $\boldsymbol{X}$ is of stable-type p then the dual of the dual space, $\boldsymbol{X}^{**}$, is also of stable-type p because $\boldsymbol{X}^{**}$ is finitely representable in $\boldsymbol{X}$ (see, Chapter 1), so that the subspaces of $\boldsymbol{X}^{**}$ are of stable-type p with the same constant as $\boldsymbol{X}$. Now, to complete the proof it is sufficient to notice that $(\boldsymbol{X}^*/\boldsymbol{Z})^*$ is a subspace of $\boldsymbol{X}^{**}$. QED

The next result is an analogue of Theorem 6.4.1 for spaces of stable-type p. For the definition of an almost continuous operator, also see Section 6.4.

Theorem 6.6.3. *If a normed space $\boldsymbol{X}$ is of stable-type p, $1 \leq p \leq 2$, then each linear continuous operator U from $\boldsymbol{X}$ into $\boldsymbol{L}_0(T,\mu)$, where μ is a finite measure, is almost continuous from $\boldsymbol{X}$ into $\boldsymbol{L}_p(T,\mu)$.*

Proof. We shall show that, for each $\epsilon \in (0, 1/8)$, there exist a measurable set $T_\epsilon \subset T$, and a constant $C > 0$, such that $\mu(T \backslash T_\epsilon) \leq 8\epsilon$, and

$$\left(\int_{T_\epsilon} |U(\boldsymbol{x})|^p d\mu\right)^{1/p} \leq c\|\boldsymbol{x}\|, \qquad \boldsymbol{x} \in \boldsymbol{X}. \tag{6.6.2}$$

For simplicity's sake, and with no restriction on generality, take $\mu(T) = 1$. Let $g_1 = U(\boldsymbol{x}_i), \ldots, g_n = U(\boldsymbol{x}_n) \in U(B_X)$, with $\boldsymbol{x}_1, \ldots, \boldsymbol{x}_n$, in the unit ball, B_X, of the space $\boldsymbol{X}$, and let $\lambda_1, \ldots, \lambda_n \in \mathbf{R}$. Then, since for each $\beta \in (0,1)$, and each $c_1, \ldots, c_n \in \mathbf{R}$,

$$\Big(\sum_i |c_i|^p\Big)^{1/p} = J_\beta^{-1}(\xi_1, \mathbf{P}) J_\beta\Big(\sum_i c_i\xi_i, \mathbf{P}\Big),$$

with (ξ_i) being, as usual, the sequence of independent and identically distributed p-stable random variables, we get, in view of the Fubini Inequality (see Chapter 1), and the assumption of stable-type p for $\boldsymbol{X}$, that

$$J_{4\epsilon}\Big(\Big(\sum_i |\lambda_i g_i|^p\Big)^{1/p}, \mu\Big) \tag{6.6.3}$$

$$\begin{aligned}
&= J_{1//2}^{-1}(\xi_1, \mathbf{P}) J_{4\epsilon}\Big(J_{1/2}\Big(\Big|\sum_i \lambda_i g_i(t)\xi_i(\omega)\Big|, \mathbf{P}(d\omega)\Big), \mu(dt)\Big) \\
&\le J_{1//2}^{-1}(\xi_1, \mathbf{P}) J_\epsilon\Big(J_\epsilon\Big(\Big|\sum_i \lambda_i g_i(t)\xi_i(\omega)\Big|, \mu(dt)\Big), \mathbf{P}(d\omega)\Big) \\
&= J_{1//2}^{-1}(\xi_1, \mathbf{P}) J_\epsilon(U(B_X), \mu) J_\alpha\Big(\Big\|\sum_i \lambda_i \boldsymbol{x}_i\xi_i\Big\|, \mathbf{P}\Big) \\
&\le J_{1//2}^{-1}(\xi_1, \mathbf{P}) J_\epsilon(U(B_X), \mu) C_1 \Big(\sum_i |\lambda_i|^p\Big)^{1/p}.
\end{aligned}$$

Now, consider the set D of functions of the form, $\sum_i |\lambda_i g_i(t)|^p$, where $g_1, \ldots, g_n \in U(B_X)$, and $\sum |\lambda_i|^p \le 1$. The set D is a convex subset of $\boldsymbol{L}_0(T, \mu)$ consisting of non-negative functions. In view of (6.6.3),

$$\begin{aligned}
J_{4\epsilon}(D, \mu) &= \sup\{J_{4\epsilon}(g, \mu) : g \in D\} \\
&\le \Big(C_1 J_{1/2}^{-1}(\xi_i, \mathbf{P}) J_\epsilon(4(B_X), \mu)\Big)^p =: C^p/2.
\end{aligned}$$

Therefore, by the Nikishin Theorem (see Chapter 1), there exists a set $T_\epsilon \subset T$ such that $\mu(T \setminus T_\epsilon) \le 8\epsilon$, and such that

$$\int_{T_\epsilon} g\, d\mu \le 2J_{4\epsilon}(D, \mu) \le C^p, \qquad g \in D,$$

and, in particular,

$$\Big(\int_{T_\epsilon} |g|^p d\mu\Big)^{1/p} \le C, \qquad g \in U(B_X),$$

which proves (6.6.2), QED

6.7 Extented basic inequalities and series of random vectors in spaces of type p

We begin this section by extending the basic inequalities defining the Rademacher-type p to a wider class of random vectors in spaces $\boldsymbol{L}_p(\boldsymbol{X})$.

Proposition 6.7.1. *The following properties of a normed space are equivalent:*

(i) The space $\boldsymbol{X}$ is of Rademacher-type p;

(ii) There exists a constant $C > 0$ such that, for all $n \in \mathbf{N}$, and all independent, zero-mean $X_1, \dots, X_n \in \boldsymbol{L}_p(\boldsymbol{X})$,

$$\Big(\mathbf{E}\Big\|\sum_i X_i\Big\|^p\Big)^{1/p} \le C\Big(\sum_i \mathbf{E}\|X_i\|^p\Big)^{1/p};$$

(iii) There exists an $\alpha \in (0, p]$ (or, for each $\alpha \in (0, p]$), and a constant $C > 0$ such that, for all $n \in \mathbf{N}$, and any independent, zero-mean $X_1, \dots, X_n \in \boldsymbol{L}_p(\boldsymbol{X})$,

$$\Big(\mathbf{E}\Big\|\sum_i X_i\Big\|^\alpha\Big)^{1/\alpha} \le C\Big(\sum_i \mathbf{E}\|X_i\|^p\Big)^{1/p}.$$

Proof. (*i*) $\Longrightarrow$ (*ii*) At the beginning let us assume that X_i's are symmetric, and that the sequence (r_i) is independent of (X_i). By Theorem 6.2.2, the space $\boldsymbol{L}_p(\boldsymbol{X})$ is of Rademacher-type p, so that

$$\Big(\mathbf{E}\Big\|\sum_i r_i X_i\Big\|^p\Big)^{1/p} \le C\Big(\sum_i \mathbf{E}\|X_i\|^p\Big)^{1/p}.$$

However, the symmetry of X_i's implies that (r_iX_i), and (X_i) are identically distributed which give *(ii)* in the symmetric case. If (X_i) are not symmetric then we proceed by symmetrization as follows: Let (X_i') be independent copies of (X_i). Then $X_i - X_i'$ are symmetric,

$$\left(\mathbf{E}\Big\|\sum_i X_i\Big\|^p\right)^{1/p} \le \left(\mathbf{E}\Big\|\sum_i (X_i - X_i')\Big\|^p\right)^{1/p}$$

$$\le C\Big(\sum_i \mathbf{E}\|X_i - X_i'\|^p\Big)^{1/p} \le 2C\Big(\sum_i \mathbf{E}\|X_i\|^p\Big)^{1/p}.$$

The implication *(i)* $\Longrightarrow$ *(iii)* can be proved exactly as the implication *(i)* $\Longrightarrow$ *(ii)*. The only additional information that is needed is contained in Remark 6.2.1. The implications *(ii)* $\Longrightarrow$ *(i)*, and *(iii)* $\Longrightarrow$ *(i)*, are evident. QED

One can further strengthen Proposition 6.7.1 by dropping the assumption of independence, and replacing it by the assumption of sign-invariance which has been defined in Section 1.1.

Theorem 6.7.1.[20] *The following properties of a normed space* $\boldsymbol{X}$ *are equivalent:*

(i) The space $\boldsymbol{X}$ *is of Rademacher-type* p*;*

(ii) There exists a sign-invariant sequence (ϕ_n) *of real random variables with* $\inf_n \mathbf{E}|\phi_n| > 0$*, and a constant* $C > 0$ *such that, for all* $(\boldsymbol{x}_i) \subset \boldsymbol{X}$*,*

$$\mathbf{E}\Big\|\sum_n \phi_n \boldsymbol{x}_n\Big\| \le C\Big(\sum_n \|\boldsymbol{x}_n\|^p\Big)^{1/p}; \tag{6.7.1}$$

(iii) For each sign-invariant sequence (ϕ_n) *of real random variables with* $\sup_n \mathbf{E}|\phi_n|^p < \infty$*, there exists a constant* $C > 0$ *such that, for each* $(\boldsymbol{x}_i) \subset \boldsymbol{X}$*, the inequality (6.7.1) holds true.*

Proof. The implication *(iii)* $\Longrightarrow$ *(ii)* is trivial. So, next, let's prove *(i)* $\Longrightarrow$ *(iii)*. By Remark 6.2.1, there exists a $C > 0$ such that for any finite $(\boldsymbol{x}_i) \subset \boldsymbol{X}$,

$$\mathbf{E}\Big\|\sum_i r_i \boldsymbol{x}_i\Big\| \le C\Big(\sum_i \|\boldsymbol{x}_i\|^{[}\Big)^{1/p},$$

[20]This results is due to G. Pisier (1973/74), Exp. 3.

so that, for a fixed $\omega \in \Omega$ (we assume (r_i), and (ϕ_i), to be independent)

$$\mathbf{E}_r\Big\|\sum_i r_i\phi_i(\omega)\boldsymbol{x}_i\Big\| \le C\Big(\sum_i \|\boldsymbol{x}_i\|^p|\phi_i(\omega)|^p\Big)^{1/p}.$$

Integrating both sides we get the inequalities,

$$\mathbf{E}\Big\|\sum_i r_i\phi_i\boldsymbol{x}_i\Big\| \le C\mathbf{E}\Big(\sum_i \|\boldsymbol{x}_i\|^p|\phi_i|^p\Big)^{1/p}.$$

$$\le C\Big(\sum_i \|\boldsymbol{x}_i\|^p\mathbf{E}|\phi_i|^p\Big)^{1/p} \le C\sup_i \mathbf{E}|\phi_i|^p\Big(\sum_n \|\boldsymbol{x}_n\|^P\Big)^{1/p},$$

which give (iii) because, for sign-invariant (ϕ_n), the sequences (ϕ_n) and $(r_n\phi_n)$ are identically distributed.

$(ii) \implies (i)$ Notice that if the sequence (ϕ_n) is sign-invariant then, for each $(\boldsymbol{x}_n) \subset \boldsymbol{X}$, and each $p \in [1.\infty)$,

$$\inf_n \mathbf{E}|\phi_n|\Big(\mathbf{E}\Big\|\sum_n r_n\boldsymbol{x}_n\Big\|^p\Big)^{1/p} \le \Big(\mathbf{E}\Big\|\sum_n \phi_n\boldsymbol{x}_n\Big\|^p\Big)^{1/p}. \tag{6.7.2}$$

Indeed, for a fixed $\omega \in \Omega$,

$$\Big\|\mathbf{E}_\phi\Big(\sum_n r_n(\omega)|\phi_n|\boldsymbol{x}_n\Big)\Big\| \le \mathbf{E}_\phi\Big\|\sum_n r_n(\omega)|\phi_n|\boldsymbol{x}_n\Big\|$$

from which it follows that

$$\Big\|\sum_n r_n(\omega)\boldsymbol{x}_n\mathbf{E}|\phi_n|\Big\|^p \le \mathbf{E}_\phi\Big\|\sum_n r_n(\omega)\boldsymbol{x}_n|\phi_n|\Big\|^p$$

so that, by integration, we get that

$$\Big(\mathbf{E}\Big\|\sum_n r_n\boldsymbol{x}_n\mathbf{E}|\phi_n|\Big\|^p\Big)^{1/p} \le \Big(\mathbf{E}\Big\|\sum_n r_n\boldsymbol{x}_n|\phi_n|\Big\|^p\Big)^{1/p}.$$

By the Contraction Principle (see Chapter 1),

$$\inf_n \mathbf{E}|\phi_n|\Big(\mathbf{E}\Big\|\sum_n r_n\boldsymbol{x}_n\Big\|^p\Big)^{1/p} \le \Big(\mathbf{E}\Big\|\sum_n r_n\boldsymbol{x}_n\mathbf{E}|\phi_n|\Big\|^p\Big)^{1/p}$$

which implies (6.7.2) because (ϕ_n), and $(r_n|\phi_n|)$ are identically distributed. Finally, from (6.7.2), and the assumption, we get that, for each $(\boldsymbol{x}_n) \subset \boldsymbol{X}$,

$$\mathbf{E}\Big\|\sum_n r_n \boldsymbol{x}_n\Big\| \le (\inf_n \mathbf{E}|\phi_n|)^{-1}\Big(\mathbf{E}\Big\|\sum_n \phi_n \boldsymbol{x}_n\Big\|^p\Big)^{1/p}$$

$$\le (\inf_n \mathbf{E}|\phi_n|)^{-1} C\Big(\mathbf{E}\Big\|\sum_n \boldsymbol{x}_n\Big\|^p\Big)^{1/p},$$

which proves that $\boldsymbol{X}$ is of Rademacher-type p. QED

Further extensions of basic inequalities can also be obtained for weakly exchangeable and exchangeable random vectors (for definitions, see Section 1.1). Let us start with the weakly exchangeable case.

Theorem 6.7.2.[21] *If $\boldsymbol{X}$ is of Rademacher-type p, then there exists a constant $C > 0$ such that, for each $n \in \mathbf{N}$, and arbitrary weakly exchangeable $X_1, \dots, X_n \in \boldsymbol{L}_p(\boldsymbol{X})$, with $X_1 + \cdots + X_n = 0$,*

$$\Big(\mathbf{E} \sup_{1\le k\le n}\Big\|\sum_{i=1}^k X_i\Big\|^p\Big)^{1/p} \le C\Big(\sum_{i=1}^n \mathbf{E}\|X_i\|^p\Big)^{1/p}. \tag{6.7.3}$$

Proof. In view of the Maximal Inequality (see Chapter 1), there exists a constant $C_1 > 0$ such that

$$\Big(\mathbf{E} \sup_{1\le k\le n}\Big\|\sum_{i=1}^k X_i\Big\|^p\Big)^{1/p} \le C_1\Big(\mathbf{E}\Big\|\sum_{i=1}^n \rho_i \boldsymbol{X}_i\Big\|^p\Big)^{1/p},$$

where $\rho_i = (1 + r_i)/2$, and the sequence (r_i) is independent of (X_i). However, $\rho_i, i = 1, \dots, n$, are independent themselves, and by Proposition 6.7.1, there exists a constant C_2, such that, for each $\omega \in \Omega$,

$$\mathbf{E}_\rho\Big\|\sum_{i=1}^n \rho_i X_i(\omega)\Big\|^p \le C_2 \sum_{i=1}^n \mathbf{E}|\rho_i|^p \|X_i(\omega)\|^p.$$

[21]This, and the next result are due to B. Maurey and G. Pisier (1974/75), Annexe I.

Integrating both sides we get the inequality,

$$\mathbf{E}\Big\|\sum_{i=1}^{n}\rho_i X_i\Big\|^p \le \frac{1}{2}C_2\sum_{i=1}^{n}\mathbf{E}\|X_i\|^p,$$

because $\mathbf{E}|\rho_i| = 1/2$. This implies (6.7.3) in view of the definition of weak exchangeability. QED

For exchangeable random vectors we get an even stronger result.

Theorem 6.7.3. *If the normed space* $\boldsymbol{X}$ *is of Rademacher-type* p, $p > 1$, *then there exists a constant* $C > 0$ *such that, for each* $n \in \mathbf{N}$, *and arbitrary exchangeable* $X_1, \ldots, X_n \in \boldsymbol{L}_p(\boldsymbol{X})$ *with* $X_1 + \cdots + X_n = 0$, *and any* $\alpha_1, \ldots, \alpha_n \in \mathbf{R}$,

$$\Big(\mathbf{E}\sup_{1\le k\le n}\Big\|\sum_{i=1}^{k}\alpha_i X_i\Big\|^p\Big)^{1/p} \le C\Big(\frac{1}{n}\sum_{i=1}^{n}|\alpha_i|^p\Big)^{1/p}\Big(\sum_{i=1}^{n}\mathbf{E}\|X_i\|^p\Big)^{1/p}. \tag{6.7.4}$$

Proof. Let $k < n$. For the sake of convenience let us introduce the notation,

$$\phi_k(\alpha_1, \ldots, \alpha_k) = \Big(\mathbf{E}\Big\|\sum_{i=1}^{k}\alpha_i X_i\Big\|^p\Big)^{1/p},$$

$$\phi_k^*(\alpha_1, \ldots, \alpha_k) = \Big(\mathbf{E}\sup_{1\le j\le k}\Big\|\sum_{i=1}^{j}\alpha_i X_i\Big\|^p\Big)^{1/p}.$$

Let $\varepsilon_1, \ldots, \varepsilon_k = \pm 1$, and let σ be a permutation of $\{1, \ldots, k\}$ such that in $\{\varepsilon_{\sigma(1)}, \ldots \varepsilon_{\sigma(k)}\}$ the plus signs precede the minus signs. If we denote by $\mathcal{A}_\rangle$ the σ-algebra in $\mathcal{F}$ spanned by $X_1, \ldots, X_i$, then, for each $j \le k$,

$$\sum_{i=1}^{j}\alpha_i X_i = \mathbf{E}\Big(\sum_{i=1}^{k}\alpha_i X_i \,\Big|\, \mathcal{A}_j\Big) + \frac{1}{n-j}\sum_{i=j+1}^{k}\alpha_i\sum_{i=1}^{j}X_i.$$

By the Doob's Martingale Inequality,

$$\phi_k^*(\alpha_1,\ldots,\alpha_k) \le \frac{p}{p-1}\phi_k(\alpha_1,\ldots,\alpha_k)+(n-k)\sum_{i=1}^{k}|\alpha_i|\phi_k^*(1,\ldots,1). \tag{6.7.5}$$

However, the exchangeability of (X_i) implies that

$$\phi_k(\alpha_1,\ldots,\alpha_k) = \phi_k(\alpha_{\sigma(1)},\ldots,\alpha_{\sigma(k)})$$

and, by the triangle inequality, we get

$$\phi_k(\alpha_{\sigma(1)},\ldots,\alpha_{\sigma(k)}) \le 3\phi_k^*(\varepsilon_{\sigma(1)}\alpha_{\sigma(1)},\ldots,\varepsilon_{\sigma(k)}\alpha_{\sigma(k)}),$$

so that (6.7.5) and the exchangeability yield again the inequalities,

$$\phi_k(\alpha_1,\ldots,\alpha_k)$$

$$\le \frac{3p}{p-1}\phi_k^*(\varepsilon_{\sigma(1)}\alpha_{\sigma(1)},\ldots,\varepsilon_{\sigma(k)}\alpha_{\sigma(k)}) + \frac{3}{n-k}\sum_{i=1}^{k}|\alpha_i|\phi_k^*(1,\ldots,1)$$

$$\le \frac{3p}{p-1}\phi_k^*(\varepsilon_1\alpha_1,\ldots,\varepsilon_k\alpha_k) + \frac{3}{n-k}\sum_{i=1}^{k}|\alpha_i|\phi_k^*(1,\ldots,1).$$

Substituting the above inequality in (6.7.5) we obtain that

$$\phi_k^*(\alpha_1,\ldots,\alpha_k) \le 3\Big(\frac{p}{p-1}\Big)^2\phi_k(\varepsilon_1\alpha_1,\ldots,\varepsilon_k\alpha_k)$$

$$+\Big(\frac{3p}{p-1}+1\Big)\frac{1}{n-k}\sum_{i=1}^{k}|\alpha_i|\phi_k^*(1,\ldots,1).$$

Averaging over possible choices of $\varepsilon_1,\ldots,\varepsilon_k = \pm 1$ (with respect to the Rademacher measure), we get

$$\phi_k^*(\alpha_1,\ldots,\alpha_k)$$

$$\le C\Big(\sum_{i=1}^{k}\mathbf{E}|\alpha_i|^p\|X_i\|^p\Big)^{1/p} + \Big(\frac{3p}{p-1}+1\Big)\frac{1}{n-k}\sum_{i=1}^{k}|\alpha_i|\phi_k^*(1,\ldots,1).$$

Now, it is easy to conclude the proof of (6.7.4) by choosing $k = \text{IntegerPart}[(n+1)/2]$, using Theorem 6.7.2 to estimate $\phi_k^*(1,\ldots,1)$, and dealing with the interval $\{k+1,\ldots,n\}$ in a similar fashion. QED

The inequalities discussed above are useful in studying the problem of convergence of random series, $\sum_i X_i$, in Banach spaces. In the remainder of this section we concentrate on the series of independent random vectors but it is easy to see that Theorems 6.7.2 and 6.7.3 make it possible, using analogous methodology, to extend the results for random series of independent random vectors to the case of weakly exchangeable and exchangeable random series.

Theorem 6.7.4.[22] *The following properties of a Banach space $\boldsymbol{X}$ are equivalent,*

(i) The space $\boldsymbol{X}$ is of Rademacher-type p;

(ii) If (X_i) is a sequence of independent, zero-mean random vectors in $\boldsymbol{X}$, then the condition

$$\sum_{i=1}^{\infty} \mathbf{E}\phi_p(\|X_i\|) < \infty, \qquad \phi_p(t) := \min\{t^p, t\},\ t \geq 0,$$

implies the almost sure convergence of the random series $\sum_i X_i$;

(iii) If (X_i) is a sequence of independent, zero-mean random vectors in $\boldsymbol{X}$, then the convergence of the series $\sum_i \mathbf{E}\|X_i\|^p$ implies the almost sure convergence of the random series $\sum_i X_i$;

(iv) For any $(\boldsymbol{x}_n) \subset \boldsymbol{X}$ with $\sum_i \|\boldsymbol{x}_i\|^p < \infty$, the random series $\sum_i r_i \boldsymbol{x}_i$ converges almost surely, and in $\boldsymbol{L}_p(\boldsymbol{X})$.

Proof. *(i)* $\Longrightarrow$ *(ii)* Denote

$$X_i' = X_i I_{[\|X_i\| \leq]}, \qquad X_i'' = X_i I_{[\|X_i\| \geq]}.$$

Clearly, $X_i = X_i' + X_i''$, and both (X_i'), and (X_i'') are sequences of independent random vectors in $\boldsymbol{X}$. Notice that

$$\mathbf{E}\Big\|\sum_{i=n}^{m} X_i''\Big\| \leq \sum_{i=n}^{m} \mathbf{E}\|X_i''\| \leq \sum_{i=n}^{m} \mathbf{E}\phi_p(\|X_i''\|),$$

[22]This result is due to W.A. Woyczynski (1973).

so that, by Cauchy's argument, the series $\sum_i X_i''$ converges in $\boldsymbol{L}_1(\boldsymbol{X})$ and, in view of the Ito-Nisio Theorem (see Chapter 1), also almost surely. Now, because $\boldsymbol{X}$ is of Rademacher-type p, by Proposition 6.7.1, and Jensen's Inequality,

$$\Big(\mathbf{E}\Big\|\sum_{I=n}^{m} X_i'\Big\|^p\Big)^{1/p} \le \Big[\mathbf{E}\Big\|\sum_{i=n}^{m}(X_i' - \mathbf{E}X_i')\Big\|^p + \Big\|\sum_{i=n}^{m}\mathbf{E}X_i'\Big\|^p\Big]^{1/p}$$

$$\le \Big(\mathbf{E}\Big\|\sum_{i=n}^{m}(X_i' - \mathbf{E}X_i')\Big\|^p\Big)^{1/p} + \mathbf{E}\Big\|\sum_{i=n}^{m} X_i'\Big\|$$

$$\le C\Big(\sum_{i=n}^{m}\mathbf{E}\|X_i' - \mathbf{E}X_i'\|^p\Big)^{1/p} + \sum_{i=n}^{m}\mathbf{E}\|X_i'\|$$

$$\le 2C\Big(2\sum_{i=n}^{m}\mathbf{E}\|X_i'\|^p\Big)^{1/p} + \sum_{i=n}^{m}\mathbf{E}\|X_i'\|$$

$$\le 2^{1+1/p}C\Big(\sum_{i=n}^{m}\mathbf{E}\phi_p(\|X_i'\|)\Big)^{1/p} + \sum_{i=n}^{m}\mathbf{E}\phi_p(\|X_i'\|).$$

Thus $\sum_i X_i'$ converges in $\boldsymbol{L}_p(\boldsymbol{X})$ and, again by the Ito-Nisio Theorem, also almost surely. This proves (*ii*).

(*ii*) $\implies$ (*iii*) This implication is immediate because $\phi_p(t) \le t^p$, $t > 0$.

(*iii*) $\implies$ (*iv*) The almost sure convergence of the series $\sum_i r_i \boldsymbol{x}_i$ is an immediate consequence of (*iii*), and the $\boldsymbol{L}_p(\boldsymbol{X})$ convergence follows from the Kahane Theorem (see Chapter 1).

(*iv*) $\implies$ (*i*) Assume (*iv*) and define two Banach spaces,

$$\boldsymbol{Y} := \Big\{(\boldsymbol{x}_n) \subset \boldsymbol{X} : \Big(\sum_n \|\boldsymbol{x}_n\|^p\Big)^{1/p} < \infty\Big\},$$

$$\boldsymbol{Z} = \Big\{(\boldsymbol{x}_n) \subset \boldsymbol{X} : \sup_n\Big(\mathbf{E}\Big\|\sum_{k=1}^{n} r_k\boldsymbol{x}_k\Big\|^p\Big)^{1/p} < \infty\Big\}.$$

By our assumption, $\boldsymbol{X} \subset \boldsymbol{Z}$. It is easy to see that the embedding $\boldsymbol{Y} \mapsto \boldsymbol{Z}$ is linear and has the closed graph, so that it is continuous in view of the Closed Graph Theorem. Therefore, there exists a constant $C > 0$ such that, for any $(\boldsymbol{x}_n) \subset \boldsymbol{X}$,

$$\sup_n \Big(\mathbf{E}\Big\|\sum_{k=1}^{n} r_k \boldsymbol{x}_k\Big\|^p\Big)^{1/p} \le C\Big(\sum_n \|\boldsymbol{x}_n\|^p\Big)^{1/p},$$

which implies that $\boldsymbol{X}$ is of Rademacher-type p. QED

Corollary 6.7.1. *Let $\boldsymbol{X}$ be a Banach space of Rademacher-type p, and let (X_n) be independent, zero mean random vectors in $\boldsymbol{X}$. If the functions $\phi_n : \mathbf{R}^+ \mapsto \mathbf{R}^+$, $n = 1, 2, \ldots$, are continuous, and such that the functions $\phi(t)/t$, and $t^p/\phi(t)$, are non-decreasing, then, for each sequence $(\alpha_i) \subset \mathbf{R}^+$, the condition*

$$\sum_n \frac{\mathbf{E}\phi_n(\|X_n\|)}{\phi(\alpha_n)} < \infty \tag{6.7.6}$$

implies the almost sure convergence of the random series $\sum_n (X_n/\alpha_n)$.

Proof. Actually it is sufficient to show that the condition,

$$\sum_n \frac{\mathbf{E}\phi_n(\alpha_n\|X_n\|)}{\phi(\alpha_n)} < \infty$$

implies the almost sure convergence of the series $\sum_n X_n$. However, by the previous Theorem it is sufficient to prove that the condition $\sum_n \mathbf{E}\Phi_n(\|X_n\|) < \infty$ implies that $\sum_n \mathbf{E}\phi_p(\|X_n\|) < \infty$, where $\Phi_n(t) := \phi_n(\alpha_n t)/\phi_n(\alpha_n)$, and that is evidently true because, for every function Φ with $\Phi(1) = 1$, and non-decreasing $\Phi(t)/t$, and $t^p/\Phi(t)$, (and all Φ_n's are such) we have the inequality $\Phi(t) \ge \phi_p(t)$, $t \ge 0$. QED

Finally, we will examine the almost sure convergence of some special random series in spaces of stable-type p.

Theorem 6.7.5.[23] *Let $1 \le p < 2$. The following properties of a Banach space $\boldsymbol{X}$ are equivalent:*

(i) The space $\boldsymbol{X}$ is of stable-type p;

[23]This result is due to B. Maurey and G. Pisier (1976).

(ii) For any sequence $(\boldsymbol{x}_i) \subset \boldsymbol{X}$, with $\sum_i \|\boldsymbol{x}_i\|^p < \infty$, and any sequence of independent and identically distributed p-stable random variables (ξ_i), the series $\sum_i \xi_i \boldsymbol{x}_i$ converges almost surely, and in $\boldsymbol{L}_q(\boldsymbol{X})$, if $q < p$;

(iii) For any bounded sequence $(\boldsymbol{x}_n) \subset \boldsymbol{X}$, the series $\sum_n n^{-1/p} r_n \boldsymbol{x}_n$ converges almost surely;

(iv) For any bounded sequence $(\boldsymbol{x}_n) \subset \boldsymbol{X}$, there exists a choice of $\varepsilon_n = \pm 1$, such that $\sum_n n^{-1/p} \varepsilon_n \boldsymbol{x}_n$ converges.

Proof. (i) $\Longrightarrow$ (ii) This implication follows directly from the definition of stable-type p and the Ito-Nisio Theorem (see Chapter 1).

(ii) $\Longrightarrow$ (i) If $\sum_i \xi_i \boldsymbol{x}_i$ converges almost surely then, by Ito-Nisio Theorem, it converges in $\boldsymbol{L}_0(\boldsymbol{X})$, and, by Hoffmann-Jorgensen Theorem, it also converges in $\boldsymbol{L}_q(\boldsymbol{X})$ for any $q < p$. Now, a routine application of the Closed Graph Theorem, along the lines of the proof of Theorem 6.7.4 (iv) $\Longrightarrow$ (i), gives the needed inequality.

(i) $\Longrightarrow$ (iii) If $\boldsymbol{X}$ is of stable-type p, then it is also of stable-type q, for some $q > p$ (see Corollary 6.5.2), and of Rademacher-type q (see Proposition 6.5.1). Furthermore, if the sequence $(\boldsymbol{x}_n)$ is bounded then $\sum_n \|n^{-1/p} \boldsymbol{x}_n\|^q < \infty$, and Theorem 6.7.4 gives the almost sure convergence of the series $\sum_n n^{-1/p} \boldsymbol{x}_n r_n$.

(iii) $\Longrightarrow$ (iv) is obvious.

(iv) $\Longrightarrow$ (i) Suppose the space $\boldsymbol{X}$ is not of stable-type p. Then, in view of Theorem 6.5.1, the canonical embedding, $\boldsymbol{l}_1 \mapsto \boldsymbol{l}_p$, is finitely factorable through $\boldsymbol{X}$. Utilizing this fact we shall construct a sequence $(\boldsymbol{x}_n) \subset \boldsymbol{X}$, and a sequence of integers $(N_k) \subset \mathbf{N}$, $N_k \to \infty$, such that $\sup_n \|\boldsymbol{x}_n\| \le 1$, and for each $\varepsilon = \pm 1$, and each k,

$$N_k^{-1/p} \Big\| \sum_{i=1}^{N_k} \varepsilon_i \boldsymbol{x}_i \Big\| > \frac{1}{2}, \tag{6.7.7}$$

so that, in view of the Kronecker Lemma, the series $\sum_n n^{-1/p} \varepsilon_n \boldsymbol{x}_n$ would diverge for all choices of $\varepsilon_i = \pm 1$.

The above mentioned construction can be accomplished as follows: Put $N_1 = 1$, and take any $\boldsymbol{x}_i \in \boldsymbol{S}_X$. Now suppose that $N_1, \ldots, N_k$, and $\boldsymbol{x}_1, \ldots, \boldsymbol{x}_{N_k}$, are chosen so that $\sup_k \{ \|\boldsymbol{x}_i\| : 1 \le$

$N_k\} \leq 1$, and for each $\varepsilon_i = \pm 1$, (6.7.7) is satisfied. Choose an integer N_{k+1} large enough to satisfy the inequality,

$$N_{k+1}^{-1/p}\Big[\frac{2}{3}(N_{k+1} - N_k)^{1/p} - N_k\Big] > \frac{1}{2}.$$

Because the canonical embedding $\boldsymbol{l}_1 \mapsto \boldsymbol{l}_p$ is finitely factorable through $\boldsymbol{X}$ one can find $\boldsymbol{x}_{N_k+1}, \ldots, \boldsymbol{x}_{N_{k+1}}$ such that, for any $(\alpha_i) \subset \mathbf{R}$,

$$\frac{2}{3}\Big(\sum_{i=N_k+1}^{N_{k+1}} |\alpha_i|^p\Big)^{1/p} \leq \Big\|\sum_{i=N_k+1}^{N_{k+1}} \alpha_i \boldsymbol{x}_i\Big\| \leq \sum_{i=N_k+1}^{N_{k+1}} |\alpha_i|,$$

so that, for any $\varepsilon_i = \pm 1$, one would have

$$N_{k+1}^{-1/p}\Big\|\sum_{I=1}^{N_{k+1}} \varepsilon_i \boldsymbol{x}_i\Big\| \geq N_{k+1}^{-1/p}\Big[\Big\|\sum_{i=N_k+1}^{N_{k+1}} \varepsilon_i \boldsymbol{x}_i\Big\| - \Big\|\sum_{i=1}^{N_k} \varepsilon_i \boldsymbol{x}_i\Big\|\Big]$$

$$\geq N_{k+1}^{-1/p}\Big[\frac{2}{3}(N_{k+1} - N_k)^{1/p} - N_k\Big] > \frac{1}{2},$$

which completes the construction and the proof of the Theorem. QED

6.8 Strong laws of large numbers and asymptotic behavior of random sums in spaces of Rademacher-type p

We begin with an analogue of the classical Kolmogorov-Chung Strong Law of Large Numbers. Its validity in a Banach space characterizes the space's Rademacher-type.

Theorem 6.8.1.[24] *Let $p \in (1, 2]$. The following properties of a Banach space $\boldsymbol{X}$ are equivalent:*

(i) The space $\boldsymbol{X}$ is of Rademacher-type p;

[24] This result is due to W.A. Woyczynski (1973), and J. Hoffmann-Jorgensen (1975).

(ii) For each sequence (X_n) of independent, zero-mean random vectors in $\boldsymbol{X}$, the convergence of the series $\sum_n n^{-p}\mathbf{E}\|X_n\|^p$ implies that

$$\lim_{n\to\infty}\frac{X_1+\cdots+X_n}{n}=0, \qquad \text{a.s.};$$

(iii) There exists a constant $C>0$ such that, for each finite sequence $(\boldsymbol{x}_i)\subset\boldsymbol{X}$,

$$\frac{1}{n}\mathbf{E}\Big\|\sum_{i=1}^n r_i\boldsymbol{x}_i\Big\|\le C\Big(\sum_{i=1}^n\frac{\|\boldsymbol{x}_i\|^p}{i^p}\Big)^{1/p};$$

(iv) There exists a constant $C>0$ such that, for each finite sequence (X_n) of independent, zero-mean random vectors in $\boldsymbol{X}$,

$$\frac{1}{n}\mathbf{E}\Big\|\sum_{i=1}^n X_i\Big\|\le C\Big(\sum_{i=1}^n\frac{\mathbf{E}\|X_i\|^p}{i^p}\Big)^{1/p}.$$

Proof. *(i)* $\Longrightarrow$ *(ii)* follows immediately from Corollary 6.7.1, with $\phi_n(t)=t^p$, and $\alpha_n=n$, and from the Kronecker's Lemma.

(ii) $\Longrightarrow$ *(iii)* Using *(ii)* in the case $X_i=r_i\boldsymbol{x}_i$ we get that $n^{-1}\sum_{i=1}^n r_i\boldsymbol{x}_i\to 0$, a.s., and also in $\boldsymbol{L}_1(\boldsymbol{X})$ (by Kahane's Lemma–see Chapter 1) as $n\to\infty$, whenever $\sum_i i^{-p}\|\boldsymbol{x}\|^p<\infty$. Thus

$$\Big\{(\boldsymbol{x}_i)\subset\boldsymbol{X};\Big(\sum_{i=1}^\infty\frac{\|\boldsymbol{x}_i\|^p}{i^p}\Big)^{1/p}<\infty\Big\}$$

$$\subset\Big\{(\boldsymbol{x}_i)\subset\boldsymbol{X}:\sup_n\frac{1}{n}\mathbf{E}\Big\|\sum_{i=1}^n r_i\boldsymbol{x}_i\Big\|<\infty\Big\},$$

and a standard application of the Closed Graph Theorem yields the desired inequality.

(iii) $\Longrightarrow$ *(i)* By the assumption, for each $(\boldsymbol{x}_i)\subset\boldsymbol{X}$,

$$\mathbf{E}\Big\|\sum_{i=1}^n in^{-1}r_i\boldsymbol{x}_i\Big\|\le C\Big(\sum_{i=1}^n\|\boldsymbol{x}_i\|^p\Big)^{1/p},$$

so that

$$\mathbf{E}\Big\|\sum_{i=1}^{n} i(2n)^{-1} r_i \boldsymbol{x}_i + \sum_{i=n+1}^{2n} r_i \boldsymbol{x}_{i-n} i(2n)^{-1}\Big\| \le C2^{1/p}\Big(\sum_{i=1}^{n} \|\boldsymbol{x}_i\|^p\Big)^{1/p}, \tag{6.8.1}$$

and, by the Contraction Principle (see Chapter 1),

$$\frac{1}{2}\mathbf{E}\Big\|\sum_{i=1}^{n} r_i \boldsymbol{x}_i\Big\| = \frac{1}{2}\mathbf{E}\Big\|\sum_{i=n+1}^{2n} r_i \boldsymbol{x}_{i-n}\Big\| \tag{6.8.2}$$

$$\le \mathbf{E}\Big\|\sum_{i=1}^{n} i(2n)^{-1} r_i \boldsymbol{x}_i + \sum_{i=n+1}^{2n} r_i \boldsymbol{x}_{i-n} i(2n)^{-1}\Big\|.$$

Now, (6.8.1) and (6.8.2) imply that $\boldsymbol{X}$ is of Rademacher-type p.

To complete the proof of the Theorem it is sufficient to notice that (*iv*) follows for (*ii*) by an application of the Closed Graph Theorem technique used above, and that (*iv*) trivially implies (*iii*). QED

Utilizing the full version of the Corollary 6.7.1 one can immediately strengthen the implication (*i*) $\implies$ (*ii*) of the above Theorem.

Proposition 6.8.1 *If* $\phi : \mathbf{R}^+ \mapsto \mathbf{R}^+$ *is continuous and such that* $\phi(t)/t$, *and* $t^p/\phi(t)$, *are non-decreasing, and if* (X_i) *is a sequence of independent, zero-mean random vectors in a Banach space* $\boldsymbol{X}$ *of Rademacher-type* p, $1 < p \le 2$, *then the convergence of the series* $\sum_n \mathbf{E}\phi(\|X_n\|)/\phi(n)$ *implies that*

$$\lim_{n\to\infty} \frac{1}{n}\sum_{i=1}^{n} X_i = 0 \qquad \text{a.s.}$$

It is possible to obtain further corollaries to the results of Section 6.7 on random series in spaces $\boldsymbol{X}$ of Rademacher -type p, and get a precise description of the asymptotic behavior of sums of independent random vectors with values in $\boldsymbol{X}$. To accomplish this task let us introduce a new classes of functions ψ.

Definition 6.8.1. The set of all functions $\psi : \mathbf{R}^+ \mapsto \mathbf{R}^+$ which do not decrease for $t > t_0$, for some $t_0 = t_0(\psi)$, and for which the series $\sum_n n^{-1}\psi^{-1}(n)$ converges, will be denoted by Ψ_c (or Ψ_d, in the case of functions ψ for which the series $\sum_n n^{-1}\psi^{-1}(n)$ diverges). The function inverse to ϕ will be denoted by $\phi^{(-1)}$.

Theorem 6.8.2.[25] *Let $\phi : \mathbf{R}^+ \mapsto \mathbf{R}^+$ be a continuous function such that $\phi(t)/t$, and $t^p/\phi(t)$, are non-decreasing, and let (X_i) be a sequence of independent, zero-mean random vectors in a Banach space $\mathbf{X}$ of Rademacher-type p. Then, if $\mathbf{E}\phi(\|X_n\|) < \infty$, and the sequence $A_n := \sum_{k=1}^n \mathbf{E}\phi(\|X_n\|) \uparrow \infty$, then*

$$\Big\|\sum_{k=1}^{n} X_k\Big\| = O\big(\phi^{(-1)}(A_n\psi(A_n))\big), \qquad n \to \infty,$$

almost surely, for each function $\psi \in \Psi_c$.

Proof. Denote $b_n = \phi^{(-1)}(A_n\psi(A_n))$. Then, of course, $b_n \uparrow \infty$, and, furthermore,

$$\sum_{n=1}^{\infty} \frac{\mathbf{E}\phi(\|X_n\|)}{A_n\psi(A_n)} < \infty. \tag{6.8.3}$$

Indeed, take n_0 such that $A_{n_0} > 0$, and $\psi(A_{n_0}) > 0$. Because the series $\sum_n n^{-1}\psi^{-1}(n)$ converges, the integral

$$I := \int_{A_{n_0}}^{\infty} \frac{dx}{x\psi(x)}$$

converges as well. Now, the Mean Value Theorem implies that

$$\int_{A_{n-1}}^{A_n} \frac{dx}{x\psi(x)} = (A_n - A_{n-1})c_n,$$

for $n > n_0$, and some c_n such that $A_n^{-1}\psi^{-1}(A_n) \leq c_n \leq A_{n-1}^{-1}\psi^{-1}(A_{n-1})$. Remembering that $A_n - A_{n-1} = \mathbf{E}\phi(\|X_n\|)$, and that

$$I = \sum_{n=n_0+1}^{\infty} \int_{A_{n-1}}^{A_n} \frac{dx}{x\psi(x)},$$

[25]This result is due to W.A. Woyczynski (1973).

we get (6.8.3). Now, $\phi(b_n) = A_n\psi(A_n)$ and, by Corollary 6.7.1, the series $\sum_n b_n^{-1}X_n$ converges almost surely, so that the Kronecker's Lemma gives the desired asymptotics of the partial sums. QED

Remark 6.8.1. The above result is, in a sense, best possible. Indeed, let $\phi : \mathbf{R}^+ \mapsto \mathbf{R}^+$ be continuous and strictly increasing with $\phi(0) = 0$, and $\phi(t) \to \infty$, as $t \to \infty$. Then, for every function $\psi \in \Psi_d$ (e.g., $\psi(t) = \log t$, or $\log t \log\log t$) there exists a sequence of independent real random variables (X_i) with $\mathbf{E}\phi(|X_i|) < \infty$, and $A_n = \sum_{k=1}^n \mathbf{E}\phi(|X_k|) \uparrow \infty$, and such that

$$\limsup_n \frac{|\sum_{i=1}^n X_i|}{\phi^{(-1)}(A_n\psi(A_n))} > 0, \qquad a.s.$$

Even without any restrictions on the moments of the random vectors (X_i) in a space of Rademacher-type p it is still possible to obtain some sort of the Strong Law of Large Numbers.

Theorem 6.8.3.[26] *Let (X_i) be a sequence of independent random vectors taking values in a Banach space $\mathbf{X}$ of Rademacher-type p, and let (ϕ_i) be a sequence of convex functions, $\phi_i : \mathbf{R}^+ \mapsto \mathbf{R}^+$, such that $\phi_i(t)/t$, and $t^p/\phi_i(t)$, are not decreasiing. If $0 < t_n \uparrow \infty$, then the convergence of the series*

$$\sum_{n=1}^{\infty} \mathbf{E}\frac{\phi_n(\|X_n\|)}{\phi_n(\|X_n\|) + \phi_n(t_n)}$$

implies that

$$\lim_{n\to\infty} \frac{1}{t_n}\sum_{k=1}^{n}(X_k - \mathbf{E}Z_k) = 0, \qquad .a.s.$$

where $Z_n := X_n I[\|X_n\| < t_n]$.

Proof. The obvious inequalities

$$\frac{\mathbf{E}\phi_n(\|Z_n\|)}{2\phi_n(t_n)} + \frac{1}{2}\mathbf{P}(\|X_n\| \geq t_n) \leq \mathbf{E}\frac{\phi_n(\|X_n\|)}{\phi_n(\|X_n\|) + \phi_n(t_n)}, \qquad n \in \mathbf{N},$$

[26] This result is due to W.A. Woyczynski (1974).

imply that

$$\sum_{n=1}^{\infty} \mathbf{P}(X_n \neq Z_n) < \infty, \tag{6.8.4}$$

and

$$\sum_{n=1}^{\infty} \frac{\mathbf{E}\phi_n(\|Z_n\|)}{\phi_n(t_n)} < \infty. \tag{6.8.5}$$

Define, again, $\phi_p(t) := \min(t, t^p)$, $t \geq 0$. Then,

$$\phi_p(t+s) \leq K\big(\phi_p(t) + \phi_p(s)\big), \qquad t, s, \geq 0, \tag{6.8.6}$$

for some constant K which, in general, may depend on p. Indeed, if $t+s \leq 1$, then (6.8.6) follows from the boundedness of the function $(1+t)^p/(1+t^p)$ on $\mathbf{R}^+$. If $t+s \geq 1$, and $t, s \leq 1$, then (6.8.6) follows from the fact that $(t+s)/(t^p+s^p) \leq 2/(t^p-(1-t)^p) < 2$. Finally, if $t+s \geq 1$, and (say) $t > 1, s < 1$, then (6.8.6) follows from the inequality $(t+s)/(t+s^p) \leq (t+1)/t \leq 2$.

In view of (6.8.6), the convexity of ϕ_n, and the fact that $\phi_n(t_n t)/\phi_n(t_n) \geq \phi_p(t)$, $t \geq 0$, we get that

$$\mathbf{E}\phi_p\big(\|(Z_n - \mathbf{E}Z_n)t_N^{-1}\|\big) \leq K\big(\mathbf{E}\phi_p\big(\|Z_n\|t_n^{-1}\big) + \phi_p\big(\mathbf{E}\|Z_n\|t_n^{-1}\big)\big)$$

$$\leq 2K\phi_n^{-1}(t_n)\mathbf{E}\phi_n(\|Z_n\|),$$

so that (6.8.5) implies that $\sum_n \mathbf{E}\phi_n(\|Z_n - \mathbf{E}Z_n)t_n^{-1}\|) < \infty$. The random vectors $(Z_n - \mathbf{E}Z_n)$ are independent and zero-mean, so that Theorem 6.7.4 implies that the series $\sum_n (Z_n - \mathbf{E}Z_n)t_n^{-1}$ converges almost surely. Now, the Kronecker's Lemma gives us that

$$\lim_{n\to\infty} t_n^{-1} \sum_{k=1}^{n} (Z_k - \mathbf{E}Z_k) = 0, \qquad a.s.$$

Furthermore, by (6.8.4), and the Borel-Cantelli Lemma, the probability that infinitely often $Z_k \neq X_k$ is equal to 0, so that also

$$\lim_{n\to\infty} t_n^{-1} \sum_{k=1}^{n} (X_k - \mathbf{E}Z_k) = 0, \qquad a.s. \qquad \text{QED}$$

To complete this Section we now turn to the case when the sequence (X_i) consists of independent, and identically distributed random vectors.

Theorem 6.8.4.[27] *Let (X_i) be a sequence of symmetric, independent, and identically distributed random vectors in a Banach space $\boldsymbol{X}$ of Rademacher-type p, and let $\phi : \mathbf{R}^+ \mapsto \mathbf{R}^+$ be a convex function such that $\phi(t)/t$, and $t^p/\phi(t)$, are non-decreasing. If the sequence (t_n), $0 < t_n \uparrow \infty$ satisfies the condition,*

$$\sum_{k=n}^{\infty} \frac{1}{\phi(t_k)} = O\Big(\frac{n}{\phi(t_n)}\Big), \tag{6.8.7}$$

then the condition

$$\sum_{n=1}^{\infty} \mathbf{P}\big(\|X_1\| \geq t_n\big) < \infty, \tag{6.8.8}$$

is necessary, and sufficient, for

$$\lim_{n\to\infty} \frac{1}{t_n} \sum_{i=1}^{n} X_i = 0. \qquad a.s.$$

Proof. Sufficiency: Let $Z_n := X_n I\big[\|X_n\| < t_n\big]$ as in Theorem 6.8.3. Then, in view of (6.8.7),

$$\sum_{n=1}^{\infty} \frac{\mathbf{E}\phi(\|Z_n\|)}{\phi(t_n)} = \sum_{n=1}^{\infty} \frac{1}{\phi(t_n)} \sum_{k=1}^{n} \mathbf{E}\phi\Big(\|X_1\| I\big[t_{k-1} \leq \|X_1\| < t_k\big]\Big)$$

$$= \sum_{k=1}^{\infty} \mathbf{E}\phi\Big(\|X_1\| I\big[t_{k-1} \leq \|X_1\| < t_k\big]\Big) \sum_{n=k}^{\infty} \frac{1}{\phi(t_n)}$$

$$\leq \text{const} \sum_{k=1}^{\infty} \frac{k}{\phi(t_k)} \mathbf{E}\phi\Big(\|X_1\| I\big[t_{k-1} \leq \|X_1\| < t_k\big]\Big)$$

$$\leq \text{const} \sum_{k=1}^{\infty} k\mathbf{P}\big(t_{k-1} \leq \|X_1\| < t_k\big) = \text{const} \sum_{k=0}^{\infty} \mathbf{P}\big(\|X_1\| \geq t_k\big).$$

[27] This result is due to W.A. Woyczynski (1974).

This inequality, together with the reasoning exactly as in the proof of Theorem 6.8.3 implies that

$$\lim_{n\to\infty}\frac{1}{t_n}\sum_{k=1}^{n}(X_k - \mathbf{E}Z_k) = 0, \qquad a.s.,$$

which completes the proof of sufficiency because $\mathbf{E}Z_k = 0$ in view of the symmetry assumption on X_n's.

Necessity: It follows from the fact that

$$\frac{X_n}{t_n} = \frac{1}{t_n}\sum_{i=1}^{n} X_i - \frac{t_{n-1}}{t_n}\cdot\frac{1}{t_{n-1}}\sum_{i=1}^{n-1} X_i \to 0, \qquad a.s.,$$

as $n \to \infty$, so that, should the series $\sum_n \mathbf{P}(\|X_1\| \geq t_n)$ diverge, by the Borel-Cantelli Lemma, the probability that $\|X_1\| \geq t_n$ happens infinitely often is equal to 1, which contradicts (6.8.8). QED

Remark 6.8.2. It is easy to check that the condition (6.8.7) is fulfilled whenever $\liminf_k \phi(t_{2k})/\phi(t_k) > 2$. This gives a handy criterion for a sequence (t_k) to satisfy (6.8.7). For instance, if $\boldsymbol{X}$ is of Rademacher-type $p, 1 < p \leq 2$, and $\phi(t) = t^q, 1 < q \leq p$, then the sequence $t_k = k$ fulfills the condition (6.8.7), but if $q = 1$, then it does not.

In the case of non-symmetric random vectors we need more restrictions on the sequence (t_k).

Theorem 6.8.5.[28] *Let (X_i) be a sequence of independent, identically distributed random vectors in a Banach space $\boldsymbol{X}$ of Rademacher-type p, and ϕ and (t_n) be as in Theorem 6.8.4. If, additionally, $\mathbf{E}X_1 = 0$, and there exists a constant C such that*

$$\frac{t_k}{t_n} \leq C\frac{k}{n}, \qquad k \geq n, \tag{6.8.9}$$

then the condition,

$$\sum_{n=1}^{\infty}\mathbf{P}\big(\|X_i\| \geq t_n\big) < \infty, \tag{6.8.10}$$

[28]This result is due to B. Maurey and G. Pisier(1976).

implies that

$$\lim_{n\to\infty} \frac{1}{t_n} \sum_{i=1}^{n} X_i = 0 \qquad a.s.$$

Proof. The proof follows the lines of the proof of Theorem 6.8.4 but to complete it (with the help of Theorem 6.8.3) we need to show that

$$\lim_{n\to\infty} \frac{1}{t_n} \sum_{k=1}^{n} \mathbf{E} Z_k = 0.$$

This can be accomplished as follows: In view of (6.8.9), and the assumption $\mathbf{E}X_1 = 0$, we get that

$$\frac{1}{t_n}\Big\|\sum_{k=1}^{n} \mathbf{E} Z_k\Big\| = \frac{1}{t_n}\Big\|\sum_{k=1}^{n} \mathbf{E}(X_k - Z_k)\Big\| \le \frac{1}{t_n} \sum_{k=1}^{n} \mathbf{E}\|X_k - Z_k\|$$

$$\le \frac{1}{t_n} \sum_{k=1}^{n} \sum_{m=k}^{\infty} \mathbf{E}\Big(\|X_1\| \cdot I\big[t_m \le \|X_1\| < t_{m+1}\big]\Big)$$

$$\le \frac{1}{t_n}\Big(\sum_{m=1}^{n}\sum_{k=1}^{m} + \sum_{m=n+1}^{\infty}\sum_{k=1}^{n}\Big)\mathbf{E}\Big(\|X_1\| \cdot I\big[t_m \le \|X_1\| < t_{m+1}\big]\Big)$$

$$\le \frac{1}{t_n} \sum_{m=1}^{n} m t_{m+1} \mathbf{P}\big(t_m \le \|X_1\| < t_{m+1}\big)$$

$$+ \frac{1}{t_n} \sum_{m=n+1}^{\infty} n t_{m+1} \mathbf{P}\big(t_m \le \|X_1\| < t_{m+1}\big).$$

Now, the first term in the above sum tends to zero because of the Kronecker's Lemma, and because of the fact that that

$$\sum_{m=0}^{\infty} (m+1)\mathbf{P}\big(t_m \le \|X_1\| < t_{m+1}\big) = \sum_{m=0}^{\infty} \mathbf{P}\big(\|X_1\| \ge t_m\big) < \infty,$$

and the second term converges to zero in view of the above inequality, and because in view of (6.8.9) we have $n t_{m+1}/t_n \le C(m+1)$. QED

6.9 Weak and strong laws of large numbers in spaces of stable-type p

The spaces of stable-type p are characterized by an infinite-dimensional analogue of the classical Kolmogorov's Weak law of Large Numbers which can be also interpreted as a Central Limit Theorem with degenerate stable limit distribution.

Theorem 6.9.1.[29] *Let $1 \leq p < 2$. The following properties of a Banach space $\boldsymbol{X}$ are equivalent:*

(i) The space $\boldsymbol{X}$ is of stable-type p;

(ii) For each sequence (X_n) of symmetric, independent, and identically distributed random vectors in $\boldsymbol{X}$,

$$\lim_{n\to\infty} \frac{1}{n^{1/p}} \sum_{i=1}^{n} X_n = 0, \tag{6.9.1}$$

in probability, if , and only if,

$$\lim_{n\to\infty} n\mathbf{P}\big(\|X_1\| > n^{1/p}\big) = 0. \tag{6.9.2}$$

Proof. (*i*) $\implies$ (*ii*) First of all notice that (6.9.1) implies (6.9.2) in any normed space. Indeed, because of the symmetry assumption,

$$\frac{1}{n^{1/p}}\Big\|\sum_{i\neq j} X_i + X_j\Big\|, \qquad \text{and} \qquad \frac{1}{n^{1/p}}\Big\|\sum_{i\neq j} X_i - X_j\Big\|,$$

are identically distributed. Thus it follows that

$$\mathbf{P}\Big(\Big\|\frac{1}{n^{1/p}}\sum_{i=1}^{n} X_i\Big\| > \epsilon\Big) \geq \frac{1}{2}\mathbf{P}\Big(\sup_{1\leq j\leq n} \|n^{-1/p} X_j\| > \epsilon\Big).$$

Hence, (6.9.1) implies the convergence $\mathbf{P}(\sup_{1\leq j\leq n} \|X_j\| > \epsilon\, n^{1/p}) \to 0$, and this is equivalent to the condition (6.9.2).

[29]This result is due to M.B. Marcus and W.A. Woyczynski (1979).

We now show that (6.9.2) implies (6.9.1). Let

$$Y_k := X_k I[\|X_k\| \le n^{1/p}], \quad Z_n := Y_1 + \cdots + Y_n,$$

and $B_n := \{X_1 + \cdots + X_n = Z_n\}$.

By assumption, and by Corollary 6.5.2(*ii*), there exists a q, $p < q < 2$, such that $\boldsymbol{X}$ is of Rademacher-type q. Now, for each $\epsilon > 0$,

$$\mathbf{P}\big(\|X_1 + \cdots + X_n\| n^{-1/p} \ge \epsilon\big)$$

$$\le \mathbf{P}(B_n) \cdot \mathbf{P}\big(\|X_1 + \cdots + X_n\| \ge \epsilon n^{1/p} \mid B_n\big)$$

$$+\mathbf{P}(B_n^c) \cdot \mathbf{P}\big(\|X_1 + \cdots + X_n\| \ge \epsilon n^{1/p} \mid B_n^c\big)$$

$$\le \mathbf{P}\big(\|Z_n\| \ge \epsilon n^{1/p}\big) + \mathbf{P}(B_n^c) \le \epsilon^{-q}\mathbf{E}\|n^{-1/p} Z_n\|^q + \sum_{k=1}^{n} \mathbf{P}\big(\|X_k\| > n^{1/p}\big)$$

$$\le \epsilon^{-q} n^{-q/p} \sum_{k=1}^{n} \mathbf{E}\|X_k\|^q I\big[\|X_k\| \le n^{1/p}\big] + \sum_{k=1}^{n} \mathbf{P}\big(\|X_k\| > n^{1/p}\big),$$

where, in the last step we employed the Proposition 6.7.1(*ii*). Thus, we now have the inequality

$$\mathbf{P}\big(\|X_1 + \cdots + X_n\| n^{-1/p} \ge \epsilon\big)$$

$$\le \epsilon^{-q} n^{1-q/p} \int_{\|X_k\| \le n^{1/p}} \|X_k\|^q d\mathbf{P} + n\mathbf{P}\big(\|X_1\| > n^{1/p}\big),$$

and we only need to show that the first term above, call it I_n, converges to zero as $n \to \infty$. Indeed,

$$I_n = \epsilon^{-q} n^{1-q/p} \sum_{k=1}^{n} \int_{k-1 \le \|X_k\|^p \le k} \|X_1\|^q d\mathbf{P}$$

$$\le \epsilon^{-q} n^{1-q/p} \sum_{k=1}^{n} k^{q/p} \mathbf{P}(k-1 \le \|X_1\|^p \le k)$$

$$\le \epsilon^{-q} n^{1-q/p} \sum_{k=1}^{n} \Big(\sum_{i=1}^{k} i^{q/p-1}\Big) \mathbf{P}(k-1 \le \|X_1\|^p \le k)$$

$$\leq \epsilon^{-q} n^{1-q/p} \sum_{i=1}^{n} i^{q/p-1} \mathbf{P}(\|X_1\|^p > i-1).$$

In view of (6.9.2), for each $\delta > 0$, there exists an $i_0 \in \mathbf{N}$ such that, for all $i > i_0$, we have the inequality $i\mathbf{P}(\|X_1\|^p \geq i-1) < \delta$. Therefore,

$$I_n \leq \epsilon^{-q} n^{1-q/p} \sum_{i=1}^{i_0} i^{q/p-1} \mathbf{P}(\|X_1\|^p > i-1) + \epsilon^q \delta_n^{1-q/p} \sum_{i=i_0+1}^{n} i^{q/p-2}.$$

The last term is less than δC, where the constant C depends only on p, and q. Since δ is arbitrary, we get that, for all $\epsilon > 0$, $\mathbf{P}(\|X_1 + \cdots + X_n\| n^{-1/p} > \epsilon) \to 0$, as $n \to \infty$. This completes the proof of the implication $(i) \implies (ii)$.

$(ii) \implies (i)$ Because of Corollary 6.5.2 (i) the space $\boldsymbol{X}$ is of stable-type p if, and only if, $\boldsymbol{l}_p$ is not finitely representable in $\boldsymbol{X}$. Thus it is sufficient to construct a counterexample in $\boldsymbol{l}_p$. Consider the random vector,

$$X(\omega) = r(\omega) \sum_{N^2(\omega) \leq j < N^2(\omega)+N} \boldsymbol{e}_j$$

where $(\boldsymbol{e}_j)$ is the canonical basis in $\boldsymbol{l}_p$, and N is a random variable with values in $\mathbf{N}$ which is independent of r, and such that $\mathbf{P}(N \geq n)$ behaves asymptotically like $Cn^{-1/p}$. Then

$$\mathbf{P}\big(\|X(\omega)\|_{l_p} > n\big) \sim Cn^{-p},$$

but $n^{-1/p} \sum_{i=1}^{n} X_i$ is not bounded in probability, which implies the existence of a random vector $Y(\omega)$ in $\mathbf{L}_0(\boldsymbol{l}_p)$ such that $n^{-p}\mathbf{P}(\|Y\| > n) \to 0$, but for which $n^{-1/p} \sum_{i=1}^{n} Y_i$ does not converge to 0 in probability. QED

In the case of sequences of special p-stable random vectors of the form $(\xi_i \boldsymbol{x}_i)$, we also have an analogue of the classical Kolmogorov-Chung's Theorem.

Theorem 6.9.2. *Let $1 \leq p \leq 2$. The following properties of a Banach space $\boldsymbol{X}$ are equivalent:*

(i) The space $\boldsymbol{X}$ *is of stable-type* p*;*

(ii) For any sequence $(\boldsymbol{x}_i) \subset \boldsymbol{X}$ *such that* $\sum_i i^{-p}\|\boldsymbol{x}_i\|^p < \infty$*, and each sequence* (ξ_i) *of independent and identically distributed* p*-stable random variables,*

$$\lim_{n\to\infty} \frac{1}{n}\sum_{i=1}^{n} \xi_i \boldsymbol{x}_i = 0, \qquad \text{a.s., and also in } \boldsymbol{L}_q,\ q < p;$$

(iii) There exists a constant C *such that, for any* $(\boldsymbol{x}_i) \subset \boldsymbol{X}$*, and* (ξ_i)*, as above ,*

$$\frac{1}{n}\Big(\mathbf{E}\Big\|\sum_{i=1}^{n} \xi_i \boldsymbol{x}_i\Big\|^{p/2}\Big)^{2/p} \le C\Big(\sum_{i=1}^{n} i^{-p}\|\boldsymbol{x}_i\|^p\Big)^{1/p}.$$

Proof. *(i)* $\implies$ *(ii)* This implication follows directly from Theorem 6.7.5, and the Kronecker's Lemma.

(ii) $\implies$ *(iii)* Assuming the almost sure convergence of the averages one gets their $\boldsymbol{L}_q$, $q < p$, convergence by the Hoffmann-Jorgensen Theorem (see Chapter 1). Then *(iii)* follows by a standard application of the Closed Graph Theorem (as in Theorem 6.7.4 *(iv)* $\implies$ *(i)*).

(iii) $\implies$ *(i)* Let $\boldsymbol{x}_1, \dots, \boldsymbol{x}_n \subset \boldsymbol{X}$, and define

$$\boldsymbol{y}_j = \begin{cases} 0, & \text{for } 1 \le j \le N, \\ \boldsymbol{x}_{j-N} & \text{for } N < j \le N+n, \end{cases}$$

for some integer N. Then, by our assumption, the inequality

$$\Big(\mathbf{E}\Big\|\sum_{i=1}^{n} \xi_i \boldsymbol{x}_i\Big\|^{p/2}\Big)^{2/p} = \Big(\mathbf{E}\Big\|\sum_{j=1}^{N+n} \xi_i \boldsymbol{y}_i\Big\|^{p/2}\Big)^{2/p}$$

$$\le C(N+n)\Big(\sum_{j=1}^{N+n} j^{-p}\|\boldsymbol{x}_{j-N}\|^p\Big)^{1/p} \le C\frac{N+n}{N+1}\Big(\sum_{j=1}^{n} \|\boldsymbol{x}_j\|^p\Big)^{1/p}$$

holds for any $N \ge 1$, so that

$$\Big(\mathbf{E}\Big\|\sum_{i=1}^{n} \xi_i \boldsymbol{x}_i\Big\|^{p/2}\Big)^{2/p} \le C\Big(\sum_{j=1}^{n} \|\boldsymbol{x}_j\|^p\Big)^{1/p}. \qquad \text{QED}$$

Also, perhaps a little surprisingly, a Strong Law of Large Numbers for Rademacher sequences (r_n) also characterizes spaces of stable-type p.

Theorem 6.9.3. *Let $1 \le p < 2$. The following properties of a Banach space $\boldsymbol{X}$ are equivalent:*

(i) The space $\boldsymbol{X}$ is of stable-type p;

(ii) For each bounded sequence $(\boldsymbol{x}_n) \subset \boldsymbol{X}$,

$$\lim_{n\to\infty} \frac{1}{n^{1/p}} \sum_{k=1}^{n} r_k \boldsymbol{x}_k = 0, \qquad a.s.;$$

(iii) For each bounded sequence $(\boldsymbol{x}_n) \subset \boldsymbol{X}$ there exists a choice of $\varepsilon_i = \pm 1$ such that

$$\lim_{n\to\infty} \frac{1}{n^{1/p}} \sum_{k=1}^{n} \varepsilon_k \boldsymbol{x}_k = 0.$$

Proof. The implication $(i) \implies (ii)$ follows directly from Theorem 6.7.5 (ii) and the Kronecker's Lemma. The implication $(ii) \implies (iii)$ is obvious, and the implication $(iii) \implies (i)$ has already been proven in the course of the proof of implication $(iv) \implies (i)$ of Theorem 6.7.5. QED

6.10 Random integrals, convergence of infinitely divisible measures and the central limit theorem

We begin with the construction of random integrals of the form $\int f dM$, where f is a deterministic function with values in a Banach space $\boldsymbol{X}$, and M is a p-stable, real-valued random measure[30]. Random integrals of this type are a natural generalization of random series of the form $\sum_i \xi_i \boldsymbol{x}_i$, $\boldsymbol{x}_i \in \boldsymbol{X}$.

[30] For a complete theory of real-valued random integrals see S. Kwapień and W.A. Woyczyński (1992).

Definition 6.10.1. Let (T, Σ) be a measurable space. A mapping $M : \Sigma \mapsto L_0(\Omega, \mathcal{F}, P)$ is said to be an (independently scattered) random measure if, for any pair-wise disjoint sets $A_1, A_2, \ldots, \in \Sigma$, the random variables $M(A_1), M(A_2), \ldots,$ are stochastically independent, and $M(\bigcup_i A_i) = \sum_i M(A_i)$, where the series on the right-hand side converges in probability (or, almost surely).

Definition 6.10.2. A random measure M on Σ is said to be p-stable with a finite, non-negative control measure m on (T, Σ), if

$$\mathbf{E} \exp[itM(A)] = \exp[-m(A)|t|^p], \qquad A \in \Sigma.$$

It follows from the Kolmogorov's Consistency Theorem that given such a control measure m one can construct the related random measure M for any value of the parameter $p, 0 < p \leq 2$.

If $f : T \mapsto \boldsymbol{X}$ is a simple function, i.e., $f = \sum_i \boldsymbol{x}_i I_{A_i}$, where $A_i \in \Sigma$ are pairwise disjoint, and $\boldsymbol{x}_i \in \boldsymbol{X}$, then we define the random integral by the obvious formula,

$$\int_T f(t)\, M(dt) := \sum_i \boldsymbol{x}_i M(A_i). \tag{6.10.1}$$

In this case, the integral $\int f\, dM$ is a p-stable $\boldsymbol{X}$-valued random vector.

For a simple function f in $\boldsymbol{L}_p(T, \Sigma, M; \boldsymbol{X})$, and for $q < p$, the map

$$f \mapsto \int_T f(t)\, M(dt),$$

is a linear operator with values in $\boldsymbol{L}_q(\Omega, \mathcal{F}, \mathbf{P}; \boldsymbol{X})$. By (6.5.2), if $\boldsymbol{X}$ is of stable-type p, and the random measure M is p-stable, we have the inequality

$$\Big\| \int_T f\, dM \Big\|_{L_p} = \Big(\mathbf{E} \Big\| \sum_i M(A_i) \Big\|^q \Big)^{1/q} \tag{6.10.2}$$

$$\leq C \Big(\sum_i m(A_i) \|\boldsymbol{x}_i\|^p \Big)^{1/p} = C \Big(\int_T \|f(t\|^p\, m(dt) \Big)^{1/p}.$$

Since the simple functions are dense in $\boldsymbol{L}_p(T,\Sigma,M;\boldsymbol{X})$, there exists a unique extension of this operator to the whole of $\boldsymbol{L}_p$. This extension will be also denoted by $\int_T f\,dM$, $f\in\boldsymbol{L}_p$, and it also satisfies (6.10.2). Summarizing the above discussion we have the following

Theorem 6.10.1.[31] *If $\boldsymbol{X}$ is a Banach space of stable-type p, and M is a p-stable random measure on (T,Σ), with control measure m, then, for each $q, 0<q<p\le 2$, there exists a linear map,*

$$\boldsymbol{L}_p(T,\Sigma,m;\boldsymbol{X})\ni f\mapsto\int_T f\,dM\in\boldsymbol{L}_q(\Omega,\mathcal{F},\mathbf{P};\boldsymbol{X}) \tag{6.10.3}$$

satisfying (6.10.1), with its values being p-stable random vectors on $\boldsymbol{X}$. The mapping, which is called the random integral of function f with respect to random measure M, satisfies the inequality

$$\Big(\mathbf{E}\Big\|\int_T f\,dM\Big\|^q\Big)^{1/q}\le C\Big(\int_T\|f\|^p dm\Big)^{1/p}, \tag{6.10.4}$$

for some constant C which is independent of f.

Remark 6.10.1. (a) The inequality (6.10.4) is also valid if the $\boldsymbol{L}_q$ norm on the left hand side is replaced by the Lorentz norm $\Lambda_p(\int f dM)$ (see Chapter 1).

(b) Of course, the fact the space $\boldsymbol{X}$ is of stable-type p is also necessary in the above Theorem.

It is quite easy to compute the characteristic functional of the random integral $\int f\,dM$. Indeed, take a sequence of simple functions $f_n=\sum_i x_i^n I(A_i^n)$ converging to f in $\boldsymbol{L}_p(m;\boldsymbol{X})$. Then, in view of (6.10.4), the distribution $\mathcal{L}(\int f_n\,dM)$ converges weakly to $\mathcal{L}(\int f\,dM)$. Therefore, for each $\boldsymbol{x}^*\in\boldsymbol{X}^*$,

$$\mathbf{E}\exp\Big[i\boldsymbol{x}^*\int f\,dM\Big]=\lim_{n\to\infty}\mathbf{E}\exp\Big[i\boldsymbol{x}^*\sum_i x_i^n M(A_i^n)\Big]$$

[31] Most of the results of this Section are due to M.B. Marcus and W.A. Woyczynski (1979).

$$= \lim_{n\to\infty} \prod_i \exp\left[-m(A_i^n)|\boldsymbol{x}^*\boldsymbol{x}_i^n|^p\right] = \lim_{n\to\infty} \exp\Big[-\int |\boldsymbol{x}^* f_n(t)|^p m(dt)\Big]$$

$$= \exp\Big[-\int |\boldsymbol{x}^* f(t)|^p m(dt)\Big].$$

Theorem 6.10.2. *Let $0 < p \le 2$. The following properties of a Banach space $\boldsymbol{X}$ are equivalent:*

(i) The space $\boldsymbol{X}$ is of stable-type p;

(ii) For any finite measure space (T, Σ, m), and any function $f \in \boldsymbol{L}_p(T, \Sigma, m; \boldsymbol{X})$, the function,

$$\boldsymbol{X}^* \ni \boldsymbol{x}^* \mapsto \exp\Big[-\int_T |\boldsymbol{x}^* f(t)|^p m(dt)\Big] \in \mathbf{R},$$

is the characteristic functional of a p-stable probability measure $\mu = \mathcal{L}(\int f\,dM)$ on $\boldsymbol{X}$.

Proof. Theorem 6.10.1, and the above computation of the characteristicfunctions of $\mathcal{L}(\int f\,dM)$, show that $(i) \implies (ii)$. We will now show that if (ii) is not satisfied then (i) is not satisfied either. Indeed, if $\boldsymbol{X}$ is not of stable-type p then, by Theorem 6.7.5, there exists a sequence $(\boldsymbol{x}_n) \subset \boldsymbol{X}$, with $\sum_i \|\boldsymbol{x}_i\|^p < \infty$, such that the random series $\sum_i \xi_i \boldsymbol{x}_i$ does not converge almost surely. Now, take $T = \boldsymbol{X}$, with Σ being the family of Borel sets of $\boldsymbol{X}$. Let m be concentrated on the set $\{\boldsymbol{x}_1, \boldsymbol{x}_2, \ldots,\}$, with $m(\{\boldsymbol{x}_i\}) = \|\boldsymbol{x}_i\|^p$. Finally, take $f(\boldsymbol{x}) = \boldsymbol{x}/\|\boldsymbol{x}\|$. For (ii) to be true, the functional

$$\exp\Big[-\int_X |\boldsymbol{x}^* f(\boldsymbol{x})|^p m(d\boldsymbol{x})\Big] = \exp\Big[-\sum_i |\boldsymbol{x}^*\boldsymbol{x}_i|^p\Big], \qquad \boldsymbol{x}^* \in \boldsymbol{X}^*,$$

would have to be a characteristic functional on $\boldsymbol{X}^*$. That is, the random vector defined by the random series $\sum_i \xi_i \boldsymbol{x}_i$ would have to exist, but it does not. QED

Remark 6.10.2. As far as representing all stable measures μ on a Banach space $\boldsymbol{X}$ of stable-type p is concerned, we can restrict our attention to measures of the form

$$\mu = \mathcal{L}\Big(\int_{S_X} \boldsymbol{x} M(d\boldsymbol{x})\Big),$$

where the control measure σ of M is concentrated on the unit sphere S_X of the space $\boldsymbol{X}$. This fact follows by a simple change of variables. Then, the characteristic functional of μ is of the form,

$$\int \exp[i\boldsymbol{x}^*\boldsymbol{x}]\mu(d\boldsymbol{x}) = \exp\Big[-\int_{S_X} |\boldsymbol{x}^*\boldsymbol{x}|^p\sigma(d\boldsymbol{x})\Big]$$

and σ will be called the *spectral measure* of the p-stable measure μ.

In spaces of stable-type p the weak convergence of the spectral measures on the unit sphere S_X implies the weak convergence of the corresponding p-stable probability distributions on $\boldsymbol{X}$. This fact will have important applications in the proof of the Central Limit Theorem to be discussed later on in this Section.

Theorem 6.10.3. *Let $0 < p \leq 2$. The following properties of a Banach space $\boldsymbol{X}$ are equivalent:*

(i) The space $\boldsymbol{X}$ is of stable-type p;

(ii) If $\sigma_1, \sigma_2, \ldots,$ are spectral measures on the unit sphere of $\boldsymbol{X}$ such that $\sigma_i \to \sigma_\infty$, weakly on S_X, and if $\mu_1, \mu_2, \ldots,$ are the corresponding p-stable probability measures on $\boldsymbol{X}$, then there exists a stable measure μ_∞ with spectral measure σ_∞, and μ_i's converge weakly to μ_∞ on $\boldsymbol{X}$.

Proof. (i) $\implies$ (ii) Let $M_n, n = 1, 2, \ldots, \infty$, be p-stable random measures on $\boldsymbol{X}$, with control measures $\sigma_n, n = 1, 2, \ldots, \infty$. By Prokhorov's Theorem, for any $\epsilon > 0$, one can find a compact set $K \subset S_X$ such that, for all $n = 1, 2, \ldots, \infty$, $\sigma_n(S_X \setminus K) < \epsilon$. Let $f = \sum_i \boldsymbol{x}_i I(A_i)$ be a simple function with a finite range such that A_i's are continuity sets of the limit spectral measure σ_∞, $\|f(\boldsymbol{x}) - \boldsymbol{x}\| \leq \epsilon$ on K, and $\|f(\boldsymbol{x})\| \leq 1$, elsewhere. Then,

$$\begin{aligned}\int_{S_X} \|\boldsymbol{x} - f(\boldsymbol{x})\|^p\sigma_n(d\boldsymbol{x}) &= \int_K \|\boldsymbol{x} - f(\boldsymbol{x})\|^p\sigma_n(d\boldsymbol{x}) \\ &\quad + \int_{S_X\setminus K} \|\boldsymbol{x} - f(\boldsymbol{x})\|^p\sigma_n(d\boldsymbol{x}) \\ &\leq \epsilon^p + 2^p\epsilon, \qquad n = 1, 2, \ldots, \infty.\end{aligned}$$

By Theorem 6.10.1, we have that if $q < p$, then

$$\left(\mathbf{E}\Big\|\int_{S_X}(\boldsymbol{x}-f(\boldsymbol{x}))M_n(d\boldsymbol{x})\Big\|^q\right)^{1/q} \le C\left(\mathbf{E}\Big\|\int_{S_X}(\boldsymbol{x}-f(\boldsymbol{x}))\sigma_n(d\boldsymbol{x})\Big\|^p\right)^{1/p}$$

$$\le C(\epsilon^p + 2^p\epsilon)^{1/p}.$$

Therefore, because $\mu_n = \mathcal{L}(\int \boldsymbol{x}M_n(d\boldsymbol{x}))$, and

$$\int \boldsymbol{x}M_n(d\boldsymbol{x}) - \int \boldsymbol{x}M_\infty(d\boldsymbol{x})$$

$$\int(\boldsymbol{x} - f(\boldsymbol{x}))M_n(d\boldsymbol{x}) + \int(f(\boldsymbol{x}) - \boldsymbol{x})M_\infty(d\boldsymbol{x})$$

$$+\int f(\boldsymbol{x})M_n(d\boldsymbol{x}) - \int f(\boldsymbol{x})M_\infty(d\boldsymbol{x}),$$

it is sufficient to show that

$$\mathcal{L}\Big(\int_{S_X} f(\boldsymbol{x})\, M_n(\boldsymbol{x})\Big) \longrightarrow \mathcal{L}\Big(\int_{S_X} f(\boldsymbol{x})\, M_\infty(\boldsymbol{x})\Big)$$

weakly, as $n \to \infty$. However, $\int f(\boldsymbol{x})M_n(d\boldsymbol{x})$ are p-stable random vectors taking values in a fixed finite-dimensional subspace of $\boldsymbol{X}$ spanned by the values of f. Therefore, to prove the weak convergence it is sufficient to prove the convergence of the characteristic functionals. And, for each $\boldsymbol{x}^* \in \boldsymbol{X}^*$, as $n \to \infty$,

$$\mathbf{E}\exp\Big[i\boldsymbol{x}^*\int f(\boldsymbol{x})M_n(d\boldsymbol{x})\Big] = \exp\Big[-\int_{S_X}|\boldsymbol{x}^*f(\boldsymbol{x})|^p\sigma_n(d\boldsymbol{x})\Big]$$

$$= \exp\Big[-\sum_j|\boldsymbol{x}^*\boldsymbol{x}_j|^p\sigma_n(A_j)\Big] \longrightarrow \exp\Big[-\sum_j|\boldsymbol{x}^*\boldsymbol{x}_j|^p\sigma_\infty(A_j)\Big]$$

$$= \mathbf{E}\exp\Big[i\boldsymbol{x}^*\int f(\boldsymbol{x})M_\infty(d\boldsymbol{x})\Big],$$

because A_j's are continuity sets for σ_∞, and $\sigma_n \to \sigma_\infty$ weakly. So, clearly, the probability measure $\mu_\infty = \mathcal{L}(\int \boldsymbol{x}M_\infty(\boldsymbol{x}))$.

(ii) $\Longrightarrow$ *(i)* As in the proof of Theorem 6.10.1, we will show that if *(i)* is not satisfied then *(ii)* is not satisfied either. So, if $\boldsymbol{X}$ is not of stable-type p, then there exists a sequence

$(\boldsymbol{x}_i) \subset \boldsymbol{X}$ with $\sum_i \|\boldsymbol{x}_i\|^p < \infty$, for which the random series $\sum_i \xi_i \boldsymbol{x}_i$ does not converge almost surely. We take for σ_∞ the measure concentrated on the set $\{\boldsymbol{x}_i/\|\boldsymbol{x}_1\|, \boldsymbol{x}_2/\|\boldsymbol{x}_2\|, \dots\}$ with values $\sigma_\infty(\{\boldsymbol{x}_i/\|\boldsymbol{x}_i\|\}) = \|\boldsymbol{x}_i\|^p, i = 1, 2, \dots,$, and for σ_n we take the measure concentrated on the set $\{\boldsymbol{x}_i/\|\boldsymbol{x}_1\|, \dots, \boldsymbol{x}_2/\|\boldsymbol{x}_n\|\}$, with values $\sigma_n(\{\boldsymbol{x}_i/\|\boldsymbol{x}_i\|\}) = \|\boldsymbol{x}_i\|^p, i = 1, \dots, n$. Clearly, as $n \to \infty$, we have the weak convergence $\sigma_n \to \sigma_\infty$ on S_X. Now, consider the corresponding p-stable measures $\mu_n, n = 1, 2, \dots$. We have $\mu_n = \mathcal{L}(\sum_{i=1}^n \xi_i \boldsymbol{x}_i)$. If μ_n converged weakly then, by Ito-Nisio Theorem (see Chapter 1) , the series $\sum_i \xi_i \boldsymbol{x}_i$ would converge almost surely, a contradiction. QED

Now, we turn to a study of Poisson random measures M, and random integrals $\int f\,dM$, where the function f takes values in a Banach space $\boldsymbol{X}$ of Rademacher-type p. This will permit us to introduce a representation of a class of infinitely divisible measures on $\boldsymbol{X}$, and will also serve as a tool in the proof of the Central Limit Theorem.

Definition 6.10.3. A random measure M on (T, Σ) is said to be (symmetric) *Poissonian*, with a σ-finite control measure m, if its characteristic function

$$\mathbf{E}\exp[itM(A)] = \exp\big[m(A)(\cos t - 1)\big], \qquad t \in \mathbf{R}, \tag{6.10.5}$$

for each set $A \in \Sigma$, with $m(A) < \infty$.

Note that if $p \le 2$, then

$$\mathbf{E}|M(A)|^p \le E(M(A))^2 = -\frac{d^2}{dt^2}\Big(\exp\big[m(A)(\cos(t-1)\big]\Big)\Big|_{t=0} = m(A), \tag{6.10.6}$$

because $M(A)$ is integer-valued.

Now, assume that $f \in \boldsymbol{L}_p(T, \Sigma, m; \boldsymbol{X})$, and $\boldsymbol{X}$ is a Banach space of Rademacher-type p. If f is a simple function, $f = \sum_i \boldsymbol{x}_i I_{A_i}, A_i \in \Sigma$, we, obviously define

$$\int f dM = \sum_i \boldsymbol{x}_i M(A_i). \tag{6.10.7}$$

Then, because $\boldsymbol{X}$ is of Rademacher-type p (see, Theorem 6.7.1),

$$\begin{aligned}\mathbf{E}\Big\|\int f\,dM\Big\|^p &= \mathbf{E}\Big\|\sum_i \boldsymbol{x}_i M(A_i)\Big\|^p \le C\sum_i \|\boldsymbol{x}_i\|^p \mathbf{E}|M(A_i)|^p \\ &\le C\sum_i \|\boldsymbol{x}_i\|^p m(A_i) = \int \|f\|^p dm,\end{aligned} \tag{6.10.8}$$

in view of (6.10.6). Thus, as has been done above for the p-stable random integrals, we can extend the integral operator $\int$ to the whole of $\boldsymbol{L}_p$ while preserving the inequality (6.10.8).

The characteristic functional, for a simple f defined above, is of the form

$$\mathbf{E}\exp\Big[i\boldsymbol{x}^*\int f\,dM\Big] = \mathbf{E}\exp\Big[i\boldsymbol{x}^*\sum_i \boldsymbol{x}_i M(A_i)\Big]$$

$$= \prod_i \mathbf{E}\exp\big[i\boldsymbol{x}^*\boldsymbol{x}_i M(A_i)\big] = \prod_i \exp\big[m(A_i)(\cos\boldsymbol{x}^*\boldsymbol{x}_i - 1)\big]$$

$$= \exp\Big[\int_T (\cos(\boldsymbol{x}^* f) - 1)\,dm\Big],$$

and, because of (6.10.8), the same formula extends to all $f \in \boldsymbol{L}_p$. Summarizing the above analysis we obtain the following,

Theorem 6.10.4. *If $\boldsymbol{X}$ is a Banach space of Rademacher-type p, and M is a Poissonian random measure on (T,Σ) with the control measure m, then there exists a linear map, called the random integral,*

$$\boldsymbol{L}_p(T,\Sigma,m;\boldsymbol{X}) \ni f \mapsto \int_T f\,dM \in \boldsymbol{L}_p(\Omega,\mathcal{F},\mathbf{P};\boldsymbol{X}), \tag{6.10.9}$$

satisfying (6.10.7). Its values are infinitely divisible $\boldsymbol{X}$-valued random vectors with the characteristic functional

$$\mathbf{E}\exp\Big[i\boldsymbol{x}^*\int f\,dM\Big] = \exp\Big[\int_T (\cos(\boldsymbol{x}*f) - 1)\,dm, \quad \boldsymbol{x}^* \in \boldsymbol{X}^*, f \in \boldsymbol{L}_p, \tag{6.10.10}$$

satisfying the inequality

$$\mathbf{E}\left\|\int f\,dM\right\|^p \leq C\int \|f\|^p dm. \tag{6.10.11}$$

Remark 6.10.3. Rademacher-type p of $\boldsymbol{X}$ is easily seen to be also the necessary condition in the above Theorem.

Remark 6.10.4. A straightforward computation shows that if $T = \boldsymbol{X}$, and $\int_X \|\boldsymbol{x}\|^p dm < \infty$, then the law of $\int \boldsymbol{x}\,dM$ is the symmetrization, call it $e_s(m)$, of the Poissonization $e(m)$ of m. Recall that

$$e(m) := e^{-m(\boldsymbol{X})}\sum_{n=0}^{\infty}\frac{m^{*n}}{n!}$$

where the inequality (6.10.11) gives the above formula its precise interpretation. In this particular case the meaning of (6.10.11) is as follows: If $\boldsymbol{X}$ is of Rademacher-type p, then

$$\int \|\boldsymbol{x}\|^p de_s(m) \leq C\int \|\boldsymbol{x}\|^p dm.$$

One can easily extend the above definition of the random integral to all Poissonian random measures M on a Banach space of Rademacher-type p for which the control measure satisfies only the condition,

$$\int_X \min(1, \|\boldsymbol{x}\|^p)\,dm < \infty. \tag{6.10.12}$$

Thus for all such m, the formula (6.10.10) also represents an infinitely divisible law on $\boldsymbol{X}$. Here is how the extension can be accomplished.

Given our previous result, we only have to be concerned about the case $\int_{\|\boldsymbol{x}\|>1}\boldsymbol{x}\,dM$. So, let us notice that

$$\int_{\|\boldsymbol{x}\|>1}\boldsymbol{x}\,dM = \sum_{k=1}^{\infty}\int_{k<\|\boldsymbol{x}\|\leq k+1}\boldsymbol{x}\,dM, \tag{6.10.13}$$

where each of the integrals on the right-hand side is well defined in view of the above Theorem, and the series converges a.s. Indeed, the space $\boldsymbol{X}$ is of Rademacher-type q, for each $q \in (0, p]$, with the same constant. Take $q_k := \log 2/\log(k+1)$. Then, by (6.10.11),

$$\mathbf{E}\Big\|\int_{k<\|\boldsymbol{x}\|\le k+1} \boldsymbol{x}\, dM\Big\|^{q_k} \le C\int_{k<\|\boldsymbol{x}\|\le k+1} \|\boldsymbol{x}\|^{q_k}\, dm$$

$$\le 2Cm(\{k < \|\boldsymbol{x}\| \le k+1\}),$$

for $k = 1, 2, \ldots$. Therefore, by Chebyshev's Inequality,

$$\mathbf{P}\Big(\Big\|\int_{k<\|\boldsymbol{x}\|\le k+1} \boldsymbol{x}\, dM\Big\| > k^{-2}\Big) \le C\frac{2m(\{k < \|\boldsymbol{x}\| \le k+1\})}{k^{-2q_k}}.$$

Because

$$\lim_{k\to\infty} k^{-2q_k} = \lim_{k\to\infty} k^{-2\log 2/\log(k+1)} = e^{-2\log 2} > 0,$$

we have the convergence of the series,

$$\mathbf{P}\Big(\Big\|\int_{k<\|\boldsymbol{x}\|\le k+1} \boldsymbol{x}\, dM\Big\| > k^{-2}\Big) \le C_1 m(\{\|\boldsymbol{x}\| > 1\}) < \infty,$$

and the Borel-Cantelli Lemma gives the desired almost sure convergence in (6.10.13).

Next, we are going to prove a continuity result analogous to the one proven for p-stable measures in Theorem 6.10.3.

Theorem 6.10.5. *Let $\boldsymbol{X}$ be a Banach space of Rademacher-type p. If m_n, $n = 1, 2, \ldots, \infty$, is a sequence of σ-finite measures on $\boldsymbol{X}$ with*

$$\int_X \min(1, \|\boldsymbol{x}\|^p) m_n(d\boldsymbol{x}) < \infty, \qquad n = 1, 2, \ldots, \infty,$$

and such that

$$\lim_{n\to\infty} m_n = m_\infty, \tag{6.10.14}$$

in weak topology outside each neighborhood of $0 \in \boldsymbol{X}$, and

$$\lim_{\epsilon\to 0} \limsup_{n\to\infty} \int_{\|x\|\le\epsilon} \|\boldsymbol{x}\|^p dm_n = 0, \tag{6.10.15}$$

then

$$\lim_{n\to\infty} \mathcal{L}\Big(\int \boldsymbol{x}\, dM_n\Big) = \mathcal{L}\Big(\int \boldsymbol{x}\, dM_\infty\Big) \tag{6.10.16}$$

in weak topology, where M_n*'s are Poissonian random measures with control measures* $m_n, n = 1, 2, \ldots, \infty$ *.*

Proof. Let $\delta > 0$. By (6.10.15), and (6.10.11), there exists an $\epsilon > 0$, and an integer N_0, such that, for each $n \geq N_0$,

$$\mathbf{E}\Big\|\int_{B_\epsilon} \boldsymbol{x}\, dM_n\Big\| \leq C \int_{B_\epsilon} \|\boldsymbol{x}\|^p dm_n < \delta. \tag{6.10.17}$$

On the other hand

$$\int_{B_\epsilon^c} \boldsymbol{x}\, dM_n - \int_{B_\epsilon^c} \boldsymbol{x}\, dM_\infty \tag{6.10.18}$$

$$= \int_{B_\epsilon^c} (\boldsymbol{x} - f(\boldsymbol{x}))\, dM_n - \int_{B_\epsilon^c} (\boldsymbol{x} - f(\boldsymbol{x}))\, dM_\infty + \int_{B_\epsilon^c} f\, dM_n - \int_{B_\epsilon^c} f\, dM_\infty,$$

for any $f \in \boldsymbol{L}_p(m; \boldsymbol{X})$. Because of (6.10.14), by Prokhorov's Theorem, there exists a compact set $K \subset B_\epsilon^c \subset \boldsymbol{X}$ such that $m_n(B_\epsilon^c \backslash K) < \delta$, for all $n = 1, 2, \ldots, \infty$. Let $f = \sum_i \boldsymbol{x}_i I_{A_i}, \boldsymbol{x}_i \in \boldsymbol{X}$, be a simple function on $(B_\epsilon^c, \mathcal{B}_X)$ with a finite range, and such that A_i's are continuity sets of m_∞, $\|f(\boldsymbol{x}) - \boldsymbol{x}\| \leq \delta$, and $\|f(\boldsymbol{x})\| \leq \|\boldsymbol{x}\|$, elsewhere. Then, by (6.10.11),

$$\mathbf{E}\Big\|\int_K (\boldsymbol{x} - f(\boldsymbol{x})) M_n(d\boldsymbol{x})\Big\|^p \leq C \int_K \|\boldsymbol{x} - f(\boldsymbol{x})\|^p m_n(d\boldsymbol{x})$$

$$\leq C\delta^p \sup_n m_n(K) \leq C_1 \delta^p, \qquad n = 1, 2, \ldots, \infty,$$

where C_1 is independent of n because, by (6.10.14), $\sup_n m_n(B_\epsilon^c) < \infty$.

Outside K we have, for each $\alpha > 0$ (as in the reasoning following Remark 6.10.4),

$$\mathbf{P}\Big(\Big\|\int_{B_\epsilon^c \backslash K} \boldsymbol{x}\, dM_n\Big\| > \alpha^2 \sum_k \frac{1}{k^2}\Big)$$

$$\leq \mathbf{P}\Big(\bigcup_{k=1}^{\infty} \Big\{\Big\|\int_{(B_\epsilon^c \backslash K) \cap \{\epsilon k < \|\boldsymbol{x}\| \leq \epsilon(k+1)\}} \boldsymbol{x}\, dM_n\Big\| > \frac{\alpha^2}{k^2}\Big)$$

$$\leq \sum_{k=1}^{\infty} \mathbf{P}\Big(\Big\| \int_{(B_\epsilon^c \setminus K)\cap\{\epsilon k<\|\boldsymbol{x}\|\leq \epsilon(k+1)\}} \boldsymbol{x}\, dM_n \Big\| > \frac{\alpha^2}{k^2}\Big)$$

$$\leq C_2 m_n(B_\epsilon^c \setminus K) < C^2\delta,$$

again, uniformly for $n = 1, 2, \ldots, \infty$. Now, in view of (6.10.17-18), and the above inequalities, to complete the proof of the Theorem it suffices to show that

$$\mathcal{L}\Big(\int_{B_\epsilon^c} f\, dM_n\Big) \longrightarrow \mathcal{L}\Big(\int_{B_\epsilon^c} f\, dM_\infty\Big),$$

weakly, as $n \to \infty$. However, $\int f\, dM_n$ are random vectors taking values in a fixed finite-dimensional subspace, $\overline{\text{span}}[\boldsymbol{x}_i] \subset \boldsymbol{X}$. Therefore, it is sufficient to check the convergence of their characteristic functionals. And, indeed, as $n \to \infty$,

$$\mathbf{E}\exp\Big[i\boldsymbol{x}^* \int f\, dM_n\Big] = \exp\Big[\sum_j (\cos(\boldsymbol{x}^*\boldsymbol{x}_j) - 1)m_n(A_j)\Big]$$

$$\longrightarrow \exp\Big[\sum_j (\cos(\boldsymbol{x}^*\boldsymbol{x}_j) - 1)m_\infty(A_j)\Big] = \mathbf{E}\exp\Big[i\boldsymbol{x}^* \int f\, dM_\infty\Big],$$

$$\boldsymbol{x}^* \in \boldsymbol{X}^*,$$

because A_j's have been chosen to be continuity sets of m_∞, and because of the assumption (6.10.14). QED

We will conclude this Section by applying the above results to the proof of the Central Limit Theorem.

Theorem 6.10.6. *Let $0 < p < 2$. The following properties of a Banach space $\boldsymbol{X}$ are equivalent:*

(i) The space $\boldsymbol{X}$ is of stable-type p;

(ii) Let X_{nk}, $n = 1, 2, \ldots, k = 1, \ldots, k_n$, be an arbitrary array of symmetric random vectors in $\boldsymbol{X}$, independent in each row, $\sigma \neq 0$ be an arbitrary finite Borel measure on $S_X = \{\|\boldsymbol{x}\| = 1\}$, and $\mu_{nk} = \mathcal{L}(X_{nk})$. If

$$\lim_{n\to\infty} \sum_{k=1}^{k_n} \mu_{nk}(\|\boldsymbol{x}\| > t, \boldsymbol{x}/\|\boldsymbol{x}\| \in A) = t^{-p}\sigma(A), \tag{6.10.19}$$

for any Borel set $A \subset S_X$, with $\sigma(\partial A) = 0$, and $t > 0$, and if, for some $q > p$,

$$\lim_{\epsilon \to 0} \limsup_{n \to \infty} \sum_{k=1}^{k_n} \int_{\|\boldsymbol{x}\| \leq \epsilon} \|\boldsymbol{x}\|^q \mu_{nk}(d\boldsymbol{x}) = 0, \tag{6.10.20}$$

then, as $n \to \infty$,

$$\mathcal{L}\Big(\sum_{k=1}^{k_n} X_{nk}\Big) \longrightarrow \mu, \tag{6.10.21}$$

weakly, where μ is the p-stable measure with the characteristic functional,

$$\int_X \exp[i\boldsymbol{x}^*\boldsymbol{x}]\,\mu(d\boldsymbol{x}) = \exp\Big[\int_{S_X \times \mathbf{R}^+} \big(\cos(\boldsymbol{x}^*\boldsymbol{x}) - 1\big)\sigma(d\boldsymbol{s})\frac{d\rho}{\rho^{1+p}}, \tag{6.10.22}$$

where $\boldsymbol{x}^* \in \boldsymbol{X}^*, \boldsymbol{x} = \boldsymbol{s} \cdot \rho,\ \boldsymbol{s} \in S_X$, and $\rho \in \mathbf{R}^+$.

Proof. (*ii*) $\Longrightarrow$ (*i*) follows immediately from the counterexample to the Weak Law of Large Numbers in the proof of Theorem 6.9.1.

(*i*) $\Longrightarrow$ (*ii*) By Theorem 6.10.2, for any finite measure σ on the unit sphere S_X, (6.10.22) is the characteristic functional of a p-stable probability measure on the Banach space $\boldsymbol{X}$ of stable-type p. Now, in view of LeCam's Theorem, it is sufficient to show that if $Y_{nk} = \int \boldsymbol{x}\, dM_{nk}$, where μ_{nk} are control measures of Poissonian M_{nk}, then the probability distributions of the array Y_{nk} converge weakly to μ, as $n \to \infty$. However,

$$\mathcal{L}\Big(\sum_{k=1}^{k_n} \int \boldsymbol{x}\, M_{nk}(d\boldsymbol{x})\Big) = \mathcal{L}\Big(\int_X \boldsymbol{x}\, M_n(d\boldsymbol{x})\Big),$$

where M_n is Poissonian with control measure

$$\mu_n = \sum_{k=1}^{k_n} \mu_{nk}$$

which is easy to check by verifying the characteristic functionals. Because $\boldsymbol{X}$ is of stable-type p, there exists a $q > p$ such that $\boldsymbol{X}$ is of Rademacher-type q (see Section 6.5). Now, both μ_n and $\sigma(d\boldsymbol{s})d\rho/\rho^{1+p}$ integrate $\min(1, \|\boldsymbol{x}\|^q)$, and hence it is sufficient to check the conditions of Theorem 6.10.5 with p replaced by q. Indeed, (6.10.19) implies evidently that $d\mu_n$ converge to $\sigma(d\boldsymbol{s})d\rho/\rho^{1+p}$ weakly outside each neighborhood of 0 because the limiting measure is a product measure. QED

Corollary 6.10.1. *Let $p < 2$. The following properties of a Banach space $\boldsymbol{X}$ are equivalent:*

(i) The space $\boldsymbol{X}$ is of stable-type p;

(ii) For each finite, non-zero Borel measure on S_X, and for any sequence (X_i) if independent, identically distributed random vectors in $\boldsymbol{X}$ with symmetric probability distributiom $\mu = \mathcal{L}(X_1)$ such that

$$\lim_{s\to\infty} \frac{\mu(\|\boldsymbol{x}\| > st)}{\mu(\|\boldsymbol{x}\| > s)}, \tag{6.10.23}$$

exists for each t, and such that for a nonnegative sequence $b_n \uparrow \infty$,

$$\lim_{n\to\infty} n\mu\Big(\{\|\boldsymbol{x}\| > tb_n, \boldsymbol{x}/\|\boldsymbol{x}\| \in A\}\Big) = t^{-p}\sigma(A), \tag{6.10.24}$$

for each Borel set $A \subset S_X$, with $\sigma(\partial A) = 0$, and any $t > 0$, we have that, as $n \to \infty$,

$$\mathcal{L}\Big(\frac{X_1 + \cdots + X_n}{b_n}\Big) \longrightarrow \mu$$

weakly, where μ is the p-stable measure determined by the characteristic functional (6.10.22).

Proof. In view of (6.10.24), the limit in (6.10.23) is exactly t^{-p}. Therefore, by the standard Karamata procedure for regularly varying functions, for each $q > p$ one can find a constant K such that

$$Kt^q\mu(\|\boldsymbol{x}\| > t) \geq \int_{\|\boldsymbol{x}\|<t} \|\boldsymbol{x}\|^q \mu(d\boldsymbol{x}). \tag{6.10.25}$$

Now we can apply the above Theorem with $X_{nk} = X_k/b_n, k = 1, \ldots, n, n = 1, 2, \ldots$. Clearly, (6.10.24) implies (6.10.19), and, on the other hand, by (6.10.23-24),

$$\lim_{\epsilon\to 0} \limsup_{n\to\infty} \sum_{k=1}^{k_n} \int_{\|\boldsymbol{x}\|\leq\epsilon} \|\boldsymbol{x}\|^q \mu_{nk}(d\boldsymbol{x})$$

$$= \lim_{\epsilon\to 0} \limsup_{n\to\infty} n^{1-q/p} \int_{\|\boldsymbol{x}\|\le \epsilon n^{1/p}} \|\boldsymbol{x}\|^q \mu(d\boldsymbol{x})$$
$$\le K \lim_{\epsilon\to 0} \limsup_{n\to\infty} n^{1-q/p} \epsilon^q n^{q/p} \mu(\|\boldsymbol{x}\| > \epsilon n^{1/p})$$
$$\le K \lim_{\epsilon\to 0} \epsilon^{q-p} = 0,$$

whenever $q > p$, so that (6.10.20) is also satisfied, which gives us our Corollary. QED

Remark 6.10.5. To conclude this Chapter we would like to mention a non-probabilistic application[32] of the above results concerning rearrangement of series in Banach spaces of Rademacher-type p. If $\boldsymbol{X}$ is of Rademacher type p, and $(\boldsymbol{x}_i) \subset \boldsymbol{X}$ is such that $\sum_i \|\boldsymbol{x}_i\|^p < \infty$, and

$$\sum_{i=1}^{n_k} \boldsymbol{x}_i \longrightarrow \boldsymbol{x} \in \boldsymbol{X}, \qquad \text{as} \qquad k \to \infty,$$

for a certain sequence $(n_k) \subset \mathbf{N}$, then there exists a rearrangement γ of positive integers such that

$$\sum_{i=1}^{\infty} \boldsymbol{x}_{\gamma(i)} = \boldsymbol{x}.$$

[32]Due to P. Assuad (1974), Exp. XVI, and B. Maurey and G. Pisier (1974/75), Annexe I. For real series it is due to Steinitz, and for $\boldsymbol{L}_p$ spaces, to Kadec.

Chapter 7

Spaces of type 2

7.1 Additional properties of spaces of type 2

For the parameter value $p = 2$, the concepts of the Rademacher-type p, and the stable-type p spaces, coincide.

Proposition 7.1.1. *A normed space $\boldsymbol{X}$ is of Rademacher-type 2 if, and only if, it is of stable-type 2.*

Proof. If $\boldsymbol{X}$ is of stable-type 2 then, by Proposition 6.5.1, it is also of Rademacher-type 2. Conversely, if $\boldsymbol{X}$ is of Rademacher-type 2, then, by Proposition 6.7.1, there exists a constant $C > 0$ such that, for any $n \in \mathbf{N}$, and any sequence $(\boldsymbol{x}_i) \subset \boldsymbol{X}$,

$$\left(\mathbf{E}\left\|\sum_{i=1}^{n} \gamma_i \boldsymbol{x}_i\right\|^2\right)^{1/2} \leq C\left(\sum_{i=1}^{n} \|\boldsymbol{x}_i\|^2\right)^{1/2},$$

where (γ_i) are independent, identically distributed, zero-mean real 2-stable (that is, Gaussian) random variables. Thus $\boldsymbol{X}$ is of stable-type 2. QED

So from this point forward we will not use the names "stable-type 2", and "Rademacher-type 2", but we will simply talk about spaces of type 2. Of course, results proven in Chapter 6 for spaces of general type p (stable, or Rademacher) apply to spaces of type 2, as well. We already know that if $2 \leq p < \infty$ then $\boldsymbol{L}_p$ is of type 2.

Here we give an elementary proof of this fact. Let, $r_1, \dots, r_n$, be, as usual, a Rademacher sequence on $(\Omega, \mathcal{F}, \mathbf{P})$, and let $\boldsymbol{x}_1, \dots, \boldsymbol{x}_n, \in \boldsymbol{L}_p(T, \Sigma, \mu)$. By Khinchine's Inequality (see Introduction), for each $p \geq 2$, there exists a constant C such that, for each $t \in T$,

$$E\big|r_1\boldsymbol{x}(t) + \cdots + r_n\boldsymbol{x}(t)\big|^p \leq C^{p/2}\Big(\sum_{j=1}^{n} |\boldsymbol{x}_j(t)|^2\Big)^{p/2}.$$

Integrating both sides with respect to μ, and applying Fubini's Theoorem to the left-hand side, we get that

$$\Big(\mathbf{E}\Big\|\sum_{j=1}^{n} r_j\boldsymbol{x}_j\Big\|^p\Big)^{2/p} \leq C\sum_{j=1}^{n} \|\boldsymbol{x}_j\|^2.$$

Furthermore, by Jensen's Inequality,

$$\Big(\mathbf{E}\Big\|\sum_{j=1}^{n} r_j\boldsymbol{x}_j\Big\|^p\Big)^{2/p} \geq \mathbf{E}\Big\|\sum_{j=1}^{n} r_j\boldsymbol{x}_j\Big\|^2,$$

so that

$$\mathbf{E}\Big\|\sum_{j=1}^{n} r_j\boldsymbol{x}_j\Big\|^2 \leq C\sum_{j=1}^{n} \|\boldsymbol{x}_j\|^2. \qquad \text{QED}$$

Many additional probabilistic properties of spaces of type 2 follow from the next Corollary and the results contained in earlier chapters.

Corollary 7.1.1.[1] *If $\boldsymbol{X}$ is of type 2 then there exists a $q < \infty$ such that $\boldsymbol{X}$ is of cotype q.*

Proof. Because of Theorem 5.4.2, and because type 2 is a superproperty, it suffices to show that if $\boldsymbol{X}$ contains a copy of $\boldsymbol{c}_0$ then it is not of type 2. Let $(\boldsymbol{e}_n)$ be the standard basis in $\boldsymbol{c}_0 \subset \boldsymbol{X}$. Consider the random series $\sum_i \gamma_i\boldsymbol{x}_i$, where $\boldsymbol{x}_i = i^{-1}\boldsymbol{e}_i$. Evidently, $\sum_i \|\boldsymbol{x}_i\|^2 < \infty$, but the series $\sum_i \gamma_i\boldsymbol{x}_i$ diverges almost surely in view of the Borel-Cantelli Lemma. Thus the above Proposition implies that $\boldsymbol{X}$ is not of type 2. QED

[1]Due to B.Maurey and G. Pisier (1976).

The next two results[2] provide sufficient conditions for a Banach space to be of type 2, and they use orthogonal rather than independent random variables.

Theorem 7.1.1. *Let (ξ_i) be a complete orthonormal system in $\boldsymbol{L}_2(\Omega, \mathcal{F}, \mathbf{P})$, and let $\boldsymbol{X}$ be a Banach space. If there exists a constant $C > 0$ such that, for all $n \in \mathbf{N}$, and any $\boldsymbol{x}_1, \ldots, \boldsymbol{x}_n \in \boldsymbol{X}$,*

$$\Big(\mathbf{E}\Big\|\sum_{i=1}^{n} \xi_i \boldsymbol{x}_i\Big\|^1\Big)^{1/2} \leq C\Big(\sum_{i=1}^{n} \|\boldsymbol{x}_i\|^2\Big)^{1/2},$$

then $\boldsymbol{X}$ is of type 2.

Proof. The standard "gliding hump" procedure implies that, for each $\epsilon > 0$, there exist increasing sequences of integers (k_j), and (m_j), and an orthonormal sequence (η_j) such that

$$\eta_j = \sum_{k=k_j}^{k_{j+1}-1} (\eta_j, \xi_j)\xi_j, \qquad \mathbf{E}|\eta_j - r_{m_j}| < \frac{\epsilon}{2j}, \quad j \in N. \tag{7.1.1}$$

Now, for each $n \in \mathbf{N}$, and any sequence $\boldsymbol{x}_1, \ldots, \boldsymbol{x}_n \in \boldsymbol{X}$,

$$\mathbf{E}\Big\|\sum_{j=1}^{n} r_j \boldsymbol{x}_j\Big\|^2 = \mathbf{E}\Big\|\sum_{j=1}^{n} r_{m_j} \boldsymbol{x}_j\Big\|^2,$$

and, by the triangle and Schwartz inequalities, we have

$$\Big(\mathbf{E}\Big\|\sum_{j=1}^{n} r_{m_j} \boldsymbol{x}_j\Big\|^2\Big)^{1/2} \leq \Big(\mathbf{E}\Big\|\sum_{j=1}^{n} (r_{m_j} - \eta_j) \boldsymbol{x}_j\Big\|^2\Big)^{1/2} + \Big(\mathbf{E}\Big\|\sum_{j=1}^{n} \eta_j \boldsymbol{x}_j\Big\|^2\Big)^{1/2}$$

$$\leq \Big(\sum_{j=1}^{n} \mathbf{E}|r_{m_j} - \eta_j|^2 \sum_{j=1}^{n} \|\boldsymbol{x}_j\|^2\Big)^{1/2} + \Big(\mathbf{E}\Big\|\sum_{j=1}^{n} \eta_j \boldsymbol{x}_j\Big\|^2\Big)^{1/2}$$

$$\leq \epsilon^{1/2}\Big(\sum_{j=1}^{n} \|\boldsymbol{x}_j\|^2\Big)^{1/2} + \Big(\mathbf{E}\Big\|\sum_{j=1}^{n} \eta_j \boldsymbol{x}_j\Big\|^2\Big)^{1/2}.$$

---

[2]Due to S. Kwapień (1972/73).

On the other hand, by (7.1.1), and orthonormality of η_j's,

$$\mathbf{E}\Big\|\sum_{j=1}^{n}\eta_j\boldsymbol{x}_j\Big\|^2=\mathbf{E}\Big\|\sum_{j=1}^{n}\Big(\sum_{k=k_j}^{k_{j+1}-1}(\eta_j,\xi_k)\xi_k\Big)\boldsymbol{x}_j\Big\|^2$$

$$\leq C\sum_{j=1}^{n}\sum_{k=k_j}^{k_{j+1}-1}|(\eta_j,\xi_k)|^2\|\boldsymbol{x}_j\|^2=C\sum_{j=1}^{n}\|\boldsymbol{x}_j\|^2.$$

Thus

$$\mathbf{E}\Big\|\sum_{j=1}^{n}r_j\boldsymbol{x}_j\Big\|^2\leq\big(\epsilon^{1/2}+C^{1/2}\big)^2\sum_{j=1}^{n}\|\boldsymbol{x}_j\|^2,$$

which concludes the proof of the fact that $\boldsymbol{X}$ is of type 2. QED

The next result provides a sufficient condition for the dual space $\boldsymbol{X}^*$ to be of type 2.

Corollary 7.1.2. *Let (ξ_i) be a complete orthonormal system in $\boldsymbol{L}_2(\Omega,\mathcal{F},\mathbf{P})$, and let $\boldsymbol{X}$ be a Banach space. If there exists a constant $C>0$ such that, for all $n\in\mathbf{N}$, and any sequence $\boldsymbol{x}_1,\dots,\boldsymbol{x}_n\in\boldsymbol{X}$,*

$$\Big(\mathbf{E}\Big\|\sum_{i=1}^{n}\xi_i\boldsymbol{x}_i\Big\|^2\Big)^{1/2}\geq C\Big(\sum_{i=1}^{n}\|\boldsymbol{x}_i\|^2\Big)^{1/2},$$

then the dual space $\boldsymbol{X}^*$ is of type 2.

Proof. In view of the above Theorem it is sufficient to show that, for each $n\in\mathbf{N}$, and any sequence $\boldsymbol{x}_1^*,\dots,\boldsymbol{x}_n^*\in\boldsymbol{X}^*$, $\mathbf{E}\Big\|\sum_{i=1}^{n}\xi_i\boldsymbol{x}_i^*\Big\|^2\leq C^{-1}\sum_{i=1}^{n}\|\boldsymbol{x}_i^*\|^2$. The completeness of the sequence (ξ_i) implies that the linear combinations of ξ_i's are dense in $\boldsymbol{L}_2$. Hence, the set V of all $\boldsymbol{X}$-valued functions of the form $\phi=\sum_{i=1}^{n}\xi_i\boldsymbol{x}_i$, $\boldsymbol{x}_i\in\boldsymbol{X}$, $i,n\in\mathbf{N}$, is dense in $\boldsymbol{L}_2(\Omega,\mathcal{F},\mathbf{P};\boldsymbol{X})$. Now, it follows from the standard duality argument that, for any $\phi^*=\sum_{i=1}^{n}\xi_i\boldsymbol{x}_i^*$, and an arbitrary $\epsilon>0$, there exists $\phi=\sum_{i=1}^{m}\xi_i\boldsymbol{x}_i\in V$, with $\mathbf{E}\|\phi\|^2=1$, such that

$$\Big(\mathbf{E}\Big\|\sum_{i=1}^{n}\xi_i\boldsymbol{x}_i^*\Big\|^2\Big)^{1/2}\leq\mathbf{E}|\phi^*\phi|+\epsilon=\sum_{i=1}^{\min(n,m)}|\boldsymbol{x}_i^*\boldsymbol{x}_i|+\epsilon$$

$$\leq \Big(\sum_{i=1}^{n} \|\boldsymbol{x}_i^*\|^2\Big)^{1/2} \cdot \Big(\sum_{i=1}^{m} \|\boldsymbol{x}_i\|^2\Big)^{1/2} + \epsilon \leq C^{-1/2}\Big(\sum_{i=1}^{m} \|\boldsymbol{x}_i\|^2\Big)^{1/2} + \epsilon.$$

QED

The sufficient conditions appearing in the following Theorem involve independence but random variables are not necessarily identically distributed.

Theorem 7.1.2. *Let $\boldsymbol{X}$ be a Banach space, and let (ξ_i) be a sequence of real independent random variables satisfying the Lindeberg-type conditions,*

$$\mathbf{E}|\xi_i|^2 = 1, \qquad i = 1, 2, \ldots, \tag{7.1.2}$$

$$\lim_{n\to\infty} \frac{1}{n} \sum_{i=1}^{n} \mathbf{E}\Big(\|\boldsymbol{x}_i\|^2 I(|\xi_i| > \epsilon\sqrt{n})\Big) = 0, \qquad \forall \epsilon > 0. \tag{7.1.3}$$

If there exists a $C > 0$ such that, for any $n \in \mathbf{N}$, and $\boldsymbol{x}_1, \ldots, \boldsymbol{x}_n \in \boldsymbol{X}$,

$$\Big\|\sum_{i=1}^{n} \xi_i \boldsymbol{x}_i\Big\| \leq C\Big(\sum_{i=1}^{n} \|\boldsymbol{x}_i\|^2\Big)^{1/2},$$

then $\boldsymbol{X}$ is of type 2.

Proof. Let $\boldsymbol{x}_1, \ldots, \boldsymbol{x}_n \in \boldsymbol{X}$. By the Central Limit Theorem the joint probability distribution of the random variables

$$\xi_{i,m} = \frac{1}{m^{1/2}}\Big(\sum_{j=1}^{m} \xi_{n(i-1)+j}\Big), \qquad i = 1, \ldots, n,$$

converges, as $m \to \infty$, to the joint distribution of independent Gaussian variables $\gamma_1, \ldots, \gamma_n$. Hence,

$$\lim_{m\to\infty} \mathbf{E}\phi(\xi_{1,m}, \ldots, \xi_{n,m}) \tag{7.1.3}$$

$$= \frac{1}{(2\pi)^{n/2}} \int_{-\infty}^{\infty} \cdots \int_{-\infty}^{\infty} \phi(s_1, \ldots, s_n) e^{-(s_1^2+\cdots+s_n^2)/2} ds_1 \ldots ds_n,$$

for any bounded and continuous function $\phi : \mathbf{R}^n \mapsto \mathbf{R}$. Consider the Banach space $\boldsymbol{Y}$ of all continuous functions $\phi : \mathbf{R}^n \mapsto \mathbf{R}$ such that

$$\lim_{\sum_{i=1}^n |s_i| \to\infty} \frac{\phi(s_1, \ldots, s_n)}{\sum_{i=1}^{n} |s_i|^2} = 0,$$

with the norm

$$\|\phi\|_Y = \max\Big(\Big(\sup_{\sum_{i=1}^n |s_i|\le 1} |\phi(s_1,\dots,s_n)|,\ \sup_{\sum_{i=1}^n |s_i|\ge 1} \frac{\phi(s_1,\dots,s_n)}{\sum_{i=1}^n |s_i|^2}\Big).$$

Let us set $\Phi_m(\phi) = \mathbf{E}\phi(\xi_{1,m},\dots\xi_{n,m})$. One can easily check that

$$|\Phi_m(\phi)| \le \|\phi\|_Y\Big(1+\sum_{i=1}^n \mathbf{E}\xi_{i,m}\Big) \le (n+1)\|\phi\|_Y, \qquad \phi\in \boldsymbol{Y}, m\in\mathbf{N}.$$

Hence, (7.1.3) also holds true for any $\phi\in\boldsymbol{Y}$ because, by (7.1.3), the limit exists on a dense subset of $\boldsymbol{Y}$. In particular, we have

$$\lim_{m\to\infty}\mathbf{E}\Big\|\sum_{i=1}^n \xi_{i,m}\boldsymbol{x}_i\Big\| = \mathbf{E}\Big\|\sum_{i=1}^n \gamma_i\boldsymbol{x}_i\Big\|,$$

because the function ϕ defined by the formula $\phi(s_1,\dots,s_n) = \|\sum_{i=1}^n s_i\boldsymbol{x}_i\|$ belongs to the space $\boldsymbol{Y}$. On the other hand, by assumption,

$$\mathbf{E}\Big\|\sum_{i=1}^n \xi_{i,m}\boldsymbol{x}_i\Big\| = \mathbf{E}\Big\|\sum_{i=1}^n\sum_{j=1}^m \xi_{m(i-1)+j}\frac{\boldsymbol{x}_i}{m^{1/2}}\Big\|$$

$$\le C\Big(\sum_{i=1}^n\sum_{j=1}^m \frac{\|\boldsymbol{x}_i\|^2}{m}\Big)^{1/2} = C\Big(\sum_{i=1}^n \|\boldsymbol{x}_i\|^2\Big)^{1/2}, \qquad m\in\mathbf{N}.$$

Therefore,

$$\mathbf{E}\Big\|\sum_{i=1}^n \gamma_i\boldsymbol{x}_i\Big\| \le C\Big(\sum_{i=1}^n \|\boldsymbol{x}_i\|^2\Big)^{1/2},$$

which proves that $\boldsymbol{X}$ is of type 2. QED

7.2 Gaussian random vectors

Spaces of type 2 are characterized by the fact that for any second order random vector in them one can find a Gaussian random vector with the same covariance functional. The following result is a special case of Theorem 6.10.2.

Theorem 7.2.1.[3] *The following properties of a Banach space* $\boldsymbol{X}$ *are equivalent:*

(i) The space $\boldsymbol{X}$ *is of type 2;*

(ii) For any random vector $Y \in \boldsymbol{L}_2(\Omega, \mathcal{F}\mathbf{P}; \boldsymbol{X})$, *with* $\mathbf{E}Y = 0$, *there exists a Gaussian random vector* X *in* $\boldsymbol{X}$ *such that its characteristic functional*

$$\mathbf{E}\exp[i\boldsymbol{x}^*X] = \exp\Big[-\frac{1}{2}\mathbf{E}(\boldsymbol{x}^*Y)^2\Big], \qquad \boldsymbol{x}^* \in \boldsymbol{X}^*.$$

Given the above result it is of interest to discuss in detail the structure of the covariance operators of Gaussian measures on spaces of type 2.

Definition 7.2.1. A linear operator $R : \boldsymbol{X}^* \mapsto \boldsymbol{X}$ is called *symmetric* if, $\boldsymbol{x}^*R\boldsymbol{y}^* = \boldsymbol{y}^*R\boldsymbol{x}^*$ for any $\boldsymbol{x}^*, \boldsymbol{y}^* \in \boldsymbol{X}^*$. The operator R is said to be *positive* if $\boldsymbol{x}^*R\boldsymbol{x}^* \geq 0$ for any $\boldsymbol{x}^* \in \boldsymbol{X}^*$. For every symmetric positive operator $R : \boldsymbol{X}^* \mapsto \boldsymbol{X}$ there exists a Hilbert space $\boldsymbol{H}$, and a continuous linear operator $A : \boldsymbol{X}^* \mapsto \boldsymbol{H}$ such that $R = A^*A$. The operator A is called the *square root* of R and will be denoted $R^{1/2}$. Any zero-mean Gaussian measure μ on $\boldsymbol{X}$ has the characteristic functional of the form,

$$\int_X \exp[i\boldsymbol{x}^*\boldsymbol{x}]\mu(d\boldsymbol{x}) = \exp\Big[-\frac{1}{2}\boldsymbol{x}^*R\boldsymbol{x}^*\Big], \qquad \boldsymbol{x}^* \in \boldsymbol{X}^*,$$

where R is symmetric and positive. R will be called the *covariance operator* of μ.

Definition 7.2.2. If (γ_i) is a sequence of independent and identically distributed Gaussian, zero-mean random variables then the closed linear span in $\boldsymbol{L}_2(\boldsymbol{X})$ of random vectors of the form $\sum_i \gamma_i\boldsymbol{x}_i$, $(\boldsymbol{x}_i) \subset \boldsymbol{X}$, will be denoted $[\gamma_i] \otimes \boldsymbol{X}$.

Proposition 7.2.1. *If* $\boldsymbol{X}$ *is of type 2 then* $[\gamma_i] \otimes \boldsymbol{X}$ *is a complemented subspace of* $\boldsymbol{L}_2(\boldsymbol{X})$ *and*

$$\boldsymbol{L}_2(\boldsymbol{X}) = ([\gamma_i] \otimes \boldsymbol{X}) \otimes ([\gamma_i] \otimes \boldsymbol{X}^*).$$

[3] Due to J. Hoffman-Jorgensen (1975). In his paper measures with Gaussian covariances are called pregaussian.

In particular, the dual space of $[\gamma_i] \otimes \boldsymbol{X}$ *may be identified with* $[\gamma_i] \otimes \boldsymbol{X}^*$.

Proof. Let $f \in \boldsymbol{L}_2(\boldsymbol{X})$. Define the operator

$$A_f : \boldsymbol{X}^* \ni \boldsymbol{x}^* \mapsto A_f \boldsymbol{x}^* = \boldsymbol{x}^* f \in \boldsymbol{L}_2(\mathbf{R}).$$

By Theorem 7.2.1, the operator $R_f = A_f^* A_f : \boldsymbol{X}^* \mapsto \boldsymbol{X}$ is the covariance operator of a Gaussian measure μ_f on $\boldsymbol{X}$. Let $\pi : \boldsymbol{L}_2(\mathbf{R}) \mapsto [\gamma_i] \otimes \mathbf{R}$ be the orthogonal projection, and consider operators $A_1 = \pi \circ A_f$, $A_2 = A_f - A_1$. Put $R_1 = A_1^* A : \boldsymbol{X}^* \mapsto \boldsymbol{X}$. Then $\boldsymbol{x}^* R_1 \boldsymbol{x}^* \leq \boldsymbol{x}^* R_f \boldsymbol{x}^*$. Thus R_1 is also the covariance operator of a Gaussian measure. Hence, the series $\sum_i A_1^* \gamma_i \gamma_i = \xi_1$ convergences almost surely, and $f = \xi_1 + \xi_2$, where $\xi_1 \in [\gamma_i] \otimes \boldsymbol{X}$, and $\xi_2 \in ([\gamma_i] \otimes \boldsymbol{X}^*)^\perp$. QED

The next theorem displays the crucial role played by absolutely summing operators. Recall that $\Pi_2(\boldsymbol{X}, \boldsymbol{Y})$ denotes the space of 2-absolutely summing operators from $\boldsymbol{X}$ to $\boldsymbol{Y}$ (see Chapter 1).

Theorem 7.2.2.[4] *The following properties of a Banach space* $\boldsymbol{X}$ *are equivalent:*

(i) The space $\boldsymbol{X}$ *is of type 2;*

(ii) A symmetric positive operator $R : \boldsymbol{X}^* \mapsto \boldsymbol{X}$ *is a covariance operator of a Gaussian measure in* $\boldsymbol{X}$ *if, and only if,* $R^{1/2} \in \Pi_2(\boldsymbol{X}^*, \boldsymbol{H})$, *where* $\boldsymbol{H}$ *is a Hilbert space.*

Proof. (*i*) $\Longrightarrow$ (*ii*) Let $R = A^* A$ with $A \in B(\boldsymbol{X}^*, \boldsymbol{H})$, and $(\boldsymbol{e}_k), k \in \mathbf{N}$, be an orthonormal basis in $\boldsymbol{H}$. In order to prove (*i*) $\Longrightarrow$ (*ii*) it suffices to show the convergence of the series $\sum_k A^* \boldsymbol{e}_k \gamma_k$. Its summands may be considered as elements of $[\gamma_i] \otimes \boldsymbol{X}$, and for our purposes it is sufficient to show that for any continuous linear functional $F : [\gamma_i] \otimes \boldsymbol{X} \mapsto \mathbf{R}$ the series $\sum_k |F(A^* \boldsymbol{e}_k \gamma_k)|$ converges (since $[\gamma_i] \otimes \boldsymbol{X}$ does not contain a subspace isomorphic to $\boldsymbol{c}_0$). According to the above Proposition this series can be written in the form

$$\sum_k |\mathbf{E}\eta^*(\mathcal{A}^* \boldsymbol{e}_k \gamma_k)|, \qquad \eta^* \in [\gamma_i] \otimes \boldsymbol{X}^*.$$

[4]Due to S.A. Chobanyan and V.I. Tarieladze (1977).

Let $R_1 : \boldsymbol{X}^{**} \mapsto \boldsymbol{X}^*$ be the covariance operator of η^*, and let $R_1 = A_1^* A_1$ be a factorization of R_1 through $\boldsymbol{H}$. Therefore $\eta^* \in [\gamma_i] \otimes \boldsymbol{X}^*$ is a Gaussian random vector in the space of cotype 2, and by Theorem 5.5.6 the operator A_1^* is 2-absolutely summing. Hence,

$$\sum_k |F(A^* \boldsymbol{e}_k \gamma_k)| = \sum_k |\mathbf{E}\eta^*(A^* \boldsymbol{e}_k \gamma_k)|$$

$$= \sum_k |(A^* \boldsymbol{e}_k)(A_1^* \boldsymbol{e}_k)| = \sum_k |(AA_1^* \boldsymbol{e}_k)\boldsymbol{e}_k| < \infty,$$

since the operator $AA_1^* : \boldsymbol{H} \mapsto \boldsymbol{H}$ is nuclear as a superposition of two 2-summing operators.

(ii) $\implies$ *(i)* Let $(\boldsymbol{x}_k) \subset \boldsymbol{X}$, with $\sum_k \|\boldsymbol{x}_k\|^2 < \infty$. We have to prove that the series $\sum_k \gamma_k \boldsymbol{x}_k$ converges almost surely. The convergence of this series will be established as soon as we show that the operator $R : \boldsymbol{X}^* \mapsto \boldsymbol{X}$ defined by the equality $R\boldsymbol{X}^* = \sum_k (\boldsymbol{x}^* \boldsymbol{x}_k)\boldsymbol{x}_k$ is the covariance operator of a Gaussian measure, i.e., the operator $A : \boldsymbol{X}^* \mapsto \boldsymbol{H}$, $A\boldsymbol{x}^* = \sum_k (\boldsymbol{x}^* \boldsymbol{e}_k)\boldsymbol{e}_k$ is 2- summing. Consider an arbitrary sequence $(\boldsymbol{x}_n^*) \subset \boldsymbol{X}^*$ such that $\sum_n (\boldsymbol{x}_n^* \boldsymbol{x})^2 < \infty$ for all $\boldsymbol{x} \in \boldsymbol{X}$. Then, by the Uniform Boundedness Principle, $\sum_n (\boldsymbol{x}_n^* \boldsymbol{x})^2 \le C\|\boldsymbol{x}\|^2$, for some constant $C > 0$, and all $\boldsymbol{x} \in \boldsymbol{X}$. Thus we have

$$\sum_n \|A\boldsymbol{x}_n^*\|^2 = \sum_n \Big\| \sum_k (\boldsymbol{x}_n^* \boldsymbol{x}_k)\boldsymbol{e}_k \Big\|^2 = \sum_n \sum_k (\boldsymbol{x}_n^* \boldsymbol{x}_k)^2 \le C \sum_k \|\boldsymbol{x}_k\|^2 < \infty,$$

that is, A is 2-absolutely summing. QED

The next result is an immediate corollary to Theorems 7.2.1 and 7.2.2.

Corollary 7.2.1. *The following properties of a Banach space* $\boldsymbol{X}$ *are equivalent:*

(i) The space $\boldsymbol{X}$ *is of type 2;*

(ii) A symmetric positive operator $R : \boldsymbol{X}^* \mapsto \boldsymbol{X}$ *is a Gaussian covariance if, and only if, for each* $B \in B(\boldsymbol{X}, \boldsymbol{H})$*, the operator* $BRB^* : \boldsymbol{H} \mapsto \boldsymbol{H}$ *is nuclear (trace class);*

(iii) Let $T \in B(\boldsymbol{H}, \boldsymbol{X})$ *and* μ_H *be a standard cylindrical Gaussian measure on* $\boldsymbol{H}$ *with the characteristic functional* $\exp[-\|\boldsymbol{x}\|^2/2]$*. A cylindrical Gaussian measure* $\mu_H \circ T^{-1}$ *on* $\boldsymbol{X}$ *is* σ*-additive if and only if* $T^* \in \Pi_2(\boldsymbol{X}^*, \boldsymbol{H})$*;*

(iv) A functional $\Phi : \boldsymbol{X}^* \mapsto \boldsymbol{C}$ *is the characteristic functional of a zero-mean Gaussian measure on* $\boldsymbol{X}$ *if, and only if,*

$$\Phi(\boldsymbol{x}^*) = \exp[-\|T\boldsymbol{x}^*\|_H^2],$$

where $T : \boldsymbol{X}^* \mapsto \boldsymbol{H}$ *is 2-absolutely summing.*

(v) There exist constants $C_1, C_2, C_3 > 0$ *such that, for each zero-mean measure* ν_R *on* $\boldsymbol{X}$ *with finite second moments,*

$$\Big(\int_X \|\boldsymbol{x}\|^2 \nu_R(d\boldsymbol{x})\Big)^{1/2} \geq C_1\pi_2(R^{1/2}) \geq C_2\Big(\int_X \|\boldsymbol{x}\|^2 \mu_R(d\boldsymbol{x})\Big)^{1/2}$$

$$\geq C_3\pi_2(R^{1/2}),$$

where μ_R *is a zero-mean Gaussian measure on* $\boldsymbol{X}$ *with the covariance operator* R.

(vi) (Under the extra assumption that $\boldsymbol{X}$ *has an unconditional basis* $(\boldsymbol{e}_k)$*). There exists a constant* $C > 0$ *such that for each masure* ν_R *on* $\boldsymbol{X}$ *with zero mean, and covariance operator* R*,*

$$\int_X \|\boldsymbol{x}\|^2 \nu_R(d\boldsymbol{x}) \geq C\Big\|\sum_k (\boldsymbol{e}_k^* R\boldsymbol{e}_k^*)^{1/2}\boldsymbol{e}_k\Big\|^2.$$

7.3 Kolmogorov's inequality and three-series theorem

Two classical results of probability theory, the Kolmogorov's inequality for the maximum of sums of independent random variables, and the three-series theorem, giving a sufficient condition for the almost sure convergence of series of independent random variables also have extensions to Banach spaces of type 2.

Theorem 7.3.1.[5] *The Banach space* $\boldsymbol{X}$ *is of type 2 if, and only if, there exists a constant* $C > 0$ *such that, for any* $n \in \mathbf{N}$*,*

[5] Due to N. Jain (1976).

and arbitrary independent random vectors $X_1, \ldots, X_n \in \boldsymbol{L}_2(\boldsymbol{X})$ *with* $\mathbf{E}X_i = 0, n = 1, 2, \ldots, n$, *we have the inequality*

$$\mathbf{P}\Big(\max_{1\le j\le n}\|X_1 + \cdots + X_j\| > \lambda\Big) \le C\lambda^{-2}\sum_{j=1}^{n}\mathbf{E}\|X_j\|^2, \qquad \forall\lambda > 0.$$

Proof. If we assume the validity of the Kolmogorov's inequality, a routine argument shows that the series $\sum_i X_i$ of zero-mean independent random vectors in $\boldsymbol{X}$ converges almost surely whenever $\sum_i \mathbf{E}\|X_i\|^2 < \infty$. This and Theorem 6.7.4 ensure that $\boldsymbol{X}$ is of type 2.

Conversely, let $\boldsymbol{X}$ be of type 2, and, as usual, define the set $\Lambda = \{\max_{1\le j\le n}\|S_j\| > \lambda\}$, where $S_j = X_1 + \cdots + X_j$, and, for $1 \le j \le n$, define the sets

$$B_j = \Big\{\|S_1\| \le \lambda, \ldots, \|S_{j-1}\| \le \lambda, \|S_j\| > \lambda\Big\}.$$

By convention, we set $S_0 = 0$. Then,

$$\int_{B_j}\|S_n\|^2 d\mathbf{P} = \mathbf{E}\Big\|S_jI_{B_j} + (S_n - S_j)I_{B_j}\Big\|^2.$$

Using Jensen's Inequality, and the independence assumption, we get that

$$\mathbf{E}''\Big\|S_jI_{B_j} + (S_n - S_j)I_{B_j}\Big\|^2 \ge \Big\|S_jI_{B_j}\Big\|^2,$$

where $\mathbf{E}''$ denotes integration on variables $X_{j+1}, \ldots, X_n$. Therefore, because $\mathbf{E} = \mathbf{E}'\mathbf{E}''$ (with $\mathbf{E}'$ standing for integration on $X_1, \ldots, X_j$) we have

$$\int_{B_j}\|S_n\|^2 d\mathbf{P} \ge \int_{B_j}\|S_j\|^2 d\mathbf{P} \ge \lambda^2\mathbf{P}(B_j),$$

and a simple addition gives the inequality

$$\int \|S_n\|^2 d\mathbf{P} \ge \lambda^2\mathbf{P}(\Lambda),$$

since B_j's are disjoint and $\bigcup_j B_j = \Lambda$. Now, the Kolmogorov's Inequality follows in view of the inequality defining type 2. QED

Employing the usual real-variable procedure we can deduce from the Kolmogorov'e inequality "one half" of the Three Series Theorem for random vectors taking values in Banach spaces of type 2.

Corollary 7.2.2. *If $\boldsymbol{X}$ is a Banach space of type 2, and (X_j) is a sequence of independent random vectors in $\boldsymbol{X}$, then the convergence of the following three series, for some $c > 0$,*

(i) $\sum_j \mathbf{P}\big[\|\boldsymbol{x}_j\| > c\big]$,

(ii) $\sum_j \mathbf{E}\big(X_j I[\|X_j\| \le c]\big)$, *and*

(iii) $\sum_j \mathbf{E}\big\|X_j I[\|X_j\| \le c] - \mathbf{E}X_j I[\|X_j\| \le c]\big\|^2$,

implies the almost sure convergence of the series $\sum_j X_j$.

7.4 Central limit theorem

The following simplest form of the Central Limit Theorem also characterizes Banach spaces of type 2.

Theorem 7.4.1.[6] *The following properties of a Banach space $\boldsymbol{X}$ are equivalent:*

(i) The space $\boldsymbol{X}$ is of type 2;

(ii) For any independent, identically distributed, zero-mean random vectors $(X_n) \subset \boldsymbol{L}_2(\boldsymbol{X})$ there exists a Gaussian measure γ on $\boldsymbol{X}$ such that

$$\mathcal{L}\Big(\frac{X_1 + \cdots + X_n}{n^{1/2}}\Big) \longrightarrow \gamma,$$

weakly, as $n \to \infty$. Moreover, the Fourier transform (characteristic functional) of γ is $\exp[-\mathbf{E}(\boldsymbol{x}^* X_1)^2/2]$.

Proof. (*i*) $\Longrightarrow$ (*ii*) By Theorems 6.10.1, 6.10.2, and 7.2.1, there exists a Gaussian random vector X in $\boldsymbol{X}$ of the form $\int_X \boldsymbol{x} M(d\boldsymbol{x})$, where M is a Gaussian random measure on $(\boldsymbol{X}, \mathcal{B}_X)$ with control measure $\mu = \mathcal{L}(X_1)$. Moreover,

$$\mathbf{E}\exp[i\boldsymbol{x}^* X] = \exp\Big[-\frac{1}{2}\mathbf{E}(\boldsymbol{x}^* X_1)^2\Big], \qquad \boldsymbol{x}^* \in \boldsymbol{X}^*.$$

[6] Due to J. Hoffmann-Jorgensen and G. Pisier (1976).

We shall show that we can take $\gamma = \mathcal{L}(X_1)$.

Since the topology induced by Lipschitzian functions on $\boldsymbol{X}$ coincides with the norm topology, it is sufficient to show that for any Lipschtzian (say, with constant K) bounded function $\phi : \boldsymbol{X} \mapsto \mathbf{R}$,

$$\lim_{n\to\infty} \mathbf{E}\phi(n^{-1/2}S_n) = \mathbf{E}\phi\Big(\int_X \boldsymbol{x}M(d\boldsymbol{x})\Big).$$

Choose a sequence $f^{(d)}, d \in \mathbf{N}$, of simple (finite range) functions from $\boldsymbol{L}_2(\boldsymbol{X}, \mathcal{B}_X, \mu; \boldsymbol{X})$ with $\int f^{(d)} d\mu = 0$, and such that

$$\int_X \|\boldsymbol{x} - f^{(d)}(\boldsymbol{x})\|^p \mu(d\boldsymbol{x}) \to 0, \qquad p = 1, 2, \quad d \to \infty. \tag{7.4.1}$$

Then, for each $d \in \mathbf{N}$, $(X_n - f^{(d)}(X_n))$ are again independent random vectors in $\boldsymbol{X}$, and we have that

$$\mathbf{E}\Big|\phi\big((X_1 + \cdots + X_n)n^{-1/2}\big) - \phi\big((f^{(d)}(X_1) + \cdots + f^{(d)}(X_n))n^{-1/2}\big)\Big|$$

$$\le K\mathbf{E}\Big\|n^{-1/2}\Big(\sum_{i=1}^{n}(X_i - f^{(d)}(X_i))\Big)\Big\|$$

$$\le K\Big(n^{-1}\mathbf{E}\Big\|\sum_{i=1}^{n}(X_i - f^{(d)}(X_i))\Big\|^2\Big)^{1/2}$$

$$\le KC^{1/2}\Big(n^{-1}\sum_{i=1}^{n}\mathbf{E}\|X_i - f^{(d)}(X_i)\|^2\Big)^{1/2}$$

$$\le KC^{1/2}\Big(\int_X \|\boldsymbol{x} - f^{(d)}(\boldsymbol{x})\|^2\mu(d\boldsymbol{x})\Big)^{1/2}.$$

The next to the last inequality used the fact that $\boldsymbol{X}$ is of type 2. Therefore, by (7.4.1),

$$\mathbf{E}\Big|\phi\big((X_1+\cdots+X_n)n^{-1/2}\big)-\phi\big((f^{(d)}(X_1)+\cdots+f^{(d)}(X_n))n^{-1/2}\big)\Big|\to 0, \tag{7.4.2}$$

as $d \to \infty$, for each $n \in \mathbf{N}$, and uniformly in n.

Now, since for each particular d, $(f^{(d)}(X_n))$ is a sequence of finite dimensional i.i.d. random vectors, by the finite-dimensional Central Limit Theorem, weakly,

$$\mathcal{L}\Big(n^{-1/2}\sum_{i=1}^{n} f^{(d)}(X_i)\Big) \longrightarrow \mathcal{L}\Big(\int_X f^{(d)}(\boldsymbol{x})M(d\boldsymbol{x})\Big), \qquad \text{as} \quad n \to \infty,$$

and, in particular, for each fixed $d \in \mathbf{N}$,

$$\mathbf{E}\Big|\phi\Big(n^{-1/2}\sum_{i=1}^{n} f^{(d)}(X_i)\Big) - \phi\Big(\int_X f^{(d)}(\boldsymbol{x})M(d\boldsymbol{x})\Big)\Big| \longrightarrow 0, \quad \text{as } n \to \infty. \tag{7.4.3}$$

In view of (7.4.2), and (7.4.3), and the definition of $f^{(d)}$, we get that

$$\mathbf{E}\Big|\phi\Big(n^{-1/2}\sum_{i=1}^{n} X_i\Big) - \phi\Big(n^{-1/2}\sum_{i-1}^{n} f^{(d)}(X_i)\Big)\Big|$$

$$+\mathbf{E}\Big|\phi\Big(n^{-1/2}\sum_{i-1}^{n} f^{(d)}(X_i)\Big) - \phi\Big(\int_X f^{(d)}(\boldsymbol{x})M(d\boldsymbol{x})\Big)\Big|$$

$$+\mathbf{E}\Big|\phi\Big(\int_X f^{(d)}(\boldsymbol{x})M(d\boldsymbol{x})\Big) - \phi\Big(\int_X \boldsymbol{x}M(d\boldsymbol{x})\Big)\Big|$$

$$\geq \mathbf{E}\Big|\phi\Big(n^{-1/2}\sum_{i=1}^{n} X_i\Big) - \phi\Big(\int_X f^{(d)}(\boldsymbol{x})M(d\boldsymbol{x})\Big)\Big| \longrightarrow 0, \quad \text{as } n \to \infty,$$

which gives the desired weak convergence. Now, the shape of the Fourier transform of γ is an immediate consequence of the formula,

$$\mathbf{E}\exp\Big[i\boldsymbol{x}^*\int_X f(\boldsymbol{x})M(d\boldsymbol{x})\Big] = \exp\Big[-\frac{1}{2}\int_X (\boldsymbol{x}^* f(\boldsymbol{x}))^2 \mu(d\boldsymbol{x})\Big), \quad \boldsymbol{x}^* \in \boldsymbol{X}^*,$$

valid for any $f \in \boldsymbol{L}_2(\boldsymbol{X}, \mathcal{B}_X, \mu; \boldsymbol{X})$ (see Theorem 6.10.2).

(ii) $\Longrightarrow$ *(i)* If for any i.i.d. zero-mean sequence $(X_n) \subset \boldsymbol{L}_2(\boldsymbol{X})$

$$\mathcal{L}((X_1 + \cdots + X_n)n^{-1/2}) \to \mathcal{L}(X), \qquad \text{as} \quad n \to \infty,$$

for some Gaussian random vector X in $\boldsymbol{X}$, then, in particular, $\mathbf{E}(\boldsymbol{x}^*X_1)^2 = \mathbf{E}(\boldsymbol{x}^*X)^2$. Therefore, for any zero-mean $X_1 \in \boldsymbol{L}_2(\boldsymbol{X})$, there exists a Gaussian X such that

$$\mathbf{E}\exp[i\boldsymbol{x}^*X] = \exp\Big[-\frac{1}{2}\mathbf{E}(\boldsymbol{x}^*X_1)^2\Big].$$

Thus, by Theorem 7.2.1, $\boldsymbol{X}$ is of type 2. QED

Having completed a discussion of the extension of the analogue of the classical Central Limit Theorem in Banach spaces of type 2, we now turn to a presentation of a different approach to the Central Limit Theorem. The next theorem works in general Banach spaces but, perhaps, gives a deeper insight into the problem.

For a sequence (X_i) of independent copies of a random vector X in $\boldsymbol{X}$, let us introduce the notation

$$CL(X) = \sup_{n\in N} \mathbf{E}n^{-1/2}\|X_1 + \cdots + X_n\|.$$

Then, we define the space $CL_\infty := \{X \in \boldsymbol{L}_0(\boldsymbol{X}) : CL(X) < \infty\}$ which is a Banach space under the norm CL. By $CL(\boldsymbol{X})$ we shall denote the closed subspace of $CL_\infty(\boldsymbol{X})$ spanned by the zero-mean simple (finite range) random vectors.

Theorem 7.4.2.[7] *Let $\boldsymbol{X}$ be a Banach space, and let (X_n) be a sequence of independent, identically distributed random vectors in $\boldsymbol{X}$. Then the laws $\mathcal{L}((X_1 + \cdots + X_n)n^{-1/2})$ converge weakly if, and only if, $X_1 \in CL(\boldsymbol{X})$.*

Proof. To prove the "if" part of the theorem let us begin by recalling the concept of the Lévy distance on the set of positive measures on $\boldsymbol{X}$:

$$d(\lambda,\mu) := \inf\big\{a > 0 : \lambda(F) \le \mu(F^a) + a, \mu(F) \le \lambda(F^a) + a,$$
$$\forall \text{ bdd } F \subset \boldsymbol{X}\big\},$$

where $F^a = \{\boldsymbol{x} \in \boldsymbol{X} : \text{dist}(\boldsymbol{x}, F) < a\}$. The metric space created with the help of this distance has the topology equivalent to the

[7]Due to G. Pisier (1975/76).

topology of weak convergence. It is easy to see that the mapping $X \mapsto \mathcal{L}(X)$ from $\boldsymbol{L}_0(\boldsymbol{X})$ into the space of measures equipped with the Lévy distance is uniformly continuous.

Now, if $X_1 \in CL(\boldsymbol{X})$, then for each $\epsilon > 0$ there exists a simple random vector Y such that $CL(X_1 - Y) < \epsilon$. The finite dimensional Central Limit Theorem implies that if (Y_i) are independent copies of Y then the laws $\mathcal{L}(n^{-1/2}(Y_1 + \cdots + Y_n))$ converge weakly. Hence, the sequence $\mathcal{L}(n^{-1/2}(X_1 + \cdots + X_n))$ is arbitrarily uniformly close in the Lévy metric to a convergent sequence of measures. Thus it converges itself in view of the completeness of the space of measures equipped with the Lévy metric. This concludes the proof of the "if" part of the Theorem.

In the proof of the "only if" part of the Theorem we shall have need of the following result which is also of independent interest.

Lemma 7.4.1.[8] *Let $\boldsymbol{X}$ be a Banach space, and let (X_n) be a sequence of independent, identically distributed random vectors in $\boldsymbol{X}$. If the laws $\mathcal{L}(n^{-1/2}(X_1 + \cdots + X_n))$ converge weakly, then*

(i) $\sup_{n \in N} \sup_{c>0} c^2 \mathbf{P}(n^{-1/2}\|X_1 + \cdots + X_n\| > c) < \infty$, *and*

(ii) $\mathbf{E}\|X_1\|^p < \infty$, *and* $\sup_{n \in N} \mathbf{E}\|(X_1 + \cdots + X_n)n^{-1/2}\|^p < \infty$, *for each* $p < 2$.

Proof. If X_1 satisfies the Central Limit Theorem then so does its symmetrization $\tilde{X}_1$. In particular,

$$\mathcal{L}\big(n^{-1/2}(\tilde{X}_1 + \cdots + \tilde{X}_n)\big) \to \mathcal{L}(Y), \qquad \text{as } n \to \infty,$$

for some Gaussian Y. Also, for each $a > 0$,

$$\overline{\lim}_{n\to\infty} \mathbf{P}\Big(n^{-1/2}\|\tilde{X}_1 + \cdots + \tilde{X}_n\| \geq a\Big) \leq \mathbf{P}(\|Y\| \geq a).$$

Therefore, for each $\epsilon < 1/2$, there exists an $a > 0$ such that

$$P\Big(n^{-1/2}\|\tilde{X}_1 + \cdots + \tilde{X}_n\| \geq a\Big) \leq \epsilon, \qquad n \in \mathbf{N}, \tag{7.4.4}$$

and by the Lévy Inequality

$$\mathbf{P}\Big(\sup_{1\leq i\leq n} \|\tilde{X}_i\| > 2an^{1/2}\Big) \leq 2\epsilon.$$

[8] Due to N. Jain (1976).

Assume that

$$\mathbf{P}\Big(\|\tilde{X}_1\| > 2an^{1/2}\Big) \le 1 - (1-2\epsilon)^{1/n},$$

which means that, for a certain constant $A(\epsilon)$

$$\mathbf{P}\Big(\|\tilde{X}_1\| > 2an^{1/2}\Big) \le A(\epsilon)n^{-1}.$$

This leads to the inequality

$$\sup_{c>0} c^2\mathbf{P}(\|\tilde{X}_1\| > c) \le B(\epsilon),$$

where $B(\epsilon)$ is another constant. To finish the proof of the Lemma it suffices to notice that if $(\tilde{X}_1 + \ldots \tilde{X}_N)N^{-1/2}$ is substituted for X_1 in (7.4.4) then we get that

$$\sup_{c>0} c^2\mathbf{P}\Big(N^{-1/2}\|\tilde{X}_1 + \cdots + \tilde{X}_N\| > c\Big) \le B(\epsilon), \qquad N \in \mathbf{N}.$$

Then the triangle inequality yields that

$$\mathbf{P}\Big(n^{-1/2}\|X_1 + \cdots + X_n\| > c + \alpha\Big)$$

$$\le \mathbf{P}\Big(n^{-1/2}\|X_1 + \cdots + X_n\| > c\Big) \Big/ \mathbf{P}\Big(n^{-1/2}\|X_1 + \cdots + X_n\| \le \alpha\Big).$$

Since X_1 satisfies the Central Limit Theorem there exists an $\alpha > 0$ such that

$$\inf_n \mathbf{P}\Big(n^{-1/2}\|X_1 + \cdots + X_n\| \le \alpha\Big) > 0,$$

from which statement (i) follows.

To obtain (ii) it is enough to use (i) and the obvious formula,

$$\mathbf{E}Y^p = \int_0^\infty pc^{p-1}\mathbf{P}(Y > c)dc, \qquad Y \ge 0.$$

This concludes the proof of the Lemma. QED

Proof of Theorem 7.4.2, continued. Now we are ready to prove the "only if" part of the Theorem. Let $X \in \boldsymbol{L}_0(\boldsymbol{X})$ and assume that X satisfies the Central Limit Theorem. By the above Lemma, $\mathbf{E}\|X\| < \infty$, and $\mathbf{E}X = 0$. One can thus consider X as a $\boldsymbol{L}_1$-limit of a martingale (X^N), $N \in \mathbf{N}$, such that, for each N, X^N takes on only finitely many values. We shall show that $\lim_{N\to\infty} CL(X - X^N) = 0$, which would complete the proof.

Notice that, by compactness argument, for each $N \in \mathbf{N}$, $X - X^N$ also satisfies the Central Limit Theorem. Now, for the sake of simplicity, assume that

$$X^N = \sum_{i=1}^{N} \phi_i \boldsymbol{x}_i,$$

where $(\boldsymbol{x}_i) \subset \boldsymbol{X}$, and (ϕ_i) is an orthonormal sequence of martingale differences. Put $Y^N = \sum_{i=1}^N \gamma_i \boldsymbol{x}_i$. Then Y^N converges almost surely to a Gaussian random vector Y. Indeed, for each $\boldsymbol{x}^* \in \boldsymbol{X}^*$, the sequence $\boldsymbol{x}^* Y^N$ converges to a random variable with the law $\gamma_X \boldsymbol{x}^{*-1}$, where $\gamma_X = \lim_{n\to\infty} \mathcal{L}\big(n^{-1/2}(X_1 + \cdots + X_n)\big)$. Thus, the Ito-Nisio Theorem (see Chapter 1) guarantees the existence of Y. By Landau-Shepp-Fernique Theorem (see Chapter 1) we also have that $\mathbf{E}\|Y - Y^N\| \to 0$, as $N \to \infty$. Checking the covariances we get that, as $n \to \infty$,

$$\mathcal{L}\big((X_1^N + \cdots + X_n^N)n^{-1/2}\big) \to \mathcal{L}(Y^N),$$

$$\mathcal{L}\big((X_1 + \cdots + X_n)n^{-1/2}\big) \to \mathcal{L}(Y),$$

$$\mathcal{L}\big(((X_1 - X_1^N) + \cdots + (X_n - X_n^N))n^{-1/2}\big) \to \mathcal{L}(Y - Y^N).$$

However, if, say $(Z_n) \subset \boldsymbol{L}_0(\boldsymbol{X})$, and $f : \boldsymbol{X} \mapsto \mathbf{R}$ is continuous and such that $f(Z_n)$ are equi-integrable, then the weak convergence of Z_n to Z implies that $\lim_{n\to\infty} \mathbf{E}f(Z_n) = \mathbf{E}f(Z)$.

Hence, in our situation, by virtue of Lemma 7.4.1, we have that

$$\lim_{n\to\infty} \mathbf{E}\Big\|n^{-1/2}\sum_{i=1}^{n}(X_i - X_i^N)\Big\| = \mathbf{E}\|Y - Y^N\|. \tag{7.4.5}$$

Let now $\epsilon > 0$, and choose N_0 such that $\mathbf{E}\|Y - Y^{N_0}\| < \epsilon$. By (7.4.5), there exists an $m \geq N_0$ such that

$$\sup_{n\geq m} \mathbf{E}\Big\|n^{-1/2}\sum_{i=1}^{n}(X_i - X_i^N)\Big\| < \epsilon. \tag{7.4.6}$$

Given m, there exists an $M \geq m$ such that

$$\sup_{N \geq M} \sup_{1 \leq n \leq m} \mathbf{E}\Big\| n^{-1/2} \sum_{i=1}^{n} (X_i - X_i^N) \Big\| < \epsilon. \tag{7.4.7}$$

Furthermore, for each $N \geq N_0$,

$$\mathbf{E}\Big\| n^{-1/2} \sum_{i=1}^{n} (X_i - X_i^N) \Big\|$$

$$\leq \mathbf{E}\Big\| n^{-1/2} \sum_{i=1}^{n} (X_i - X_i^{N_0}) \Big\| + \mathbf{E}\Big\| n^{-1/2} \sum_{i=1}^{n} (X_i^N - X_i^{N_0}) \Big\|$$

$$\leq 2\mathbf{E}\Big\| n^{-1/2} \sum_{i=1}^{n} (X_i - X_i^{N_0}) \Big\|,$$

because $\sum_{i=1}^n (X_i^N - X_i^{N_0})$ can be obtained by taking a conditional expectation of $\sum_{i=1}^n (X_i - X_i^{N_0})$. Hence, for each $N \geq N_0$, and $n \geq m$, by (7.4.6)

$$\mathbf{E}\Big\| n^{-1/2} \sum_{i=1}^{n} (X_i - X_i^N) \Big\| \leq 2\epsilon,$$

so that, by (7.4.7), for $N \geq N_0$, we have $CL(X - X^N) \leq 2\epsilon$. QED

The Central Limit Theorem for triangular arrays can also be proven for random vectors taking values in spaces of type 2. Below, we discuss two sample results in this area which, however, also rely on the explicit compactness assumptions.

Theorem 7.4.3.[9] *Suppose that* $(Y_{nj}), n \in \mathbf{N}, j = 1, \ldots, j_n$, *is an array of row-wise independent random vectors in a Banach space* $\boldsymbol{Y}$ *of type 2, and satisfying the following three conditions:*

(i) $\mathbf{E}Y_{nj} = 0$;

(ii) $0 \leq \sigma_{nj}^2 := \mathbf{E}\|Y_{nj}\|^2 < v_{nj}^2$, *with* $\sum_{j=1}^{j_n} v_{nj}^2 = 1$;

[9]Due to D.J.H. Garling (1976). That paper also contains a functional Central Limit Theorem (invariance principle) written in a similar spirit.

(iii) For each $\eta > 0$

$$\sum_{j=1}^{j_n} \int_{|y^* Y_{nj}| > \eta} (\boldsymbol{y}^* Y_{nj})^2 d\mathbf{P} \longrightarrow 0, \quad \text{as} \quad n \to \infty,$$

for each $\boldsymbol{y}^* \in \boldsymbol{Y}^*$.

Suppose that T *is a compact linear operator from* $\boldsymbol{Y}$ *into a Banach space* $\boldsymbol{X}$, *and let* $X_{nj} = TY_{nj}$. *Then, weakly,*

$$\mathcal{L}\Big(\sum_{j=1}^{j_n} X_{nj}\Big) \longrightarrow \gamma, \quad \text{as} \quad n \to \infty, \tag{7.4.8}$$

where γ *is a Gaussian measure on* $\boldsymbol{X}$, *provided*

$$\lim_{n\to\infty} \mathbf{E}\Big(\boldsymbol{x}^* \sum_{j=1}^{j_n} X_{nj}\Big) \quad \textit{exists} \quad \forall \boldsymbol{x}^* \in \boldsymbol{X}^*.$$

Proof. As usual, it is sufficient to show that the sequence in (7.4.8) is uniformly tight. By (i), and the fact that $\boldsymbol{Y}$ is of type 2,

$$\mathbf{E}\Big\|\sum_{j=1}^{j_n} Y_{nj}\Big\|^2 \le C \sum_{j=1}^{j_n} \mathbf{E}\|Y_{nj}\|^2 \le C, \qquad n \in \mathbf{N},$$

so that

$$\mathbf{P}\Big(\Big\|\sum_{j=1}^{j_n} Y_{nj}\Big\|^2 > C/\epsilon\Big) \le \epsilon, \qquad n \in \mathbf{N}.$$

Therefore, since $K = \overline{TB_Y}$ is a compact set, we have

$$\mathbf{P}\Big(\sum_{j=1}^{j_n} Y_{nj} \notin (C/\epsilon)^{1/2} K\Big) \le \epsilon, \qquad n \in \mathbf{N},$$

which gives the desired uniform tightness. QED

The next result also deals with arrays of random vectors but has the advantage of involving no compact operators.

Theorem 7.4.4.[10] *Suppose that* (X_{nj}) *is a row-wise independent array of random vectors satisfying conditions (i), (ii), and (iii), of Theorem 7.4.3, and taking values in a Banach space* $\boldsymbol{X}$ *of type 2. If*

$$\lim_{n\to\infty} \mathbf{E}\Big(\boldsymbol{x}^* \sum_{j=1}^{j_n} X_{nj}\Big)^2$$

exists for each $\boldsymbol{x}^* \in \boldsymbol{X}^*$, *and for each* $\epsilon > 0$ *there exists a finite-dimensional subspace* $\boldsymbol{Y} \subset \boldsymbol{X}$ *such that*

$$\sum_{j=1}^{j_n} \mathbf{E}\big[\mathrm{dist}(X_{nj}, \boldsymbol{Y})\big]^2 < \epsilon, \qquad n \in \mathbf{N}, \tag{7.4.9}$$

then, there exists a Gaussian measure γ *on* $\boldsymbol{X}$ *such that, as* $n \to \infty$, *weakly,*

$$\mathcal{L}\Big(\sum_{j=1}^{j_n} X_{nj}\Big) \longrightarrow \gamma.$$

Proof. By Proposition 6.2.4, the quotient space $\boldsymbol{X}/\boldsymbol{Y}$ is of type 2 whenever $\boldsymbol{X}$ is of type 2, and the constant C is the same in both cases. Given $\epsilon > 0$, let $\eta = \epsilon^3/2C^2$, and let $\boldsymbol{Y}$ be a finite dimensional subspace corresponding to η in view of (7.4.9). Then,

$$\mathbf{E}\Big[\mathrm{dist}\Big(\sum_{j=1}^{j_n} X_{nj}, \boldsymbol{Y}\Big)\Big]^2 \le C^2\eta, \qquad n \in \mathbf{N},$$

so that

$$\mathbf{P}\Big[\mathrm{dist}\Big(\sum_{j=1}^{j_n} X_{nj}, \boldsymbol{Y}\Big) > \epsilon\Big] \le C^2\eta/\epsilon^2 = \epsilon/2.$$

Also,

$$\mathbf{P}\Big(\Big\|\sum_{j=1}^{j_n} X_{nj}\Big\| > 2C/\epsilon^{1/2}\Big) \le \epsilon/4,$$

[10]Due to D.J.H. Garling (1976).

and, furthermore, if $K = \{\boldsymbol{x} \in \boldsymbol{Y} : \|\boldsymbol{x}\| \le 2C/\epsilon^{1/2}\}$,

$$\mathbf{P}\Big[\mathrm{dist}\Big(\sum_{j=1}^{j_n} X_{nj}, K\Big) > \epsilon\Big] \le \epsilon/2, \qquad n \in \mathbf{N},$$

so that, by a standard argument[11] the sequence $\mathcal{L}\Big(\sum_{j=1}^{j_n} X_{nj}\Big)$ is uniformly tight. QED

7.5 Law of iterated logarithm

Definition 7.5.1. We shall say that $X \in \boldsymbol{L}_0(\boldsymbol{X})$ satisfies the Law of the Iterated Logarithm if, for a sequence of independent copies (X_n) of X, the sequence

$$\Big((2n \log\log n)^{-1/2} \sum_{i=1}^{n} X_i\Big), \qquad n = 3, 4, \dots,$$

is almost surely a conditionally compact set in $\boldsymbol{X}$.

For other, but equivalent formulations of the Law of Iterated Logarithm see Section 5.6.

In what follows we shall use the following notation: By $IL_\infty(\boldsymbol{X})$ we will denote that set of random vectors in $\boldsymbol{X}$ for which

$$IL(X) := \mathbf{E} \sup_n (2n \log\log n)^{-1/2} \Big\|\sum_{i=1}^{n} X_i\Big\| < \infty.$$

Note that $IL_\infty(\boldsymbol{X})$ is a Banach space equipped with the norm IL. Denote by $IL(\boldsymbol{X})$ the closure in $IL_\infty(\boldsymbol{X})$ of the set of zero-mean, simple random vectors in $IL_\infty(\boldsymbol{X})$. In view of the real-valued Law of the Iterated Logarithm, the latter set is contained in $IL_\infty(\boldsymbol{X})$.

Theorem 7.5.1.[12] *Let $\boldsymbol{X}$ be a Banach space, and let X be a random vector in $\boldsymbol{X}$. Then X satisfies the Law of the Iterated Logarithm if, and only if, $X \in IL(\boldsymbol{X})$.*

[11] See, e..g., K.R. Parthasarathy (1967), p. 49.
[12] Due to G. Pisier (1975/76).

Proof. *"If"* If $X \in IL(\boldsymbol{X})$ then, for every $\epsilon > 0$, there exists a finite range random vector Y such that

$$IL(X - Y) < \epsilon. \tag{7.5.1}$$

Now, consider the random variable α_X defined by the formula,

$$\alpha_X := \lim_{N\to\infty} \sup_{n\in\mathbf{N}} \inf_{1\le j\le N} \Big\| (2n\log\log n)^{-1/2} \sum_{i=1}^{n} X_i \\ - (2j\log\log j)^{-1/2} \sum_{i=1}^{j} X_i \Big\|.$$

Clearly, it suffices to show that $\alpha_X = 0$ almost surely. So, notice that

$$\alpha_X \le \alpha_Y + 2\sup_n (2n\log\log n)^{-1/2} \Big\| \sum_{i=1}^{n} (X_i - Y_i) \Big\|.$$

Since Y satisfies the Law of the Iterated Logarithm, $\alpha_Y = 0$ almost surely, so that it follows from (7.5.1) that $\mathbf{E}\alpha_X \le 2\epsilon$, for any $\epsilon > 0$. Therefore $\alpha_X = 0$ almost surely.

"Only if" Assume that X satisfies the Law of the Iterated Logarithm. First, we shall prove that $X \in IL_\infty(\boldsymbol{X})$. By assumption,

$$\sup_{n\in\mathbf{N}} (2n\log\log n)^{-1/2} \Big\| \sum_{i=1}^{n} X_i \Big\| < \infty, \qquad a.s., \tag{7.5.2}$$

so that, in view of the Hoffmann-Jorgensen Theorem (see Chapter 1), it is sufficient to prove that

$$\sup_{n\in\mathbf{N}} (2n\log\log n)^{-1/2} \|X_n\| \in \boldsymbol{L}_1.$$

Actually, we shall show that it even belongs to $\boldsymbol{L}_p$, for every $p < 2$. It follows from (7.5.2) that $\sup_{n\in\mathbf{N}} (2n\log\log n)^{-1/2}\|X_n\| < \infty$ almost surely, and by Borel-Cantelli Lemma, there exists a $d > 0$ such that

$$\Lambda = \sum_{n=1}^{\infty} \mathbf{P}\Big(\|X\| > d(2n\log\log n)^{1/2} \Big) < \infty.$$

Therefore, for each $k \in \mathbf{N}$,

$$\sum_{n=1}^{\infty} k\mathbf{P}\Big(\|X\| > d(2nk \log\log nk)^{1/2}\Big) \leq \Lambda.$$

Furthermore, for some constant $B > 0$,

$$(2nk \log\log nk)^{1/2} \leq B(2n \log\log n)^{1/2}(2k \log\log k)^{1/2},$$

so that

$$\sum_{n=1}^{\infty} k\mathbf{P}\Big(\|X\| > Bd(2n \log\log n)^{1/2}(2k \log\log k)^{1/2},\Big) \leq \Lambda.$$

Hence, for each $c > 0$

$$\begin{aligned}\Phi(c) &:= \mathbf{P}\Big(\sup_n (2n \log\log n)^{-1/2}\|X_n\| > c\Big)\\ &\leq \sum_n \mathbf{P}\Big(\|X\| > c(2n \log\log n)^{1/2}\Big),\end{aligned}$$

so that

$$\mathbf{E}\Big(\sup_n (2n \log\log n)^{-1/2}\|X_n\|\Big)^p = \iint_0^{\infty} pc^{p-1}\Phi(c)dc$$

$$\leq (Bd)^p + \int_{Bd}^{\infty} pc^{p-1}\sum_n \mathbf{P}\Big(\|X\| > c(2n \log\log n)^{1/2}\Big)dc$$

$$\leq (Bd)^p + \sum_k \int_{Bd(2k \log\log k)^{1/2}}^{Bd(2(k+1)\log\log(k+1))^{1/2}} pc^{p-1}$$

$$\times \sum_n \mathbf{P}\Big(\|X\| > Bd(2n \log\log n)^{1/2}(2k \log\log k)^{1/2}\Big)dc$$

$$\leq (Bd)^p + \sum_k \int_{Bd(2k \log\log k)^{1/2}}^{Bd(2(k+1)\log\log(k+1))^{1/2}} pc^{p-1}\Lambda k^{-1}dc$$

$$\leq (Bd)^p + (Bd)^p\Lambda \sum_{k=1}^{\infty} k^{-1}$$

$$\times\Big(\big(2(k+1)\log\log(k+1)\big)^{p/2} - (2k\log\log k)^{p/2}\Big),$$

and, if $p < 2$, the last series converges because, asymptotically, the numerator behaves like $(\log\log k)^{p/2}k^{p/2-1}$.

Now, as in the proof of Theorem 7.4.2, one can define a martingale X^N of simple random vectors such that

$$\mathbf{E}\|X - X^N\| \longrightarrow 0, \quad \text{as} \quad N \to \infty.$$

Evidently, for each N, the random vector $X - X^N$ satisfies the Law of the Iterated Logarithm. We shall show that $IL(X - X^N) \to 0$.

By Kuelbs' Theorem[13]

$$\overline{\lim}_{n\to\infty}(2n\log\log n)^{-1/2}\Big\|\sum_{i=1}^{n}(X_i - X_i^N)\Big\| \le \sup\{\|\boldsymbol{x}\| : \boldsymbol{x} \in K_{X-X^N}\}, \tag{7.5.3}$$

where K_{X-X^N} is the compact set for $X - X^N$ as defined in Definition 7.5.1. Denote by λ_N the right-hand side of (7.5.3). By the method of the gliding hump, in view of the compactness of K_X, one can deduce immediately that $\lambda_N \to 0$ as $N \to \infty$. Now, let N_0 be such that $\lambda_{N_0} < \epsilon$. By (7.5.3), there exists an $m \ge N_0$ such that

$$\mathbf{E}\sup_{n\ge m}(2n\log\log n)^{-1/2}\Big\|\sum_{i=1}^{n}(X_i - X_i^{N_0})\Big\| < \epsilon.$$

Then, for each $N \ge N_0$,

$$\mathbf{E}\sup_{n\ge m}(2n\log\log n)^{-1/2}\Big\|\sum_{i=1}^{n}(X_i - X_i^{N})\Big\|$$

$$\le 2\mathbf{E}\sup_{n\ge m}(2n\log\log n)^{-1/2}\Big\|\sum_{i=1}^{n}(X_i - X_i^{N_0})\Big\| < 2\epsilon.$$

Now, for a fixed m, there exists an $M \ge m$ such that

$$\sup_{n\ge M}\mathbf{E}\sup_{1\le n\le m}(2n\log\log n)^{-1/2}\Big\|\sum_{i=1}^{n}(X_i - X_i^{N})\Big\| < \epsilon.$$

[13]See, J. Kuelbs (1977).

Finally,

$$IL(X - X^N) \le \mathbf{E} \sup_{n\le m} (2n \log\log n)^{-1/2} \Big\| \sum_{i=1}^n (X_i - X_i^N) \Big\|$$

$$+\mathbf{E} \sup_{n\ge m} (2n \log\log n)^{-1/2} \Big\| \sum_{i=1}^n (X_i - X_i^N) \Big\|,$$

so that if $N \ge M$, we have $IL(X - X^N) < \epsilon + 2\epsilon$. QED

As a corollary to the above theorem we have the following analogue of the Law of the Iterated Logarithm in spaces of type 2.

Theorem 7.5.2.[14] *Let (X_i) be a sequence of independent and identically distributed random vectors in a Banach space $\mathbf{X}$ of type 2. If $\mathbf{E}X_1 = 0$, and $\mathbf{E}\|X_1\|^2 < \infty$, then there exists a constant $C > 0$ such that*

$$IL(X_1) \le C\big(\mathbf{E}\|X_1\|^2\big)^{1/2},$$

and, in particular, (X_i) satisfies the Law of the Iterated Logarithm.

Proof. Take the Rademacher sequence (r_n) independent of (X_n) and assume (without loss of generality) that the random vectors (X_n) are symmetric. Then, $(r_n X_n)$ and (X_n) are identically distributed and

$$IL(X_1) \le \Big[\mathbf{E} \sup_n (2n \log\log n)^{-1} \Big\| \sum_{i=1}^n r_i X_i \Big\|^2 \Big]^{1/2}.$$

By Kwapień's Theorem (see Chapter 1) the random variable $\exp \| \sum_i r_i \boldsymbol{x}_i \|^2$ is integrable if $\sum_i r_i \boldsymbol{x}_i$ converges almost surely, where $(\boldsymbol{x}_n) \subset \boldsymbol{X}$. Then, by a standard real-line argument based on Lévy's Inequality, we get that, for a constant D,

$$IL(X_1) \le D \sup_n n^{-1/2} \Big(\mathbf{E}_r \Big\| \sum_{i=1}^n r_i X_i \Big\|^2 \Big)^{1/2}.$$

[14]Due to G. Pisier (1975/76).

Since $\boldsymbol{X}$ is of type 2,

$$IL(X_1) \leq CD\,\mathbf{E}\sup_n\Big(n^{-1}\sum_{i=1}^n \|X_i\|^2\Big)^{1/2},$$

so that, by the maximal inequality associated with the Strong Law of Large Numbers for real-valued random variables $(\|X_i\|^2)$, for a constant M, $IL(X_1) \leq CDM(\mathbf{E}\|X_1\|^2)^{1/2}$. QED

Remark 7.5.1. G. Pisier (1975/76) has also demonstrated that if $\boldsymbol{X}$ is a Banach space then each zero-mean random vector $X \in \boldsymbol{L}_2(\boldsymbol{X})$ satisfies the Law of the Iterated Logarithm if, and only if, there exists a constant C such that for all $n \in \mathbf{N}$, and every sequence $(\boldsymbol{x}_i) \subset \boldsymbol{X}$,

$$\mathbf{E}\Big\|\sum_{i=1}^n r_i\boldsymbol{x}_i\Big\| \leq C(\log\log n)^{1/2}\Big(\sum_{i=1}^n \|\boldsymbol{x}_i\|^2\Big)^{1/2}.$$

This theorem depends on a result of J. Kuelbs (1977) which says that if $\mathbf{E}\|X\|^2 < \infty$ then $X \in IL(\boldsymbol{X})$ if, and only if,

$$(2n\log\log n)^{-1/2}\sum_{i=1}^n X_i \longrightarrow 0$$

in probability, as $n \to \infty$.

7.6 Spaces of type 2 and cotype 2

Theorem 7.6.1.[15] *If a Banach space is of both type 2 and cotype 2, then it is isomorphic to a Hilbert space.*

Proof. Step 1. We first prove that if $\boldsymbol{X}$ is of type 2, and cotype 2, then there exists a constant C such that, for each $n \in \mathbf{N}$, each sequence $(\boldsymbol{x}_i) \subset \boldsymbol{X}$, and each matrix $(a_{ij}) \subset \mathbf{R}$,

$$\sum_{i=1}^n\Big\|\sum_{j=1}^n a_{ij}\boldsymbol{x}_j\Big\|^2 \leq C^2\|a\|^2\sum_{i=1}^n \|\boldsymbol{x}_i\|^2, \tag{7.6.1}$$

[15]Due to S. Kwapień (1972/73). The paper also contains other characterizations of Banach spaces isomorphic to a Hilbert space $\boldsymbol{H}$. In particular, $\boldsymbol{X}$ is isomorphic to $\boldsymbol{H}$ iff the Fourier transform is a bounded operator from $\boldsymbol{L}_2(\mathbf{R};\boldsymbol{X})$ into itself.

where $\|a\|$ denotes the norm of the operator $a : \boldsymbol{l}_2^{(n)} \mapsto \boldsymbol{l}_2^{(n)}$ defined by the matrix (a_{ij}).

From Remark 5.5.1 (b), and Proposition 7.1.1, it follows that, in view of our assumptions, there exists a constant C such that, for each $n \in \mathbf{N}$, and each sequence $(\boldsymbol{x}_i) \subset \boldsymbol{X}$,

$$C^{-1}\sum_{i=1}^{n} \|\boldsymbol{x}_i\|^2 \leq \mathbf{E}\Big\|\sum_{i=1}^{n} \gamma_i \boldsymbol{x}_i\Big\|^2 \leq C\sum_{i=1}^{n} \|\boldsymbol{x}_i\|^2, \tag{7.6.2}$$

where (γ_i) is the sequence of standard independent Gaussian random variables. Now, it is known that the extreme points of the unit ball of the bounded operator space $B(\boldsymbol{l}_2^{(n)}, \boldsymbol{l}_2^{(n)})$ are exactly linear isometries. Hence, by the Krein-Milman Theorem, any $a \in B(\boldsymbol{l}_2^{(n)}, \boldsymbol{l}_2^{(n)})$ with $\|a\| \leq 1$ is a convex combination of matrices of isometries. Thus it is clear that it suffices to prove (7.6.1) in the case when the matrix a is an isometry. However, by (7.6.2), in this situation

$$\begin{aligned}\sum_{i=1}^{n}\Big\|\sum_{j=1}^{n} a_{ij}\boldsymbol{x}_j\Big\|^2 &\leq C\mathbf{E}\Big\|\sum_{i=1}^{n}\gamma_i\Big(\sum_{j=1}^{n} a_{ij}\boldsymbol{x}_j\Big)\Big\|^2 \\ &= C\mathbf{E}\Big\|\sum_{j=1}^{n}\boldsymbol{x}_j\Big(\sum_{i=1}^{n}\gamma_i a_{ij}\Big)\Big\|^2 = C\mathbf{E}\Big\|\sum_{j=1}^{n}\boldsymbol{x}_j\gamma_j\Big\|^2 \\ &\leq C^2\sum_{j=1}^{n}\|\boldsymbol{x}_j\|^2,\end{aligned}$$

because the isometry does not change the distribution of a Gaussian symmetric measure on $\boldsymbol{l}_2^{(n)}$.

Step 2. Let $u : \boldsymbol{l}_1(I; \mathbf{R}) \mapsto \boldsymbol{X}$ be a bounded linear operator onto $\boldsymbol{X}$ (which can be achieved if $|I|$ is sufficiently large). We shall show first that $u \in \Pi_2(\boldsymbol{l}_1(I); \boldsymbol{X})$.

Let $(\boldsymbol{y}_i) \in \boldsymbol{l}_1(I)$ be such that, for each $\boldsymbol{y}^* \in \boldsymbol{l}_\infty(I)$, the series $\sum_i (\boldsymbol{y}^*\boldsymbol{y}_i) < \infty$. Then also $\sum_i \|\boldsymbol{y}_i\|^2 < \infty$, and there exists a linear bounded operator, $v : \boldsymbol{l}_2(I) \mapsto \boldsymbol{l}_1(I)$, such that $\boldsymbol{y}_i = v(\boldsymbol{e}_i^2)$, where $(\boldsymbol{e}_i)$ is the canonical basis in $\boldsymbol{l}_2(I)$. By Grothendieck's Factorization Theorem,

$$v : \boldsymbol{l}_2(I) \overset{d}{\mapsto} \boldsymbol{l}_2(I) \overset{a}{\mapsto} \boldsymbol{l}_1(I),$$

where $d = (d_i)$ is diagonal and Hilbert-Schmidt, and $a = (a_{ij})$ is bounded. Therefore, by Step 1,

$$\sum_{i\in I} \|u\boldsymbol{y}_i\|^2 = \sum_{i\in I} \|uv(\boldsymbol{e}_i^2)\|^2 \leq \sum_{i\in I} \|u\|^2 \|ad(\boldsymbol{e}_i^2)\|^2$$

$$\leq \sum_{i\in I} \|u\|^2 \|a\|^2 |d_i|^2 < \infty,$$

so that $u \in \Pi_2$.

Now, by the Pietsch Factorization Theorem (see Chapter 1)

$$u : \boldsymbol{l}_1(I) \overset{u_1}{\mapsto} \boldsymbol{H} \overset{u_2}{\mapsto} \boldsymbol{X},$$

where $\boldsymbol{H}$ is a Hilbert space, and u_1, and u_2, are bounded. Since u was an operator onto $\boldsymbol{X}$, the operator u_2 maps $\boldsymbol{H}$ onto $\boldsymbol{X}$ as well, so that $\boldsymbol{X}$ is isomorphic to a Hilbert space. QED

Remark 7.6.1. Almost the same proof as above shows that if $u : \boldsymbol{X} \mapsto \boldsymbol{Y}$, and $\boldsymbol{X}$ is of type 2, and $\boldsymbol{Y}$ is of cotype 2, then u can be factorized through a Hilbert space.

Remark 7.6.2. Tail probabilities of sums of random vectors in Banach spaces of type 2 can also be controlled. Here are two sample results[16]:

(a) Let (X_n) be independent, identically distributed, zero-mean random vectors in a Banach space $\boldsymbol{X}$ of type 2 with $\mathbf{E}\|X_1\| < \infty$. If $t > 0$, and $r > \max(t, 2)/2$ then $\mathbf{E}\|X_1\|^2 < \infty$ if, and only if, for each $\epsilon > 0$,

$$\sum_j j^{r-2} \mathbf{P}\big(\|X_1 + \cdots + X_j\| > \epsilon j^{r/t}\big) < \infty.$$

(b) Let (X_n) be independent, identically distributed, random vectors in a Banach space $\boldsymbol{X}$ of type 2. Then $\mathbf{E}\|X_1\| < \infty$, and $\mathbf{E}X_1 = 0$, if, and only if, for each $\epsilon > 0$,

$$\sum_j j^{-1} \mathbf{P}\big(\|X_1 + \cdots + X_j\| > \epsilon j\big) < \infty.$$

[16]Due to N. Jain (1975).

Remark 7.6.3. If $\boldsymbol{X}$ is of type 2 then it is also possible to prove the following extension of the Salem and Zygmund Theorem on exponential moments for lacunary trigonometric series with coefficients in $\boldsymbol{X}$[17] :

If G is a compact Abelian group and $\Delta = (\chi_j)$ is a Sidon set in the dual group of G then, if $\sum_j \|\boldsymbol{x}_j\|^2 < \infty$ then, for all $\beta > 0$, and any bounded sequence $(\alpha_j) \subset \mathbf{C}$,

$$\int_G \exp\left[\beta \sup_n \left\| \sum_{j=1}^{n} \alpha_j \chi_j(g) \boldsymbol{x}_j \right\|^2 \right] dg < \infty,$$

where dg denotes the integration with respect to the Haar measure on G.

Remark 7.6.4. Also, $\boldsymbol{X}$ is of type 2 if, and only if, the class of Gaussian covariance operators coincides with the class of positive, symmetric and nuclear operators.[18]

[17]Due to J. Kuelbs and W.A. Woyczyński (1978).
[18]Due to S. Chevet, S.A. Chobanyan, W. Linde and V.I. Tarieladze (1977).

Chapter 8

Beck convexity

8.1 General definitions and properties and their relationship to types of Banach spaces

We begin this chapter with the basic definition of the Beck-convexity[1], which for the sake of brevity will be called here B-convexity. We shall also briefly discuss its relationship to other geometric properties of Banach spaces which were introduced and discussed in depth in the previous chapters.

Definition 8.1.1. Let $k \in \mathbf{N}^+$, and $\epsilon \in (0,1)$. A real, normed space $\boldsymbol{X}$ is said to be (k,ϵ)-*convex* if, for each sequence $\boldsymbol{x}_1, \ldots, \boldsymbol{x}_k \in S_X$, where S_X denotes the unit sphere in $\boldsymbol{X}$,

$$\inf_{\epsilon_i = \pm 1} \|\epsilon_1 \boldsymbol{x}_1 + \cdots + \epsilon_k \boldsymbol{x}_k\| \leq k(1-\epsilon).$$

The space $\boldsymbol{X}$ is said to be *B-convex* if it is (k,ϵ)-convex for some $k \in \mathbf{N}^+$, and some $\epsilon \in (0,1)$.

The following simple Proposition shows certain relationship between (k,ϵ)- and (j,δ)- convexity.

---

[1]Introduced by A. Beck (1962), and (1963).

Proposition 8.1.1.[2] *(i) If $\boldsymbol{X}$ is (k,ϵ)-convex, $2 \leq j < k$, and $k\epsilon > k-j$, then $\boldsymbol{X}$ is also (j,δ)-convex with $\delta = (k\epsilon - k - j)/j$.*

(ii) If $\boldsymbol{X}$ is (k,ϵ)-convex and $\boldsymbol{Y} \subset \boldsymbol{X}$ then $\boldsymbol{Y}$ is also (k,ϵ)-convex.

(iii) If $\boldsymbol{X}$ is (k,ϵ)-convex and $\delta < \epsilon$, then $\boldsymbol{X}$ is (k,δ)-convex.

(iv) If a normed space $\boldsymbol{X}$ is (k,ϵ)-convex then its completion is also (k,ϵ)-convex.

Proof. (i) Take an arbitrary sequence $\boldsymbol{x}_1,\dots,\boldsymbol{x}_j \in S_X$, and put $\boldsymbol{x}_{j+1} = \cdots = \boldsymbol{x}_k = 0$. Since $\boldsymbol{X}$ is (k,ϵ)-convex, there exist $\varepsilon_1,\dots,\varepsilon_k = \pm 1$ such that

$$\|\varepsilon_1\boldsymbol{x}_1 + \cdots + \varepsilon_j\boldsymbol{x}_j\| = \|\varepsilon_1\boldsymbol{x}_1 + \cdots + \varepsilon_k\boldsymbol{x}_k\| \leq k(1-\epsilon) = j(1-\delta),$$

so that $\boldsymbol{x}$ is (j,δ)-convex.

The proofs of (ii) – (iv) are obvious and we omit them. QED

In infinite-dimensional normed space $\boldsymbol{X}$ there exists also a less trivial relation between k, and ϵ.

Theorem 8.1.1. *If $\boldsymbol{X}$ is an infinite-dimensional (k,ϵ)-convex Banach space then $1-\epsilon \geq k^{-1/2}$.*

Proof. Dvoretzky's Theorem (see Chapter 1) assures the existence of finite-dimensional spaces of arbitrarily high dimension, say k, approximating Hilbert spaces $\boldsymbol{l}_2^{(k)}$ with any prescribed accuracy. Therefore, it is sufficient to check the inequality in $\boldsymbol{l}_2^{(k)}$. However, the inequality is evident in $\boldsymbol{l}_2^{(k)}$ because, for an orthonormal basis $\boldsymbol{e}_1,\dots,\boldsymbol{e}_k \in \boldsymbol{l}_2^{(k)}$ we have the equality $\|\varepsilon_e e_i + \cdots + \varepsilon_k\boldsymbol{e}_k\| = k^{1/2}$. QED

Remark 8.1.1. The above estimate cannot be improved in general because in the Hilbert space $\boldsymbol{H}$, for any $\boldsymbol{x}_1,\dots,\boldsymbol{x}_k \in S_H$, there are $\varepsilon_1,\dots,\varepsilon_k = \pm 1$ such that

$$k^{-1}\|\varepsilon_1\boldsymbol{x}_1 + \cdots + \varepsilon_k\boldsymbol{x}_k\| \leq k^{-1/2},$$

in view of the generalized parallelepiped equality

$$\sum_{\varepsilon_1,\dots,\varepsilon_k=\pm 1} \|\varepsilon_1\boldsymbol{x}_1 + \cdots + \varepsilon_k\boldsymbol{x}_k\|^2 = 2^k \sum_{i=1}^{k} \|\boldsymbol{x}_i\|^2.$$

[2]This Proposition, Theorem 8.1.1, and Propositions 8.1.3, and 8.1.5 are due to D.P. Giesy (1966).

The concept of (k, ϵ)-convexity is a property of normed spaces $\boldsymbol{X}$ that imposes restrictions only on the k-dimensional subspaces and, similarly, B-convexity is a local property of normed spaces in the sense that it imposes restrictions only on the structure of the finite-dimensional subspaces of $\boldsymbol{X}$. In particular, we have the following statement.

Proposition 8.1.2. *If a normed space* $\mathbf{Y}$ *is finitely representable in a B-convex space* $\boldsymbol{X}$*, then* $\mathbf{Y}$ *is also B-convex. In other words, B-convexity is a super-property of a normed space.*

Also, the B-convexity of the following classes of normed spaces is not difficult to check.

Theorem 8.1.2. *(i) If* $\boldsymbol{X}$ *is less than k-dimensional, then it is* (k, k^{-1})*-convex, and thus B-convex.*

(ii) If $\boldsymbol{X}$ *is uniformly convex then it is* $(2, \epsilon)$*-convex for some* $\epsilon > 0$*, and thus B-convex.*

Proof. (i) For any $\boldsymbol{x}_1, \dots, \boldsymbol{x}_k \in S_X$ there exist $\alpha_1, \dots, \alpha_k \in \mathbf{R}$, with $\max_i |\alpha_i| = 1$, such that $\alpha_1 \boldsymbol{x}_1 + \cdots + \alpha_k \boldsymbol{x}_k = 0$. Define $\lambda_i = \alpha_i / |\alpha_i|$, if $\alpha_i \neq 0$, and $\lambda_i = 1$, otherwise. Then, $|\lambda_i - \alpha_i| \leq 1$, for $i = 1, \dots, k$, and for some i, $\lambda_i = \alpha_i$. Then

$$k^{-1} \|\lambda_1 \boldsymbol{x}_1 + \cdots + \lambda_k \boldsymbol{x}_k\| = k^{-1} \left\| \sum_{i=1}^{k} \alpha_i \boldsymbol{x}_i + \sum_{i=1}^{k} (\lambda_i - \alpha_i) \boldsymbol{x}_i \right\|$$

$$\leq k^{-1} \sum_{i=1}^{k} |\lambda_i - \alpha_i| \leq k^{-1}(k-1) = 1 - k^{-1}.$$

(ii) The uniform convexity implies existence of a $\delta > 0$ such that, for any $\boldsymbol{x}, \boldsymbol{y} \in S_X$ with $\|\boldsymbol{x} - \boldsymbol{y}\| > 1$, we have $\|\boldsymbol{x} + \boldsymbol{y}\| \leq 2(1 + \delta)$. Now, the space $\boldsymbol{X}$ is $(2, \min((2^{-1}, \delta)$-convex because , for any $\boldsymbol{x}, \boldsymbol{y} \in S_X$, either

$$\|\boldsymbol{x} - \boldsymbol{y}\| \leq 1 = 2(1 - 1/2) \leq 2(1 - \min(2^{-1}, \delta)),$$

or

$$\|\boldsymbol{x} + \boldsymbol{y}\| \leq 2(1 - \delta) \leq 2(1 - \min(2^{-1}, \delta)). \qquad \text{QED}$$

Below we list examples of space that are B-convex and that are not B-convex.

Example 8.1.1. (a) By Part (*ii*) of the above Theorem the spaces $\boldsymbol{L}_p(T, \Sigma, \mu), 1 < p < \infty$, over any measure space (T, Σ, μ) are B-convex.

(b) The space $\boldsymbol{l}_1$ is not B-convex. Indeed, if $(\boldsymbol{e}_i)$ is the standard basis in $\boldsymbol{l}_1$ then, for each $n \in \mathbf{N}, n \geq 2,$, and arbitrary $\varepsilon_i = \pm 1$, we have the equality $\|\varepsilon_1 \boldsymbol{e}_i + \cdots + \varepsilon_n \boldsymbol{e}_n\| = n$.

(c) The space $\boldsymbol{c}_0$, and thus also $\boldsymbol{l}_\infty$, are not B-convex. Indeed, take an arbitrary $n \in \mathbf{N}, n \geq 2$, and define

$$\boldsymbol{x}_1 = (+1, -1, +1, -1, \ldots, +1, -1, 0, 0, 0, \ldots),$$

$$\boldsymbol{x}_2 = (+1, +1, -1, -1, \ldots, -1, -1, 0, 0, 0, \ldots),$$

....

$$\boldsymbol{x}_n = (+1, +1, +1, +1, \ldots, -1, -1, 0, 0, 0, \ldots),$$

where in each vector there are 2^n non-zero terms, and the non-zero terms in $\boldsymbol{x}_i$ consist of alternating blocks of $+1$'s and -1's, each block of length 2^{i-1}.

By the very construction, for any sequence $\varepsilon_1, \ldots, \varepsilon_n = \pm 1$, there exists a $j, 1 \leq j \leq 2^n$, such that the j's coordinates of $\boldsymbol{x}_1, \ldots, \boldsymbol{x}_n$, are exactly $\varepsilon_1, \ldots, \varepsilon_n$, so that $\|\varepsilon_1 \boldsymbol{x}_1 + \cdots + \varepsilon_n \boldsymbol{x}_n\| = n$.

(d) If $\boldsymbol{X}$, and $\boldsymbol{Y}$ are infinite-dimensional normed spaces, and if $\boldsymbol{X}$ is a dual space, then the space $B(\boldsymbol{X}, \boldsymbol{Y})$ of bounded operators from $\boldsymbol{X}$ into $\boldsymbol{Y}$ is not B-convex.

To verify this fact let $n \in \mathbf{N}, n \geq 2$, and $m = 2^n$. Then, for each k, let $(\beta_i^{(k)})$ be the periodic sequence with period 2^k which starts with 2^{k-1} terms equal to $+1$ followed by 2^{k-1} terms equal to -1. By Dvoretzky's Theorem (see Chapter 1), for each $\epsilon > 0$ there exists $\boldsymbol{x}_1, \ldots, \boldsymbol{x}_m \in S_X$ such that, for all $\alpha_1, \ldots, \alpha_m \in \mathbf{R}$,

$$(1 - \epsilon)\Big\|\sum_{i=1}^{m} \alpha_i \boldsymbol{x}_i\Big\| \leq \Big(\sum_{i=1}^{m} \alpha_i^2\Big)^{1/2}.$$

Now, for each $j = 1, \ldots, n$, let us define a linear, continuous operator $T_j : \boldsymbol{l}_2 \mapsto \boldsymbol{X}$, determining its values on elements of the standard basis $(\boldsymbol{e}_k) \subset \boldsymbol{l}_2$ as follows:

$$T_j(\boldsymbol{e}_i) = (1-\epsilon)\beta_i^{(j)}\boldsymbol{x}_i, \quad \text{if} \quad 1 \le i \le m,$$

and 0 otherwise. Then $T_j \in B(\boldsymbol{l}_2, \boldsymbol{X})$, and $\|T_j\| \le 1$. Again, as in the preceding example ($B(\boldsymbol{X}, \boldsymbol{Y})$ has the sup-type norm !), for arbitrary $\varepsilon_1, \ldots, \varepsilon_n = \pm 1$,

$$\|\varepsilon_1 T_1 + \cdots + \varepsilon_n T_n\| \ge n(1-\epsilon),$$

so that $B(\boldsymbol{l}_2, \boldsymbol{X})$ is not (n, ϵ)-convex. Since $n \ge 2$, and $\epsilon > 0$, were arbitrary, we conclude that $B(\boldsymbol{l}_2, \boldsymbol{X})$ is not B-convex.

The adjoint mapping of $B(\boldsymbol{l}_2, \boldsymbol{X})$ into $B(\boldsymbol{X}^*, \boldsymbol{l}_2)$ is an isometry. Hence, the latter space is not B-convex either, so that, for any $n \in \mathbf{N}, n \ge 2$, and $\epsilon > 0$, one can find $T_1, \ldots, T_n \in S_{B(X^*, l_2)}$ such that, for arbitrary $\varepsilon_1, \ldots, \varepsilon_n = \pm 1$, we have the inequality $\|\varepsilon_1 T_1 + \cdots + \varepsilon_n T_n\| \ge n(1-\epsilon)$. By considering the image of points where these 2^n linear combinations of T_i's nearly achieve their norms, we find a projection P of $\boldsymbol{l}_2$ onto a finite-dimensional subspace such that, for arbitrary $\varepsilon_1, \ldots, \varepsilon_n = \pm 1$,

$$\|\varepsilon_1 PT_1 + \cdots + \varepsilon_n PT_n\| \ge n(1-\epsilon).$$

Utilizing again Dvoretzky's Theorem we find a linear map $S; P(\boldsymbol{l}_2) \mapsto \boldsymbol{Y}$, of norm 1, which is so nearly an isometry that, for all $\varepsilon_1, \ldots, \varepsilon_n = \pm 1$,

$$\|\varepsilon_1 SPT_1 + \cdots + \varepsilon_n SPT_n\| \ge n(1-3\epsilon).$$

Since SPT_j is an element of $B(\boldsymbol{X}^*, \boldsymbol{Y})$ of norm at most 1, we see that $B(\boldsymbol{X}^*, \boldsymbol{Y})$ is not $(n, 3\epsilon)$-convex, for any $n \ge 2$, and $\epsilon > 0$. This proves our original assertion.

In the next step we will indicate the relationship between the B-convexity and the infratype of a Banach space that was studied in Section 6.1. Let us begin by recalling the definition of the numerical constant $a_k^\infty(\boldsymbol{X})$ that were essential in investigation of the cotype of a Banach space $\boldsymbol{X}$:

$$a_n^\infty(\boldsymbol{X}) = \inf\Big\{a \in \mathbf{R}^+ : \forall \boldsymbol{x}_1, \ldots, \boldsymbol{x}_n \in \boldsymbol{X}, \inf_{\varepsilon_i = \pm 1}\Big\|\sum_{i=1}^n \varepsilon_i \boldsymbol{x}_i\Big\| \le a \max_{1 \le i \le n} \|\boldsymbol{x}_i\|\Big\}.$$

It is easy to see that

$$a_n^\infty(\boldsymbol{X}) = \inf\Big\{a \in \mathbf{R}^+ : \forall \boldsymbol{x}_1, \dots, \boldsymbol{x}_n \in S_X, \inf_{\varepsilon_i=\pm 1}\Big\|\sum_{i=1}^n \varepsilon_i \boldsymbol{x}_i\Big\| \le a\Big\}.$$

The following Proposition provides a characterization of B-convexity in terms of the constants a_n^∞.

Proposition 8.1.3. *The following properties of a normed space $\boldsymbol{X}$ are equivalent:*
(i) The space $\boldsymbol{X}$ is B-convex;
(ii) The constants $a_n^\infty(\boldsymbol{X}) < n$, for some $n \ge 2$;
(iii) Asymptotically, $n^{-1}a_n^\infty(\boldsymbol{X}) \to 0$, as $n \to \infty$;
(iv) The constants $a_n^\infty(\boldsymbol{X}) = O(n^\gamma)$, for some $\gamma \in [1/2, 1)$.

Proof. Implications $(iv) \implies (iii) \implies (ii) \implies (i)$ being obvious it is sufficient to prove that $(i) \implies (iv)$. The latter implication is an immediate corollary to the Lemma 6.1.3 and Proposition 6.1.1. QED

The above Proposition permits us to prove the following general statement about the relationship between B-convexity and the infratype of a normed space.

Theorem 8.1.3. *A normed space $\boldsymbol{X}$ is B-convex if, and only if, it is of infratype p for some $p \in (1, 2]$.*

Proof. If $\boldsymbol{X}$ is B-convex then, by Proposition 8.1.3 (iv), we can find a $\gamma \in [1/2, 1)$ such that $a_n^\infty(\boldsymbol{X}) = O(n^\gamma)$. Therefore, by Theorem 6.1.4,

$$p_{\text{inf}} = \lim_{n\to\infty}[\log n/\log a_n^\infty(\boldsymbol{X})] = (1-\gamma)^{-1} > 1,$$

so that $\boldsymbol{X}$ is of infratype p for some $p > 1$. Conversely, if $\boldsymbol{X}$ is of infratype p for some $p \in (1, 2]$ then Proposition 6.1.1(iii) gives the statement (iv) of the preceding Proposition and, hence, also B-convexity of $\boldsymbol{X}$. QED

In a similar fashion one can relate the concepts of Rademacher and stable type to B-convexity. Again, let us recall the definitions of the constants that played important roles in the study of

Rademacher and stable types in Chapter 6:

$$b_n^2(\boldsymbol{X}) = \inf\Big\{b \in \mathbf{R}^+ : \forall \boldsymbol{x}_1, \dots, \boldsymbol{x}_n \in \boldsymbol{X}, \Big(\mathbf{E}\Big\|\sum_{i=1}^n r_i \boldsymbol{x}_i\Big\|^2\Big)^{1/2}$$
$$\leq b\Big(\sum_{i=1}^n \|\boldsymbol{x}_i\|^2\Big)^{1/2}\Big\}.$$

Proposition 8.1.5. *The following properties of a normed space* $\boldsymbol{X}$ *are equivalent:*
(i) The space $\boldsymbol{X}$ *is B-convex;*
(ii) $b_n^2(\boldsymbol{X}) < n^{1/2}$, *for some* $n \geq 2$;
(iii) $n^{-1/2} b_n^2(\boldsymbol{X}) \to 0$, *as* $n \to \infty$;
(iv) $b_n^2(\boldsymbol{X}) = O(n^\gamma)$, *for some* $\gamma \in [0, 1/2)$.

Proof. Implications $(iv) \implies (iii) \implies (ii)$ are evident. The implication $(ii) \implies (i)$ follows from the fact that $a_n^\infty(\boldsymbol{X}) \leq n^{1/2} b_m^2(\boldsymbol{X})$ in view of Proposition 6.2.2(i), and from Proposition 8.1.4. So, we only need to prove the implication $(i) \implies (iv)$.

Assume to the contrary that $b_n^2(\boldsymbol{X}) = n^{1/2}$, for all n. Then, for each $n \in \mathbf{N}$, and each $\epsilon > 0$, there exist $\boldsymbol{x}_1, \dots \boldsymbol{x}_n \in \boldsymbol{X}$, with $\sum_i \|\boldsymbol{x}_i\|^2 = n,$, such that

$$(1-\epsilon)n^2 \leq \mathbf{E}\Big\|\sum_{i=1}^n r_i \boldsymbol{x}_i\Big\|^2 \leq \Big(\sum_{i=1}^n \|\boldsymbol{x}_i\|\Big)^2,$$

so that

$$\frac{1}{2}\sum_{1\leq i,j\leq n}\big(\|\boldsymbol{x}_i\| - \|\boldsymbol{x}_j\|\big)^2 = n\sum_{i=1}^n \|\boldsymbol{x}_i\|^2 - \Big(\sum_{i=1}^n \|\boldsymbol{x}_i\|\Big)^2 \leq \epsilon k^2.$$

In particular, if $i_0, 1 \leq i_0 \leq n$, is such that

$$\|\boldsymbol{x}_{i_0}\| = \sup_{1\leq i\leq n} \|\boldsymbol{x}_i\|, \qquad \text{then} \qquad \sum_{j=1}^n \big(\|\boldsymbol{x}_{i_0}\| - \|\boldsymbol{x}_j\|\big)^2 \leq 2\epsilon n^2,$$

so that

$$n^{1/2} = \Big(\sum_{i=1}^n \|\boldsymbol{x}_i\|^2\Big)^{1/2} \geq n^{1/2}\|\boldsymbol{x}_{i_0}\| - \Big(\sum_{j=1}^n \big(\|\boldsymbol{x}_{i_0}\| - \|\boldsymbol{x}_j\|\big)^2\Big)^{1/2}$$

$$\geq n^{1/2}\|\boldsymbol{x}_{i_0}\| - n(2\epsilon)^{1/2},$$

and $\|\boldsymbol{x}_{i_0}\| \leq 1 + (2\epsilon n)^{1/2}$. Finally, we have that

$$\left(\mathbf{E}\left\|\sum_{i=1}^{n} r_i \boldsymbol{x}_i\right\|^2\right)^{1/2} \geq \frac{1-\epsilon}{1+(2\epsilon n)^{1/2}} n \sup_{1\leq i\leq n} \|\boldsymbol{x}_i\|,$$

from which we get that

$$\frac{1-\epsilon}{1+(2\epsilon n)^{1/2}} n \leq \sup_{x_1+\cdots+x_n \in S_X} \left(\mathbf{E}\left\|\sum_{i=1}^{n} r_i \boldsymbol{x}_i\right\|^2\right)^{1/2}$$

$$\leq \sup_{x_1+\cdots+x_n \in S_X} 2^{-n/2}\left(\inf_{\varepsilon_i = \pm 1}\left\|\sum_{i=1}^{n} \varepsilon_i \boldsymbol{x}_i\right\| + n^2(2^n - `1)\right)^{1/2}$$

$$= \left(\frac{a_n^{\infty}(\boldsymbol{X}) + n^2(2^n - 1)}{2^n}\right)^{1/2} \leq n.$$

Therefore, for each $n \in \mathbf{N}$, we have $a_n^{\infty}(\boldsymbol{X}) = n$, and by Proposition 8.1.4, $\boldsymbol{X}$ is not B-convex, a contradiction. QED

Finally, the next result shows the connection between spaces that are B-convex and spaces of Rademacher and stable type.

Theorem 8.1.4.[3] *The following properties of a normed space* $\boldsymbol{X}$ *are equivalent:*

(i) The space $\boldsymbol{X}$ *is B-convex;*

(ii) $\boldsymbol{X}$ *is of Rademacher type* p*, for some* $p \in (1, 2]$*;*

(iii) $\boldsymbol{X}$ *is of stable type* p*, for some* $p \in (1, 2]$*;*

(iv) $\boldsymbol{X}$ *is of stable type 1.*

Proof. We prove the implications $(i) \implies (ii) \implies (iii) \implies (i)$.

$(ii) \implies (ii)$ If X is B-convex then, by Proposition 8.1.5 (iv), there exists a $\gamma \in [0, 1/2)$ such that $b_n^2(\boldsymbol{X}) = O(n^{\gamma})$. Therefore, by Theorem 6.2.1, the supremum of those p for which $\boldsymbol{X}$ is of Rademacher-type p,

$$p_{\text{Rtype}}(\boldsymbol{X}) = \lim_{n\to\infty} \frac{\log n}{\log n^{1/2} b_n^2(\boldsymbol{X})} \geq \frac{1}{\gamma + 1/2} \geq 1,$$

[3]Due to G. Pisier (1973/73).

and (*ii*) is satisfied.

(*ii*) $\Longrightarrow$ (*iii*) This implication follows directly from Proposition 6.5.1 (*ii*) which says that if $\boldsymbol{X}$ is of Rademacher-type p then it is of stable-type q for each $q < p$.

(*iii*) $\Longrightarrow$ (*iv*) This implication is obvious.

(*iv*) $\Longrightarrow$ (*i*) Assume that $\boldsymbol{X}$ is not B-convex. We first prove that in this case the space $\boldsymbol{l}_1$ is not finitely representable in $\boldsymbol{X}$. Indeed, for each $n \in \mathbf{N}$, and each $\epsilon > 0$, there exist $\boldsymbol{x}_1, \ldots, \boldsymbol{x}_n \in S_X$ such that, for any $\varepsilon_1, \ldots, \varepsilon_n = \pm 1$ $\|\varepsilon_1 \boldsymbol{x}_1 + \cdots + \varepsilon_n \boldsymbol{x}_n\| > n(1-\epsilon)$. This implies that, for all real sequences (α_i), if i_0 is chosen in such a way that $|\alpha_{i_0}| = \max\{|\alpha_i| : 1 \leq i \leq n\}$ then

$$\Big\|\sum_{i=1}^n \alpha_i \boldsymbol{x}_i\Big\| = \Big\|\sum_{i=1}^n [\operatorname{sgn}(\alpha_i)|\alpha_i|]\boldsymbol{x}_i - \sum_{i=1}^n [\operatorname{sgn}(\alpha_{i_0})|\alpha_{i_0}| - \alpha_i]\boldsymbol{x}_i\Big\|$$

$$\geq |\alpha_n|\Big\|\sum_{i=1}^n \operatorname{sgn}(\alpha_i)\boldsymbol{x}_i\Big\| - \sum_{i=1}^n \Big||\alpha_{i_0}| - |\alpha_i|\Big|\|\boldsymbol{x}_i\| \tag{8.1.1}$$

$$\geq n|\alpha_{i_0}|(1-\epsilon) + \sum_{i=1}^n (|\alpha_i| - |\alpha_{i_0}|) = \sum_{i=1}^n |\alpha_i| - n\epsilon|\alpha_{i_0}| \geq (1-n\epsilon)\sum_{i=1}^n |\alpha_i|,$$

so that $\boldsymbol{l}_1$ is finitely representable in $\boldsymbol{X}$.

Now, if $\boldsymbol{X}$ were of stable-type 1 there would exist a constant $C > 0$ such that, for all $n \in \mathbf{N}$, and all sequences $\boldsymbol{x}_1, \ldots, \boldsymbol{x}_n \in \boldsymbol{X}$,

$$\Big(\mathbf{E}\Big\|\sum_{i=1}^n \theta_i \boldsymbol{x}_i\Big\|^{1/2}\Big)^2 \leq C \sum_{i=1}^n \|\boldsymbol{x}_i\|,$$

where (θ_i) are i.i.d. stable random variables of exponent 1. Then (8.1.1) would imply that, for any $n \in \mathbf{N}$, $(\alpha_i) \subset \mathbf{R}$, $(\boldsymbol{x}_i) \subset S_X$,

$$\Big(\mathbf{E}\Big(\sum_{i=1}^n |\alpha_i||\theta_i|\Big)^{1/2}\Big)^2 \leq C_1\Big(\mathbf{E}\Big\|\sum_{i=1}^n \alpha_i\theta_i\boldsymbol{x}_i\Big\|^{1/2}\Big)^2 \leq C_1 C \sum_{i=1}^n |\alpha_i|.$$

However, this inequality may not be true because, by Schwartz' Theorem (see Chapter 1), the series $\sum_i \alpha_i\theta_i$ converges absolutely almost surely if, and only if, $\sum_i |\alpha_i|(1 + \log(1/|\alpha_i|)) < \infty$. QED

8.2 Local structure of B-convex spaces and preservation of B-convexity under standard operations

Proposition 8.2.1.[4] *A normed linear space $\boldsymbol{X}$ is (k,ϵ)-convex [B-convex] if, and only if, each k-dimensional [separable] subspace of $\boldsymbol{X}$ is (k,ϵ)-convex [B-convex].*

Proof. The "only if" part is obvious because every subspace of a (k,ϵ)-convex space $\boldsymbol{X}$ is (k,ϵ)-convex. Conversely, if $\boldsymbol{X}$ is not (k,ϵ)-convex then one can find $\boldsymbol{x}_1,\ldots,\boldsymbol{x}_k \in S_X$ such that, for any $\varepsilon_1,\ldots,\varepsilon_k = \pm 1$,

$$\|\varepsilon \boldsymbol{x}_1 + \cdots + \varepsilon_k \boldsymbol{x}_k\| > k(1-\epsilon),$$

so that $\overline{\text{span}}[\boldsymbol{x}_1,\ldots,\boldsymbol{x}_k] \subset \boldsymbol{X}$ is an – at most k-dimensional – subspace of $\boldsymbol{X}$ which is not (k,ϵ)-convex.

Similarly, if $\boldsymbol{X}$ is not B-convex, then $\boldsymbol{X}$ is not (k,n^{-1})-convex for any $k \geq 2, n \geq 1$, and, as above, one can find a sequence $\boldsymbol{X}_{n,k} \subset \boldsymbol{X}$ of finite-dimensional subspaces that are not (k,n^{-1})-convex. The span of $\boldsymbol{X}_{n.k}, k \geq 2, n \geq 1$, in $\boldsymbol{X}$ is a separable subspace of $\boldsymbol{X}$ which is not B-convex. QED

Geometrically speaking, the following important theorem states that a normed space which is not B-convex must contain arbitrarily good approximations of $\boldsymbol{l}_1^{(k)}$, for any $k \in \mathbf{N}$. It is an immediate corollary of Theorem 8.1.6, and the nontrivial Corollary 6.5.2(i). However, we provide here an independent elementary proof.

Theorem 8.2.1.[5] *The following properties of a normed space $\boldsymbol{X}$ are equivalent:*

(i) The space $\boldsymbol{X}$ is B-convex;

(ii) The space $\boldsymbol{l}_1$ is not crudely finitely representable in $\boldsymbol{X}$;

(iii) The space $\boldsymbol{l}_1$ is not finitely representable in $\boldsymbol{X}$.

[4]Due to D.P. Giesy (1966).

[5]Due to D.P. Giesy and R.C. James (1973).

Proof. $(i) \implies (ii)$ Assume to the contrary that $\boldsymbol{l}_1$ is crudely finitely representable in $\boldsymbol{X}$. Then, in particular, there would exist a constant $\lambda \in (0,1]$ such that, for each $k \in \mathbf{N}$, there exist $\boldsymbol{x}_1, \ldots, \boldsymbol{x}_k \in S_X$ such that, for any sequence $\varepsilon_1, \ldots, \varepsilon_k = \pm 1$,

$$k\lambda \leq \|\varepsilon_1 \boldsymbol{x}_1 + \cdots + \varepsilon_k \boldsymbol{x}_k\|.$$

This implies that $k^{-1} a_k^{\infty}(\boldsymbol{X}) \geq \lambda > 0$, for all $k \geq 1$. Hence, by Proposition 8.1.5(iii), the space $\boldsymbol{X}$ is not B-convex. A contradiction.

$(ii) \implies (iii)$ This implication is obvious.

$(iii) \implies (i)$ Proof of this implication is contained in the proof of the implication $(iv) \implies (i)$ of Theorem 8.1.6. QED

Corollary 8.2.1. *If a normed space $\boldsymbol{X}$ is B-convex and $\boldsymbol{Y}$ is isomorphic to $\boldsymbol{X}$ in the sense of normed spaces then $\boldsymbol{Y}$ is B-convex as well.*

Now we are turning to the problem of preservation of B-convexity under standard operations on normed spaces such as selection of a subspace, the closure, completion, mapping via linear bounded operators, and taking duals and preduals. The following Proposition follows directly from the definitions so we omit the proofs.

Proposition 8.2.2. *(i) If a normed space $\boldsymbol{Y} \subset \boldsymbol{X}$, and $\boldsymbol{X}$ is (k,ϵ)-convex, then $\boldsymbol{Y}$ is also (k,ϵ)-convex;*

(ii) If a normed space $\boldsymbol{Y} \subset \boldsymbol{X}$, and $\boldsymbol{Y}$ is (k,ϵ)-convex, then the closure of $\boldsymbol{Y}$ in $\boldsymbol{X}$ is also (k,ϵ)-convex;

(iii) The completion of a B-convex normed space is B-convex.

The following result shows that B-convexity is preserved under some linear mappings.

Theorem 8.2.2.[6] *If $\boldsymbol{X}$ is a B-convex normed space and $T : \boldsymbol{X} \mapsto \boldsymbol{Y}$ is a continuous, linear and open mapping into a normed space $\boldsymbol{Y}$, then the image $T\boldsymbol{X}$ is B-convex.*

Proof. Since T is open, one can find a $\delta > 0$ such that

$$\{\boldsymbol{y} \in T\boldsymbol{X} : \|\boldsymbol{y}\| < \delta\} \subset T(S_X).$$

[6] Due to D.P. Giesy (1966).

Now, let $\boldsymbol{y}_1, \ldots, \boldsymbol{y}_k \in S_{TX}$. Because $\delta \boldsymbol{y}_i \in TS_X$, there exist $\boldsymbol{x}_i \in S_X$ such that $T\boldsymbol{x}_i = \delta \boldsymbol{y}_i$. In view of B-convexity of $\boldsymbol{X}$, and Proposition 8.1.5, one can find a $k \in \mathbf{N}$, and $\varepsilon_1, \ldots, \varepsilon_k = \pm 1$, such that

$$\|\varepsilon_1 x_1 + \cdots + \varepsilon_k \boldsymbol{x}_k\| \leq k\delta \|T\|/2.$$

Therefore,

$$\Big\|\sum_{i=1}^{k} \varepsilon_i \boldsymbol{y}_i\Big\| = \frac{1}{\delta}\Big\|\sum_{i=1}^{k} \varepsilon_i T\boldsymbol{x}_i\Big\| \leq \frac{1}{\delta}\|T\|\Big\|\sum_{i=1}^{k} \varepsilon_i \boldsymbol{x}_i\Big\| \leq \frac{1}{2},$$

so that $T\boldsymbol{X}$ is $(k, 1/2)$-convex. QED

Remark 8.2.1. In the above Theorem the openness of the operator T is essential. Indeed, define

$$T : \boldsymbol{l}_1 \ni (\alpha_1, \alpha_2, \ldots) \mapsto \Big(\alpha_1, \frac{\alpha_2}{2}, \ldots, \frac{\alpha_n}{n}, \ldots\Big) \in \boldsymbol{l}_1.$$

The operator T is linear, continuous and $T\boldsymbol{l}_2$ is dense in $\boldsymbol{l}_1$. But $\boldsymbol{l}_2$ is B-convex, and $T\boldsymbol{l}_2$ is not . If it were B-convex then its completion, $\boldsymbol{l}_1$, would be B-convex and it is not.

From Proposition 6.5.4, Theorem 6.5.6, and Theorem 8.1.6(*iv*) we immediately get a result concerning the factoring operation.

Theorem 8.2.3.[7] *Let $\boldsymbol{X}$ be a normed space and $\boldsymbol{Y}$ be its closed subspace. Then $\boldsymbol{X}$ is B-convex if, and only if, both $\boldsymbol{Y}$ and $\boldsymbol{X}/\boldsymbol{Y}$ are B-convex.*

Corollary 8.2.2. *Let $\boldsymbol{X}$, and $\boldsymbol{Y}$, be normed spaces, and $T : \boldsymbol{X} \mapsto \boldsymbol{Y}$ be linear, continuous and open map. The $\boldsymbol{X}$ is B-convex if, and only if,* $\operatorname{Ker} T$, *and* $\operatorname{Im} T$ *are B-convex.*

Proof. If $\boldsymbol{X}$ is B-convex then $\operatorname{Ker} T$ is its linear subspace and also B-convex by Proposition 8.2.2. At the same time $\operatorname{Im} T$ is B-convex by Theorem 8.2.2. Conversely, if $\operatorname{Ker} T$, and $\operatorname{Im} T$, are both B-convex then $\boldsymbol{X}$ is also B-convex by the above Theorem because $\boldsymbol{X}/\operatorname{Ker} T$ is also B-convex as an image of $\operatorname{Im} T = T\boldsymbol{X}$ by a continuous linear open map $U(T(\boldsymbol{x})) = \boldsymbol{x} + \operatorname{Ker} T$.

[7]Theorems 8.2.3 and 8.2.4 are due to D.P. Giesy (1966).

Theorem 8.2.4. *A normed space $\boldsymbol{X}$ is B-convex if, and only if, its dual $\boldsymbol{X}^*$ is B-convex.*

Proof. Assume that $\boldsymbol{X}$ is not B-convex. Then, by Theorem 8.2.2, the space $\boldsymbol{l}_1$ is crudely finitely representable in $\boldsymbol{X}$, i.e., there exists a $\lambda > 1$, and subspaces $\boldsymbol{X}_n \subset \boldsymbol{X}$, such that, for every $n \in \mathbf{N}$, the distance $d(\boldsymbol{X}_n, \boldsymbol{l}_1^{(n)}) < \lambda$. By duality, for each $n \in \mathbf{N}$, we have the distance $d(\boldsymbol{X}_n^*, \boldsymbol{l}_\infty^{(n)}) < \lambda$.

On the other hand, $\boldsymbol{X}_n^* \sim \boldsymbol{X}^*/\boldsymbol{X}_n^0$, where $E_n^0 = \{\boldsymbol{X}^* \in \boldsymbol{X}^* : \boldsymbol{x}^*\boldsymbol{x} = 0, \forall \boldsymbol{x} \in \boldsymbol{X}_n\}$. In particular, for every $n \in \mathbf{N}$,

$$d\Big(\boldsymbol{X}^*/\boldsymbol{X}_{2^n}^0, \boldsymbol{l}_\infty^{(2^n)}\Big) < \lambda.$$

However, $\boldsymbol{l}_1^{(n)}$ may be embedded isometrically into $\boldsymbol{l}_\infty^{(2^n)}$ so that, for each $n \in \mathbf{N}$, each sequence $(\bar{\boldsymbol{x}}_i^{(n)})_{1 \le i \le n}$ in $\boldsymbol{X}_{2^n}$, and each $(\alpha_i) \in \mathbf{R}^n$,

$$\sum_{i=1}^n |\alpha_i| \le \Big\|\sum_{i=1}^n \alpha_i \bar{\boldsymbol{x}}_i^{(n)}\Big\|, \qquad \text{and} \qquad \sup_{1 \le i \le n} \|\bar{\boldsymbol{x}}_i^{(n)}\| < \lambda.$$

Now, let $(\boldsymbol{x}_i^{(n)})_{1 \le i \le n}$ be in $\boldsymbol{X}_n$, and such that $\boldsymbol{x}_i^{(n)}$ represents $\bar{\boldsymbol{x}}_i^{(n)} \bmod \boldsymbol{X}_{2^n}^0$ with $\|\boldsymbol{x}_i^{(n)}\| < \lambda$. Then, for every $(\alpha_i) \in \mathbf{R}^n$,

$$\sum_{i=1}^n |\alpha_i| \le \Big\|\sum_{i=1}^n \alpha_i \boldsymbol{x}_i^{(n)}\Big\| \le \lambda \sum_{i=1}^n |\alpha_i|,$$

and this means that $\boldsymbol{l}_1$ is crudely finitely representable in $\boldsymbol{X}^*$, so that, again by Theorem 8.2.2, $\boldsymbol{X}^*$ is not B-convex.

Conversely, if $\boldsymbol{X}^*$ is not B-convex then $\boldsymbol{l}_1$ is crudely finitely representable in $\boldsymbol{X}^*$. However, $\boldsymbol{l}_1^{(n)}$'s are necessarily duals of quotients of $\boldsymbol{X}$ which are close to $\boldsymbol{l}_\infty^{(n)}$'s, so that one can finish the proof proceeding as above. QED

The (k, ϵ)-convexity is also inherited by the biduals, and the result also offers an alternative proof of the "only if" part of the above Theorem 8.2.4.

Theorem 8.2.5.[8] *A Banach space $\boldsymbol{X}$ is (k, ϵ)-convex if, and only if, the bidual space $\boldsymbol{X}^{**}$ is (k, ϵ)-convex.*

[8] Due to D.P. Giesy (1966).

Proof. Because $\boldsymbol{X}$ embeds isometrically into $\boldsymbol{X}^{**}$, we immediately get that the (k,ϵ)-convexity of $\boldsymbol{X}^{**}$ implies (k,ϵ)-convexity of $\boldsymbol{X}$.

Conversely, assume that $\boldsymbol{X}$ is (k,ϵ)-convex and choose $\boldsymbol{x}_1^{**},\dots,\boldsymbol{x}_k^{**}\in S_{X^{**}}$, and $\varepsilon_1,\dots,\varepsilon_k=\pm1$. Choose $\boldsymbol{x}^*_{\varepsilon_1,\dots,\varepsilon_k}\in\boldsymbol{X}^*$, with $\|\boldsymbol{x}^*_{\varepsilon_1,\dots,\varepsilon_k}\|=1$ in such a way that

$$\Big\|\sum_{i=1}^k\varepsilon_i\boldsymbol{x}_i^{**}\Big\|-\delta<\Big|\Big(\sum_{i=1}^k\varepsilon_i\boldsymbol{x}_i^{**}\Big)(\boldsymbol{x}^*_{\varepsilon_1,\dots,\varepsilon_k})\Big|.$$

For each i consider the $\boldsymbol{X}^*$ neighborhood N of $\boldsymbol{x}_i^{**}$ determined by δ, and the 2^k functionals $\boldsymbol{x}^*_{\varepsilon_1,\dots,\varepsilon_k},\varepsilon,\dots,\varepsilon_k=\pm1$. Since the canonical mapping $U:\boldsymbol{X}\mapsto\boldsymbol{X}^{**}$ maps S_X into a dense set of $S_{X^{**}}$, one can find an $\boldsymbol{x}_i\in S_X$ such that $U\boldsymbol{x}_i\in N$. This means that

$$\Big|\boldsymbol{x}_i^{**}(\boldsymbol{x}^*_{\varepsilon_1,\dots,\varepsilon_k})-(U\boldsymbol{x}_i)(\boldsymbol{x}^*_{\varepsilon_1,\dots,\varepsilon_k})\Big|=\Big|\boldsymbol{x}_i^{**}(\boldsymbol{x}^*_{\varepsilon_1,\dots,\varepsilon_k})-\boldsymbol{x}^*_{\varepsilon_1,\dots,\varepsilon_k}(\boldsymbol{x}_i)\Big|<\delta,$$

for each $(\varepsilon_1,\dots,\varepsilon_k)$. In view of the (k,ϵ)-convexity of $\boldsymbol{X}$ there exist $\delta_1,\dots,\delta_k=\pm1$ such that

$$\Big\|\sum_{i=1}^k\delta_i\boldsymbol{x}_i\Big\|<k(1-\epsilon+\delta).$$

Then

$$\Big\|\sum_{i=1}^k\delta_i\boldsymbol{x}_i^{**}\Big\|\le\Big|\Big(\sum_{i=1}^k\delta_i\boldsymbol{x}_i^{**}\Big)(\boldsymbol{x}^*_{\delta_1,\dots,\delta_k})\Big|+k\delta$$

$$\le\Big|\sum_{i=1}^k\delta_i\Big(\boldsymbol{x}_i^{**}\boldsymbol{x}^*_{\delta_1,\dots,\delta_k}-\boldsymbol{x}^*_{\delta_1,\dots,\delta_k}\boldsymbol{x}_i\Big)\Big|+\Big|\sum_{i=1}^k\delta_i\boldsymbol{x}^*_{\delta_1,\dots,\delta_k}(\boldsymbol{x}_i)\Big|+k\delta$$

$$\le\sum_{i=1}^k|\delta_i|\Big|\boldsymbol{x}_i^{**}\boldsymbol{x}^*_{\delta_1,\dots,\delta_k}-\boldsymbol{x}^*_{\delta_1,\dots,\delta_k}\boldsymbol{x}_i\Big|+\Big|\boldsymbol{x}^*_{\delta_1,\dots,\delta_k}\Big(\sum_{i=1}^k\delta_i\boldsymbol{x}_i\Big)\Big|+k\delta$$

$$<k\delta+\|\boldsymbol{x}^*_{\delta,\dots,\delta_K}\|\cdot\Big\|\sum_{i=1}^k\delta_i\boldsymbol{x}_i\Big\|+k\delta\le k(1-\epsilon+\delta),$$

which implies the (k, ϵ)-convexity of $\boldsymbol{X}^{**}$ in view of arbitrariness of $\delta > 0$, and the above sharp inequality. QED

It is also elementary to check that B-convexity is preserved under finite direct sums and span operations.

Proposition 8.2.3. *(i) Let $\boldsymbol{X}_1, \ldots, \boldsymbol{X}_n$ be normed spaces and $\boldsymbol{X} = \boldsymbol{X}_1 \oplus \cdots \oplus \boldsymbol{X}_n$ be the direct sum under component-wise arithmetic and (say) $\boldsymbol{l}_1^{(n)}$-norm. Then $\boldsymbol{X}$ is B-convex if, and only if, all of $\boldsymbol{X}_1, \ldots, \boldsymbol{X}_n$ are B-convex.*

(ii) Let $\boldsymbol{X}$ be a Banach space and $\boldsymbol{X}_1, \ldots, \boldsymbol{X}_n$ be its linear subspaces such that $\boldsymbol{X} = \overline{\text{span}}(\boldsymbol{X}_1, \ldots, \boldsymbol{X}_n)$. Then $\boldsymbol{X}$ is B-convex if, and only if, all of $\boldsymbol{X}_1, \ldots, \boldsymbol{X}_n$ are B-convex.

Proof. (i) Because $\boldsymbol{X}_i$ may be identified with a subspace of $\boldsymbol{X}$, Proposition 8.2.2 implies that if $\boldsymbol{X}$ is B-convex then, for each $i = 1, \ldots, n$, the space $\boldsymbol{X}_i$ is B-convex. The converse may be proven by induction on n. For $n = 1$ the result is trivial. Suppose it is true for $n - 1$, and that $\boldsymbol{X}_1, \ldots, \boldsymbol{X}_n$ are B-convex. The projection $T : \boldsymbol{X} \mapsto \boldsymbol{X}_n$ is continuous, linear, and open, with both $\operatorname{Im} T = \boldsymbol{X}_n$, and the $\operatorname{Ker} T = \boldsymbol{X}_1 \oplus \cdots \oplus \boldsymbol{X}_{n-1} \oplus 0$ being B-convex. Thus, by Corollary 8.2.2, $\boldsymbol{X}$ is B-convex.

(ii) Again, by Proposition 8.2.2, if $\boldsymbol{X}$ is B-convex then for each $i = 1, \ldots, n$ the space $\boldsymbol{X}_i$ is B-convex. To prove the converse, by Proposition 8.2.2, we may assume that $\boldsymbol{X}_i$ are closed and B-convex. Then, by (i), the space $\boldsymbol{Y} = \boldsymbol{X} = \boldsymbol{X}_1 \oplus \cdots \oplus \boldsymbol{X}_n$ is B-convex, and of course complete, and the linear operator,

$$T : \boldsymbol{Y} \ni (\boldsymbol{x}_1, \ldots, \boldsymbol{x}_n) \mapsto \boldsymbol{x}_1 + \cdots + \boldsymbol{x}_n \in \boldsymbol{X},$$

is onto, and continuous, because

$$\|T(\boldsymbol{x}_1, \ldots, \boldsymbol{x}_n)\|_X = \|\boldsymbol{x}_1 + \cdots + \boldsymbol{x}_n\|$$

$$\leq \|\boldsymbol{x}_1\| + \cdots + \boldsymbol{x}_n\| = \|(\boldsymbol{x}_1, \ldots, \boldsymbol{x}_n)\|^Y.$$

Therefore, T is open by the Banach Open Mapping Theorem, and an application of Theorem 8.2.2 completes the proof. QED

Finally, we shall verify that some spaces of functions with values in B-convex spaces are also B-convex.

Theorem 8.2.6.[9] *Let $1 < p < \infty$, and let (T, Σ, μ) be a measure space. Then, a Banach space $\boldsymbol{X}$ is B-convex if, and only if, the space $\boldsymbol{L}_p(T, \Sigma, \mu; \boldsymbol{X})$ is B-convex.*

Proof. If $\boldsymbol{L}_p(\boldsymbol{X})$ is B-convex then $\boldsymbol{X}$ is also B-convex as a subspace of $\boldsymbol{L}_p(\boldsymbol{X})$. Conversely, if $\boldsymbol{X}$ is B-convex then, by Theorem 8.1.4, $\boldsymbol{X}$ is of stable-type 1 and, by Theorem 6.5.3, $\boldsymbol{L}_p(\boldsymbol{X})$ is also of stable type 1, i.e. B-convex. QED

8.3 Banach lattices and reflexivity of B-convex spaces

We begin by characterizing subspaces of the space of integrable functions which are B-convex.

Theorem 8.3.1.[10] *Let (T, Σ, μ) be a measure space, and let $\boldsymbol{X}$ be a closed subspace of $\boldsymbol{L}_1(T, \Sigma, \mu)$. Then the following properties of $\boldsymbol{X}$ are equivalent:*

(i) The space $\boldsymbol{X}$ is B-convex;

(ii) $\boldsymbol{X}$ is reflexive;

(iii) $\boldsymbol{X}$ does not contain an isomorphic copy of $\boldsymbol{l}_1$;

(iv) $\boldsymbol{X}$ does not contain an isomorphic copy of $\boldsymbol{l}_1$ complemented in $\boldsymbol{L}_1$.

Proof. Because $\boldsymbol{X}$ is B-convex if, and only if, $\boldsymbol{X}$ is of stable-type 1 (see Theorem 8.1.4), Theorem 6.5.2 gives the equivalence of $(i), (iii)$, and (iv). The implication $(ii) \implies (iii)$ is evident. Therefore it is sufficient to show that if $\boldsymbol{X}$ is a close non-reflexive subspace of $\boldsymbol{L}_1$ then it contains a basic sequence which is equivalent to the standard basis of $\boldsymbol{l}_1$. In the course of the proof we shall have need of the following Lemma:

Lemma 8.3.1. *Let $\boldsymbol{e}_n, n \in \mathbf{N}$ be the standard basis in $\boldsymbol{l}_1$, and let $k_j, j \in \mathbf{N}$, and $n_k, k \in \mathbf{N}$, be two sequences of increasing*

[9] Due to G. Pisier (1973/74). An alternative proof can be found in H.P. Rosenthal (1976).

[10] Due to G. Pisier (1972/73), Exp. XVIII and XIX.

integers. Denote

$$\boldsymbol{z}_j = \sum_{i=k_j+1}^{k_{j+1}} a_i^{(j)} \left(\boldsymbol{e}_{n_{2i}} - \boldsymbol{e}_{n_{2i+1}}\right),$$

with $a_i^{(j)} \geq 0$*, for* $k_j < i \leq k_{j+1}$*, and* $\sum_i a_i^{(j)} = 1$*. Then,* $\boldsymbol{z}_j, j \in N$*, is a basic sequence in* $\boldsymbol{l}_1$ *which is equivalent to* $\mathbf{e}_n, n \in \mathbf{N}$*, and for which the coordinate functional* $\boldsymbol{z}_j^*(\sum_i \alpha_i \boldsymbol{z}_i) = \alpha_j$*, has the norm 1/2, for each* j*.*

Proof. The proof of the Lemma is obvious in view of the equality

$$\Big\|\sum_{j=1}^{m} t_j \boldsymbol{z}_j\Big\|_{l_1} = 2\sum_{j=1}^{m} |t_j|.$$

Proof of Theorem 8.3.1, continued. Now, assume that $\boldsymbol{X}$ is non-reflexive. By the Dunford-Pettis compactness criterion, the unit sphere S_X is not equiintegrable in $\boldsymbol{L}_1$, so that

$$\lim_{a\to\infty} \sup_{x\in S_X} \int_{|x|>a} |\boldsymbol{x}| d\mu = \delta > 0.$$

Hence, one can find $a_n, n \in \mathbf{N}, a_n \uparrow \infty$, such that

$$\delta - \frac{\delta}{2n} < \sup_{x\in S_X} \int_{|x|>a_n} |\boldsymbol{x}| d\mu < \delta + \frac{\delta}{2n}, \qquad n \in \mathbf{N}, \tag{8.3.1}$$

from which it follows that there exists a sequence $(\boldsymbol{x}_n) \subset S_X$, such that, for each $n \in \mathbf{N}$,

$$\delta - \frac{\delta}{2n} < \int_{|x_n|>a_n} |\boldsymbol{x}_n| d\mu < \delta + \frac{\delta}{2n}, \qquad n \in \mathbf{N}, \tag{8.3.2}$$

Let us define

$$\boldsymbol{x}_n' = \boldsymbol{x}_n I[|\boldsymbol{x}_n| > a_n], \qquad \text{and} \qquad \boldsymbol{x}_n'' = \boldsymbol{x}_n - \boldsymbol{x}_n'.$$

Then, for each $\epsilon > 0$,

$$\mu\big\{|\boldsymbol{x}_n'| \geq \epsilon\|\boldsymbol{x}_n'\|\big\} \leq \mu\{|\boldsymbol{x}_n'| \geq 0\} \leq \mu\{|\boldsymbol{x}_n| > a_n\} \leq 1/a_n,$$

and, because $1/a_n \to 0$, as $n \to \infty$, for each $\epsilon > 0$,

$$\{\boldsymbol{x}'_n : n \in N\} \subset M_{p_\epsilon} := \{\boldsymbol{x} \in \boldsymbol{L}_p : \mu\{|\boldsymbol{x}| \geq \epsilon \|\boldsymbol{x}\|_{L_p}\} \geq \epsilon\}.$$

Therefore, by Kadec-Pelczynski's Theorem[11], one can find a basic subsequence $(\boldsymbol{x}'_{n_i}) \subset (\boldsymbol{x}'_n)$ which is equivalent to the standard basis of $\boldsymbol{l}_1$ because, in view of (2), we have $\delta/2 \leq \|\boldsymbol{x}_{n_i}\| \leq 3\delta/2$.

On the other hand, the sequence $(\boldsymbol{x}''_n)$ is equi-integrable because

$$\sup_{\boldsymbol{x} \in (\boldsymbol{x}''_i)} \int_{|\boldsymbol{x}|>a} |\boldsymbol{x}| d\mu = \sup_{p>n} \int_{|\boldsymbol{x}''_p|>a_n} |\boldsymbol{x}''_p| d\mu$$

$$= \sup_{p>n} \Big(\int_{|\boldsymbol{x}_p|>a_n} |\boldsymbol{x}_p| d\mu - \int_{|\boldsymbol{x}_p|>a_p} |\boldsymbol{x}_p| d\mu \Big)$$

$$\leq \sup_{p>n} \Big(\delta - \frac{\delta}{2n} - \Big(\delta - \frac{\delta}{2p} \Big) \Big) \leq \frac{\delta}{n} \to 0,$$

as $n \to \infty$. Hence, one can find a subsequence $(\boldsymbol{x}''_{n_k}) \subset (\boldsymbol{x}''_{n_i})$ which converges weakly so that

$$(\boldsymbol{x}_{n_{2k}} - \boldsymbol{x}_{n_{2k+1}}) \to 0,$$

weakly. Zero is a strong accumulation point of the convex envelope of the latter sequence. So, there exists an increasing sequence (k_j), and $a_i^{(j)} > 0$, with

$$\sum_{i=k_j+1}^{k_{j+1}} a_i^{(j)} = 1,$$

such that, if

$$\boldsymbol{z}_j = \sum_{i=k_j+1}^{k_{j+1}} a_i^{(j)} (\boldsymbol{x}_{n_{2i}} - \boldsymbol{x}_{n_{2i+1}})$$

then

$$\lim_{j \to \infty} \|\boldsymbol{z}_j - \boldsymbol{z}'_j\| = \lim_{j \to \infty} \|\boldsymbol{z}''_j\| = 0,$$

[11]See *Studia Mathematatica* 21(1962), 161-176.

where

$$\boldsymbol{z}'_j = \sum_{i=k_j+1}^{k_{j+1}} a_i^{(j)} \left(\boldsymbol{x}'_{n_{2i}} - \boldsymbol{x}'_{n_{2i+1}}\right), \quad \text{and} \quad \boldsymbol{z}''_j = \boldsymbol{z}_j - \boldsymbol{z}'_j.$$

Now, by the above Lemma, $(\boldsymbol{z}'_j)$ is a basic sequence equivalent to the standard basis of $\boldsymbol{l}_1$, and one can choose $(\boldsymbol{z}_j)$ such that

$$\sum_{j=1}^{\infty} \|\boldsymbol{z}^*_j\| \|\boldsymbol{z}''_j\| = \sum_{j=1}^{\infty} \|\boldsymbol{z}^*_j\| \|\boldsymbol{z}_j - \boldsymbol{z}'_j\| < 1.$$

By Bessaga-Pelczynski's Theorem[12], $(\boldsymbol{z}_j)$ is equivalent as a basic sequence to $(\boldsymbol{z}'_j)$. Thus we have obtained a basic sequence in $\boldsymbol{X}$ which is equivalent to the standard basis of $\boldsymbol{l}_1$. QED

In the next Theorem we turn to a characterization of B-convex Banach lattices and Banach spaces with unconditional bases.

Theorem 8.3.2. *Let $\boldsymbol{X}$ be either a Banach lattice, or a Banach space with unconditional basis. Then the following properties of $\boldsymbol{X}$ are equivalent:*

(i) The space $\boldsymbol{X}$ is B-convex;

(ii) $\boldsymbol{X}$ is reflexive;

(iii) $\boldsymbol{X}$ is superreflexive;

(iv) $\boldsymbol{X}$ does not have subspaces isomorphic to either $\boldsymbol{c}_0$, or to $\boldsymbol{l}_1$.

Proof. It is sufficient to prove the above theorem for Banach lattices because all isomorphically invariant properties of a Banach lattice are shared by Banach spaces with an unconditional basis. This follows from the fact that for every such space one can find an equivalent norm which makes the space isometrically isomorphic to a Banach lattice[13].

Now, the implication $(i) \implies (iv)$ is trivial because neither $\boldsymbol{c}_0$, not $\boldsymbol{l}_1$ are B-convex, and because B-convexity is preserved by subspaces. The implications $(iv) \implies (ii)$, and $(iv) \implies (iii)$ follow from James' Theorem (see Chapter 1). If $\boldsymbol{X}$ is superreflexive,

[12]See *Studia Mathematica* 17 (1958), 151-164.

[13]See, e.g., M.M. Day (1973), p. 73, Theorem 1.

then, by Enflo's Theorem (see Chapter 1), $\boldsymbol{X}$ admits an equivalent uniformly convex norm. Therefore, by Theorem 8.1.2, $\boldsymbol{X}$ is B-convex. QED

Not all B-convex spaces are reflexive. In particular, there exists a $(3,\epsilon)$-convex space which is not reflexive whenever $\epsilon < 1-(3^{-1/2}+2^{-1}(2/3)^{1/2})$[14]. However, we have a positive result for $(2,\epsilon)$-convex Banach spaces.

Theorem 8.3.3.[15] *If $\boldsymbol{X}$ is a $(2,\epsilon)$-convex Banach space for some $\epsilon>0$, then $\boldsymbol{X}$ is reflexive (and even superreflexive).*

Proof. Suppose that $\boldsymbol{X}$ is non-reflexive. For each sequence $(\boldsymbol{x}_j^*)\subset S_{X^*}$, and each $p_1,\ldots,p_{2n}\in\mathbf{N}$, define

$$S(p_1,\ldots,p_{2n},(\boldsymbol{x}_k^*)) := \Big\{\boldsymbol{x} : \forall k,i\in\mathbf{N},\ \frac{3}{4}\le(-1)^{i-1}\boldsymbol{x}_k^*(\boldsymbol{x})\le 1\Big\},$$

if $p_{2i-1}\le k\le p_{2i}$, and $1\le i\le n$. Let

$$K(n,(\boldsymbol{x}_j^*)) := \liminf_{p_1\to\infty}\big[\ldots\liminf_{p_{2n}\to\infty}\big[\inf\{\|\boldsymbol{z}\| : \boldsymbol{z}\in S(p_1,\ldots,p_{2n},(\boldsymbol{x}_k^*))\}\big]\ldots\big],$$

and

$$K_n := \inf\big\{K(n,(\boldsymbol{x}_j^*));\, \|\boldsymbol{x}_j^*\|=1,\ \forall j\big\}.$$

To show that $K_n<\infty$ let us suppose that $\{p_1,\ldots,p_{2n}\}$ is an increasing sequence of integers. It is known[16] that for each $r<1$ there exist $(\boldsymbol{z}_i)\subset B_X$, and $(\boldsymbol{x}_j^*)\subset S_{X^*}$, such that $\boldsymbol{x}_n^*(\boldsymbol{z}_i)>r$ if $n\le i$, and $\boldsymbol{x}_n^*(\boldsymbol{z}_i)=0$, if $n>i$. Let

$$\boldsymbol{w}=\sum_{j=1}^{n}(-1)^{j-1}\Big(-\boldsymbol{z}_{p_{2j-1}-1}+\boldsymbol{z}_{p_{2j}}\Big).$$

Then

$$\boldsymbol{x}_k^*(\boldsymbol{w})=\sum_{j=1}^{n}(-1)^{j-1}\Big(-\boldsymbol{x}_k^*(\boldsymbol{z}_{p_{2j-1}-1})+\boldsymbol{x}_k^*(\boldsymbol{z}_{p_{2j}})\Big)=\sum_{j=1}^{n}(-1)^{j-1}A_{k_j},$$

[14] See R.C. James (1974). Other versions can be found in R.C. James and J. Lindenstrauss (1975), and J. Farahat (1974/75).

[15] Due to R.C. James (1974). For another proof see A. Brunel and L. Sucheston (1974).

[16] See R. C. James, *Studia Mathematica* 23(1964), 205-216.

where $A_{kj} = 0$, if $k > p_{2j}$,

$$A_{kj} = \boldsymbol{x}_k^*(\boldsymbol{z}_{p_{2j}}), \quad \text{if} \quad p_{2j-1} \leq k \leq p_{2j},$$

and $|A_{kj}| < 1 - r$, if $k < p_{2j-1}$. Therefore, if $p_{2i-1} \leq k \leq p_{2i}$, then

$$(-1)^{i-1}\boldsymbol{x}_k^*(\boldsymbol{w}) = \boldsymbol{x}_k^*(\boldsymbol{z}_{p_{2i}}) + A_k, \quad \text{with} \quad |A_k| < n(1-r).$$

Therefore, to have $\boldsymbol{w} \in S(p_1, \ldots, p_{2n}, (\boldsymbol{x}_i^*))$ it is sufficient to have $\boldsymbol{x}_k^*(\boldsymbol{z}_{p_{2i}}) > 7/8$, and $n(1-r) < 1/8$. To accomplish this we can choose r as the larger of the $7/8$, and $1-1/(8n)$. Since $\|\boldsymbol{w}\| \leq 2n$, it follows that $K(n, (\boldsymbol{x}_i^*)) \leq 2n$, and, hence, $K_n \leq 2n$.

Now, we shall show that $\boldsymbol{X}$ is not $(2, \epsilon)$-convex for any $\epsilon > 0$. Since K_n is positive and monotonically increasing, for any number r satisfying $1 > r > 1 - \delta$, there is a $\delta > 0$, and an N, such that

$$(K_n - \delta)/(K_n + 2\epsilon) > r > 1 - \epsilon, \quad \text{if} \quad n > N. \tag{8.3.3}$$

Since $K_n - K_{n-1} \to 0$ if $\lim K_n$ is finite, and the conditions $K_n \leq 2n$, and $\lim K_n = +\infty$, imply that $\liminf K_n/K_{n-1} = 1$, it follows from (8.3.3) that there is an $m > N$ such that

$$(K_{m-1} - \delta)/(K_m + 2\epsilon) > 1 - \epsilon. \tag{8.3.4}$$

In view of the definition of the sequence $K - N$, there is a sequence $(\boldsymbol{x}_i^*) \subset S_{X^*}$ such that

$$K(m, (\boldsymbol{x}_i^*)) < K_m + \epsilon. \tag{8.3.5}$$

Also, by choosing $p_1, \ldots, p_{2m}$, and $q_1, \ldots, q_{2m}$ successively in the followiing order

$$p_1, q_1, p_2, p_3, q_2, q_3, p_4, p_5, \ldots, q_{2m-2}, q_{2m-1}, p_{2m}, q_{2m},$$

and in such a way that the above sequence is increasing, we can get increasing sequences $(p_1, \ldots, p_{2m})$, and $(q_1, \ldots, q_{2m})$, with the following three properties:

(a) $S(p_1, \ldots, p_{2m}; (\boldsymbol{x}_i^*))$, and $S(q_1, \ldots, q_{2m}; (\boldsymbol{x}_i^*))$, have elements $\boldsymbol{u}$, and $\boldsymbol{v}$, respectively, such that

$$\|\boldsymbol{u}\| \leq K(m, (\boldsymbol{x}_i^*)) + \delta, \quad \text{and} \quad \|\boldsymbol{v}\| \leq K(m, (\boldsymbol{x}_i^*)) + \delta;$$

(b) $K(m, (\boldsymbol{x}_i^*)) - \delta \leq \|\boldsymbol{z}\|$, if

$$z \in S(q_1, p_2, q_3, p_4, q_5, p_6, \ldots, q_{2m-1}, p_{2m}; (\boldsymbol{x}_i^*));$$

(c) $K(m-1, (\boldsymbol{x}_i^*)) - \delta \leq \|\boldsymbol{z}\|$, if

$$z \in S(p_3, q_2, p_5, q_4, p_7, q_6 \ldots, p_{2m-1}, q_{2m-2}; (\boldsymbol{x}_i^*)).$$

Then

$$(\boldsymbol{u}+\boldsymbol{v})/2 \in S(q_1, p_2, q_3, p_4, q_5, p_6, \ldots, q_{2m-1}, p_{2m}; (\boldsymbol{x}_i^*)),$$

and from (b) we have

$$\|\boldsymbol{u}+\boldsymbol{v}\|/2 > K(m, (\boldsymbol{x}_i^*)) - \delta. \tag{8.3.6}$$

Also,

$$(\boldsymbol{u}-\boldsymbol{v})//2 \in S(p_3, q_2, p_5, q_4, p_7, q_6 \ldots, p_{2m-1}, q_{2m-2}; (\boldsymbol{x}_i^*)),$$

and

$$\|\boldsymbol{u}-\boldsymbol{v}\|/2 > K(m-1, (\boldsymbol{x}_i^*) - \delta. \tag{8.3.7}$$

Finally, let $\boldsymbol{x} = \boldsymbol{u}/(K_m + 2\delta)$, and $\boldsymbol{y} = \boldsymbol{v}/(K_m + 2\delta)$. Then, from (a) and (8.3.5), we have $\|\boldsymbol{x}\|, \|\boldsymbol{y}\| \leq 1$. Since $K(m, (\boldsymbol{x}_i^*)) \geq K_m$, it follows from (8.3.6) and (8.3.3) (and (8.3.7), and (8.3.4), respectively) that

$$\|\boldsymbol{x}+\boldsymbol{y}|/2 > (K_m - \delta)/(K_m + 2\delta) > 1 - \epsilon,$$

and

$$\|\boldsymbol{x}-\boldsymbol{y}|/2 > (K_{m-1} - \delta)/(K_m + 2\delta) > 1 - \epsilon,$$

so that $\boldsymbol{X}$ is not $(2, \epsilon)$-convex, for any $\epsilon > 0$. QED

Corollary 8.3.1. *If $\boldsymbol{X}$ is a $(3, \epsilon)$-convex Banach space for some $\epsilon > 1/3$, then it is reflexive.*

Proof. The result is an immediate consequence of the above Theorem, Proposition 8.1.1, and Theorem 8.1.1. QED

On the other hand, there are reflexive (even locally uniformly convex) Banach spaces that are not B-convex.

Example 8.3.1. Let $n_i \uparrow \infty$, $p_i \downarrow 1$, and let $\boldsymbol{X}_i = \boldsymbol{l}_{p_1}^{(n_i)}$. Then $\boldsymbol{X}_i$ are uniformly convex, and hence locally uniformly convex. The space $\boldsymbol{X}_i$ is finite-dimensional and thus reflexive. Define $\boldsymbol{X} = (\boldsymbol{X}_i \oplus \boldsymbol{X}_2 \oplus \dots)_{l_2}$. Note that the $\boldsymbol{l}_2$ sum of reflexive spaces is reflexive, and the $\boldsymbol{l}_2$ sum of locally uniformly convex spaces is such as well[17]. Thus $\boldsymbol{X}$ is reflexive and locally uniformly convex. Pick $k \geq 2$, and $\epsilon > 0$. As $p \downarrow 1$, $k^{(1/p-1)} \uparrow 1$. Obviously, $\boldsymbol{X}_i$ is embedded isometrically in $\boldsymbol{X}$. In $\boldsymbol{X}_i$ take $\boldsymbol{x}_j = (0, \dots, 0, 1, 0, \dots, 0)$, with 1 in the j-th position. Then, let $\boldsymbol{y}_j = (0, \dots, 0, \boldsymbol{x}_j, 0, \dots, 0) \in \boldsymbol{X}$. Then $\|\boldsymbol{y}_j\| = \|\boldsymbol{x}_j\| = 1$, and for $\varepsilon_j = \pm 1$,

$$\left\|\sum_{j=1}^{k} \varepsilon_j \boldsymbol{y}_j\right\| = \left\|\sum_{j=1}^{k} \varepsilon_j \boldsymbol{x}_j\right\| = \left(\sum_{j=1}^{k} |\varepsilon_j|^{p_i}\right)^{1/p_i} = k^{1/p_i} > k(1-\epsilon),$$

so that $\boldsymbol{X}$ is not (k, ϵ)-convex for any $k \in \mathbf{N}, \epsilon > 0$, and, by definition, it is not B-convex.

Remark 8.3.1. Other characterizations of Banach spaces which are not B-convex are also possible[18]:

(a) The Banach space $\boldsymbol{X}$ is not B-convex if, and only if, there exists an $\epsilon \in (0,1)$, such that for each $k \geq 2$ there exist $\boldsymbol{x}_1^*, \dots, \boldsymbol{x}_k^* \in B_{X^*}$ such that, for each $\varepsilon_1, \dots, \varepsilon_k = \pm 1$ there exists an $\boldsymbol{x} \in B_X$ such that for each $j = 1, \dots, k$, $\boldsymbol{x}_j^*(\varepsilon_j \boldsymbol{x}) > \epsilon$.

(b) The Banach space $\boldsymbol{X}$ is not B-convex if, and only if, there exists an $\epsilon \in (0,1)$, such that for each $k \geq 2$ there exist $\boldsymbol{x}_1, \dots, \boldsymbol{x}_k \in B_X$ such that, for each $\varepsilon_1, \dots, \varepsilon_k = \pm 1$ there exists an $\boldsymbol{x}^* \in B_{X^*}$ such that for each $j = 1, \dots, k$, $\boldsymbol{x}^*(\varepsilon_j \boldsymbol{x}_j) > \epsilon$.

(c) If $\boldsymbol{X}$ is not B-convex then it contains a subspace with basis that is not B-convex either.

8.4 Classical weak and strong laws of large numbers in B-convex spaces

In view of Theorems 8.1.4 and 6.9.1, the validity of the classical Kolmogorov's Weak Law of Large Numbers characterizes B-convex

[17] See M.M. Day (1973), p. 31.
[18] See D.R. Brown (1974).

Banach spaces.

Theorem 8.4.1.[19] *A Banach space is B-convex if, and only if, for each sequence* (X_n) *of symmetric, independent, and identically distributed random vectors in* $\boldsymbol{X}$,

$$\lim_{n\to\infty} \frac{X_1+\cdots+X_n}{n} = 0, \qquad \text{in probability,}$$

if, and only if,

$$\lim_{N\to\infty} n\mathbf{P}(\|X_1\| > n) = 0.$$

A study of the following Strong Law of Large Numbers initiated investigation of B-convex spaces.

Theorem 8.4.2.[20] *The following properties of a Banach space* $\boldsymbol{X}$ *are equivalent:*

(i) The space $\boldsymbol{X}$ *is B-convex;*

(ii) For any sequence (X_n) *of independent, zero-mean random vectors in* $\boldsymbol{X}$ *with* $\sup_n \|X_n\|^2 < \infty$,

$$\lim_{n\to\infty} \frac{X_1+\cdots+X_n}{n} = 0, \qquad \text{with probability 1,}$$

(iii) For any bounded sequence $(\boldsymbol{x}_n) \subset \boldsymbol{X}$,

$$\lim_{n\to\infty} \frac{r_1\boldsymbol{x}_1+\cdots+r_n\boldsymbol{x}_n}{n} = 0, \qquad \text{in} \quad \boldsymbol{L}_1(\Omega,\mathcal{F},\mathbf{P};\boldsymbol{X}).$$

Proof. $(i) \implies (ii)$ By Theorem 8.1.4, the space $\boldsymbol{X}$ is of Rademacher-type p, for some $p \in (1,2]$. The boundedness of (X_n) in $\boldsymbol{L}_2(\boldsymbol{X})$ implies its boundedness in $\boldsymbol{L}_p(\boldsymbol{X})$. Therefore, the series $\sum_n \mathbf{E}\|X_n\|^p/n^p$ converges, and by Theorem 6.8.1, we obtain (ii).

$(ii) \implies (iii)$ This implication is obvious in view of the Lebesgue Dominated Convergence Theorem.

[19]Due to M.B. Marcus and W.A. Woyczyński (1977), and (1979).
[20]Due to A. Beck (1962), and (1963).

(iii) $\Longrightarrow$ *(i)* Suppose that $\boldsymbol{X}$ is not B-convex. Then, for each $n \in \mathbf{N}$, there exist $\boldsymbol{x}_1^n, \ldots, \boldsymbol{x}_{n^n}^n \in B_X$ such that

$$n^{-n}\mathbf{E}\Big\|\sum_{i=1}^{n^n} r_i \boldsymbol{x}_i^n\Big\| \geq 1/2.$$

Now, let us construct a sequence $(\boldsymbol{x}_n) \subset B_X$, as follows:

$$\boldsymbol{x}_j = \boldsymbol{x}_i^n, \text{ when } j = k_n + i,$$

where

$$i \in \{1, 2, \ldots, (n+1)^{n+1}\}, \text{ and } k_n = 1 + 2^2 + 3^3 + \cdots + n^n.$$

Then,

$$\mathbf{E}\Big\|\sum_{i=1}^{k_n} r_i \boldsymbol{x}_i\Big\| \geq \frac{1}{2}n^n - k_{n-1} \geq \frac{1}{2}n^n - (n-1)^n,$$

from which it follows that

$$\limsup_n \frac{1}{k_n}\mathbf{E}\Big\|\sum_{i=1}^{k_n} r_i \boldsymbol{x}_i\Big\| \geq \limsup_n \frac{1}{2n^n}\mathbf{E}\Big\|\sum_{i=1}^{k_n} r_i \boldsymbol{x}_i\Big\|$$

$$\geq \lim_n \frac{1}{2}\Big[\frac{1}{2} - \Big(1 - \frac{1}{n}\Big)^n\Big] > 0.$$

A contradiction. QED

It is possible to prove the equivalence *(i)* $\Longleftrightarrow$ *(ii)* using only basic definitions[21] Here we include only the proof of the implication *(ii)* $\Longrightarrow$ *(i)* because it gives an additional insight into the structure of B-convex spaces.

Alternative proof of (ii) $\Longrightarrow$ *(i)*. Assume, a contrario, that $\boldsymbol{X}$ is not B-convex, i.e., for each $k \in \mathbf{N}$, and each $\epsilon > 0$, there exist $\boldsymbol{x}_1, \ldots \boldsymbol{x}_k \in S_X$ such that, for all $\varepsilon_1, \ldots, \varepsilon_k = \pm 1$,

$$\|\varepsilon_1 \boldsymbol{x}_1 + \cdots + \varepsilon_k \boldsymbol{x}_k\| > k(1 - \epsilon).$$

[21] Due to A. Beck (1963).

Pick arbitrary sequences (β_i), and (δ_i), of real numbers converging to zero and proceed as follows: Choose $k_1 \in \mathbf{N}$ with $k_1 > (1 - \delta_1)/\delta_1$, and $\boldsymbol{x}_i^{(1()}, \ldots, \boldsymbol{x}_{k_1}^{(1)} \in S_X$ such that, for each $\varepsilon_1, \ldots, \varepsilon_{k_1} = \pm 1$,

$$\|\varepsilon_1 \boldsymbol{x}_1^{(1)} + \cdots + \varepsilon_{k_1} \boldsymbol{x}_{k_1}^{(1)}\| \geq k_1(1 - \beta_1).$$

Then, for each $n \in \mathbf{N}$, set $m_n = k_1 + \cdots + k_{n-1}$, and choose $k_n > m_n(1 - \delta_n)/\delta_n$, and $\boldsymbol{x}_1^{(n)}, \ldots \boldsymbol{x}_{k_n}^{(n)} \in \mathcal{S}_X$ such that, for every $\varepsilon_1, \ldots, \varepsilon_{k_n} = \pm 1$,

$$\|\varepsilon_1 \boldsymbol{x}_i^{(n)} + \cdots + \varepsilon_{k_n} \boldsymbol{x}_{k_n}^{(n)}\| \geq k_n(1 - \beta_n).$$

This gives us the inequalities,

$$\frac{k_n}{m_{n+1}} > 1 - \delta_n, \qquad \text{and} \qquad \frac{m_n}{m_{n+1}} < \delta_n.$$

For any integer i, we have the bounds $m_j < i \leq m_{j+1}$, for some value of j, i.e., $i = m_j + r$, where $1 \leq r \leq k_j$. Define $\boldsymbol{y}_i = \boldsymbol{x}_r^{(j)}$. This gives us a sequence $(\boldsymbol{y}_i) \subset \boldsymbol{X}$ which is uniformly bounded. On the other hand,

$$\left\|\frac{1}{m_{j+1}} \sum_{i=1}^{m_j+1} r_i \boldsymbol{x}_i\right\| \geq \left\|\frac{1}{m_{j+1}} \sum_{i=m_j+1}^{m_{j+1}} r_i(\omega) \boldsymbol{x}_i\right\| - \left\|\frac{1}{m_j} \sum_{i=1}^{m_j+1} r_i(\omega) \boldsymbol{x}_i\right\|$$

$$\geq \frac{1}{m_{j+1}} \|\varepsilon_1 \boldsymbol{x}_1^{(j)} + \cdots + \varepsilon_{k_j}^{(j)}\| - \frac{m_j}{m_{j+1}}$$

$$\geq k_j(1 - \beta_j)/m_{j+1} - m_j/m_{j+1} > (1 - \delta_j)(1 - \beta_j) - \delta_j.$$

Thus,

$$\limsup_n \left\|\sum_{i=1}^{n} X_i(\omega)\right\| n^{-1} = 1, \qquad \forall \omega \in \Omega.$$

A contradiction. QED

The following result demonstrates that there is a uniformity in the Law of Large Numbers in B-convex Banach spaces.

Theorem 8.4.3.[22] *If $\boldsymbol{X}$ is a B-convex Banach space, then there exists a sequence $(p_n), 0 < p_n \uparrow 1$, such that, for any $\epsilon \in (0,1)$, and any sequence $(\boldsymbol{x}_n) \subset B_X$,*

$$\mathbf{P}\Big(\sup_{m>n}\Big\|\frac{r_1\boldsymbol{x}_1 + \cdots + r_m\boldsymbol{x}_m}{m}\Big\| \le \epsilon\Big) > p_n, \qquad \forall n \in \mathbf{N}.$$

Proof. We proceed by contradiction and will show that if this result is false then the Strong Law of Large Numbers of Theorem 8.4.2 fails in $\boldsymbol{X}$ as well.

Suppose that there exist $\epsilon, \eta > 0$ such that, for each $n \in \mathbf{N}$, there exist independent random vectors $(X_i^{(n)}), i, n \in \mathbf{N}$ such that

$$\mathbf{P}\Big(\sup_{m>n}\Big\|\frac{X_1^{(n)} + \cdots + X_m^{(n)}}{m}\Big\| \le \epsilon\Big) < 1 - \eta.$$

We will construct inductively a pair of sequences of integers, (n_i), and (m_j), as follows: Choose $n_1 = 0$. Since

$$\mathbf{P}\Big(\sup_{m>n}\Big\|X_1^{(n_1)} + \cdots + X_m^{(n_1)}\Big\| \,/\, m \le \epsilon\Big) < 1 - \eta,$$

we can find $m_1 > n_1$ for which

$$\mathbf{P}\Big(\sup_{n_1<m<m_1}\Big\|X_1^{(n_1)} + \cdots + X_m^{(n_1)}\Big\| \,/\, m \le \epsilon\Big) < 1 - \eta,$$

Once $n_1, \ldots, n_{j-1}$, and $m_1, \ldots, m_{j-1}$ have been chosen, we define $p_{j-1} = m_1 + \cdots + m_{j-1}$, and pick $n_j = 1p_{j-1}/\epsilon$. Then

$$\mathbf{P}\Big(\sup_{m>n_j}\Big\|\sum_{i=1}^{m} X_i^{(n_j)}\Big\| \,/\, m \le \epsilon\Big) < 1 - \eta.$$

Choose $m_j > n_j$ so that

$$\mathbf{P}\Big(\sup_{n_j<m<m_j}\Big\|\sum_{i=1}^{m} X_i^{(n_j)}\Big\| \,/\, m \le \epsilon\Big) < 1 - \eta.$$

[22]Due to A. Beck (1976).

Now, we define a sequence of random vectors, (Y_k), as follows: Every integer $k \in \mathbf{N}$ lies between p_{j-1}, and p_j, for some value of j (take $p_0 = 0$). Thus we define $j(k)$, and $t(k)$, so that $k = p_{j(k)-1} + t(k)$, and $0 < t(k) \leq n_{j(k)}$. Then we set

$$Y_k = X_{t(k)}^{(n_{j(k)})},$$

and

$$A_j = \Big\{\omega : \sup_{n_j < m < m_j} \Big\| \sum_{i=1}^{m} X_i^{(n_j)} \Big\| \,/\, m \leq \epsilon \Big\}, \qquad \forall j \in \mathbf{N}.$$

By our construction, the events of the sequence (A_j) are independent, and $\mathbf{P}(A_j) > \eta, \forall j \in \mathbf{N}$. Thus, almost every ω lies in infinitely many of A_j's. For each $\omega \in A_j$, and for each $m \in (p_{j-1}, p_j]$, we have

$$\frac{1}{n}\Big\| X_1(\omega) + \cdots + X_m(\omega) \Big\| \geq \frac{1}{m}\Big\| \sum_{i=p_{j-1}+1}^{m} X_i(\omega) \Big\| - \frac{1}{m}\Big\| \sum_{i=1}^{p_{j-1}} X_i(\omega) \Big\|$$

$$\geq \frac{m - p_{j-1}}{m}\Big\| \frac{1}{m - p_{j-1}} \sum_{i=1}^{m-p_{j-1}} X_i^{(n_j)}(\omega) \Big\| - \frac{p_{j-1}}{m}$$

$$\geq \Big(1 - \frac{p_{j-1}}{p_{j-1} + n_j}\Big)\Big\| \frac{1}{m - p_{j-1}} \sum_{i=1}^{m-p_{j-1}} X_i^{(n_j)}(\omega) \Big\| - \frac{p_{j-1}}{p_{j-1} + n_j}$$

$$\geq \Big(1 - \frac{\epsilon}{3}\Big)\Big\| \frac{1}{m - p_{j-1}} \sum_{i=1}^{m-p_{j-1}} X_i^{(n_j)}(\omega) \Big\| - \frac{\epsilon}{3},$$

since $n_j > 3p_{j-1}/\epsilon$, so that $p_{j-1}/(p_{j-1} + n_j).\epsilon/3$. Since

$$\sup_{n_j < m - p_{j-1} < m_j} \Big\| \frac{1}{m - p_{j-1}} \sum_{i=1}^{m-p_{j-1}} X_i^{(n_j)}(\omega) \Big\| > \epsilon,$$

we obtain the inequality,

$$\sup_{p_{j-1} + n_j < m \leq p_j} \Big\| \frac{1}{m} \sum_{i=1}^{m} Y_i(\omega) \Big\| \geq \frac{2}{3}\epsilon - \frac{\epsilon}{3} = \frac{\epsilon}{3}.$$

It follows that, for almost all ω's (i.e., those which lie in infinitely many A_j's) we have

$$\frac{1}{m}\|Y_i(\omega)+\cdots+Y_m(\omega)\| \geq \frac{\epsilon}{3},$$

for infinitely many m's. A contradiction. QED

Finally, in the case of (k,ϵ)-convex spaces it is possible to obtain more precise uniform Laws of Large Numbers for convergence in $\boldsymbol{L}_p(\boldsymbol{X})$.

Theorem 8.4.4.[23] *Let $1 \leq p < q \leq \infty$, $2 \leq q$, $k \geq 2$, and $\epsilon > 0$. Then there exists a sequence $b_n = b_n(p,q,k,\epsilon) \in \mathbf{R}$, $n \in \mathbf{N}$, with $\lim_n b_n = 0$, and such that, if $\boldsymbol{X}$ is (k,ϵ)-convex, and $X_1,\ldots,X_n$ are independent, zero-mean random vectors in $\boldsymbol{X}$ with $\sup_i \mathbf{E}\|X_i\|^q \leq M < \infty$, then*

$$\left(\mathbf{E}\Big\|\frac{1}{n}\sum_{i=1}^n X_i\Big\|^p\right)^{1/p} \leq Mb_n.$$

Proof. By assumption, $\mathbf{E}\|X_n\|^2 \leq 1 + M^q$, if $q < \infty$, and $\mathbf{E}\|X_n\|^2 \leq M^2$, if $q = \infty$. Hence, by the Strong Law of Large Numbers (Theorem 8.4.2),

$$\lim_{n\to\infty}\Big\|\frac{1}{n}\sum_{i=1}^n X_i\Big\| = 0, \qquad \text{a.s.}$$

Also,

$$\left(\mathbf{E}\Big\|\frac{1}{n}\sum_{i=1}^n X_i\Big\|^q\right)^{1/q} \leq \frac{1}{n}\left(\sum_{i=1}^n \mathbf{E}\|X_i\|^q\right)^{1/q} \leq M,$$

and by the Lebesgue Dominated Convergence Theorem,

$$\lim_{n\to\infty}\mathbf{E}\Big\|\frac{1}{n}\sum_{i=1}^n X_i\Big\|^p = 0. \tag{8.4.1}$$

[23] Due to D.P. Giesy (1966).

Now, define

$$b_n = \sup\Big(\mathbf{E}\Big\|\frac{1}{n}\sum_{i=1}^{n} X_i\Big\|^p\Big)^{1/p},$$

where the supremum extends over all (k,ϵ)-convex Banach spaces $\boldsymbol{X}$, all probability spaces $(\Omega, \mathcal{F}, \mathbf{P})$, and all sequences $X_1, \dots, X_n$ of independent random vectors in $\boldsymbol{X}$ with $\mathbf{E}X_i = 0$, and $\mathbf{E}\|X_i\|^q \leq 1, 1 \leq i \leq n$. Then b_n depends only on p, q, k, ϵ, and $0 \leq b_n \leq \infty$. We shall show that $\alpha = \limsup_n b_n = 0$.

Suppose that $\alpha > 0$. Choose $\beta > 0$ such that $2\beta < \alpha$. We will now construct a sequence (X_i) of independent, zero-mean random vectors in $\boldsymbol{X}$, with $\mathbf{E}\|X_i\|^{\leq}1$, for which

$$\limsup_n \mathbf{E}\Big\|\frac{1}{n}\sum_{i=1}^{n} X_i\Big\|^p \geq \beta > 0.$$

This will contradict (8.4.1)

Since $\limsup_n b_n > 2\beta$, for each $n \in \mathbf{N}$, there exists $m(n) \geq n$, the (k,ϵ)-convex spaces $\boldsymbol{X}_n$, probability spaces $(\Omega_n, \mathcal{F}_n, \mathbf{P}_n)$, and zero-mean, independent random vectors $Y_1^{(n)}, \dots, Y_{m(n)}^{(n)}$ on Ω_n, with $\mathbf{E}\|Y_i^{(n)}\|^q \leq 1$, such that

$$\Big\|\frac{1}{m(n)}\sum_{i=1}^{m(n)} Y_i^{(n)}\Big\| \geq 2\beta.$$

Let $\boldsymbol{X} = (\boldsymbol{X}_1 \oplus \boldsymbol{X}_2 \oplus \dots)_{l_2}$, and $(\Omega, \mathcal{F}, \mathbf{P}) = \prod(\Omega_n, \mathcal{F}_n, \mathbf{P}_n)$. Then, by Theorem 8.2.6, $\boldsymbol{X}$ is B-convex. For all $n \in \mathbf{N}$, and $1 \leq i \leq m(n)$, define random vectors $X_i^{(n)} : \Omega \mapsto \boldsymbol{X}$, by the formula

$$X_i^{(n)}(\omega_1, \omega_2, \dots) = (\boldsymbol{x}_1, \boldsymbol{x}_2, \dots),$$

where $\boldsymbol{x}_n = Y_i(\omega_n)$, and $\boldsymbol{x}_i = 0$, for $j \neq n$. Then, for all $n \in \mathbf{N}$, and $1 \leq i \leq m(n)$, we have $\mathbf{E}(X_i^{(n)}) = 0$, and $\mathbf{E}\|X_i^{(n)}\|^q \leq 1$, and

$$\mathbf{E}\Big\|\frac{1}{m(n)}\sum_{i=1}^{m(n)} X_i^{(n)}\Big\|^p = \mathbf{E}\Big\|\frac{1}{m(n)}\sum_{i=1}^{m(n)} Y_i^{(n)}\Big\|^p \geq (2\beta)^p.$$

The random vectors $\{X_i^{(n)} : 1 \le i \le m(n), n \ge 1\}$ form a set of independent $\boldsymbol{X}$-valued random vectors.

Let $n_1 = 1, p_1 = m(n_1)$, and $X_i = X_i^{(n_i)}$, for $1 \le i \le m(n_1)$. Then

$$\mathbf{E}\Big\|\frac{1}{p_1}\sum_{i=1}^{p_1} X_i\Big\|^p \ge (2\beta)^p > \beta^p.$$

Suppose we have chosen $p_1 < p_2 < \cdots < p_s$, and a sequence $X_i, 1 \le i \le p_s$, such that

$$\mathbf{E}\Big\|\frac{1}{p_j}\sum_{i=1}^{p_j} X_i\Big\|^p \ge \beta^p, \qquad 1 \le j \le s.$$

Let

$$\gamma_s = \mathbf{E}\Big\|\frac{1}{p_s}\sum_{i=1}^{p_s} X_i\Big\|^p$$

Choose $n_{s=1}$ so large that if we let $m_{s+1} = m(n_{s=1})$ then

$$\big(2m_{s+1} - p_s(\gamma_s/\beta)\big) \;/\; (m_{s+1} + p_s) \ge 1. \tag{8.4.2}$$

For $1 \le i \le m(n_{s+1})$ define $X_{p_s+i} = X_i^{(n_{s+1})}$. Let $p_{s+1} = m(n_{s+1}) + p_s$. Then, from (8.4.2), we have

$$\big(2m_{s+1} - p_s(\gamma_s/\beta)\big) \;/\; p_{s+1} \ge 1. \tag{8.4.3}$$

Finally, in view of (8.4.3), we obtain the inequality,

$$\begin{aligned}\left(\mathbf{E}\Big\|\frac{1}{p_{s+1}}\sum_{i=1}^{p_{s+1}} X_i\Big\|^p\right)^{1/p} &\ge \frac{1}{p_{s+1}}\left(\mathbf{E}\Big\|\sum_{i=p_s+1}^{p_{s+1}} X_i\Big\|^p\right)^{1/p} \\ &\quad - \frac{1}{p_{s+1}}\left(\mathbf{E}\Big\|\sum_{i=1}^{p_s} X_i\Big\|^p\right)^{1/p} \\ = \frac{m_{s+1}}{p_{s+1}}\frac{1}{m_{s+1}}\left(\mathbf{E}\Big\|\sum_{i=1}^{m_{s+1}} X_i^{(n_{s+1})}\Big\|^p\right)^{1/p} &- \frac{p_s}{p_{s+1}}\frac{1}{p_s}\left(\mathbf{E}\Big\|\sum_{i=1}^{p_s} X_i^{(n_{s+1})}\Big\|^p\right)^{1/p} \\ \ge \frac{m_{s+1}}{p_{s+1}}(2\beta) - \frac{p_s}{p_{s+1}}\gamma_s &= \beta\big(2m_{s+1} - p_s(\gamma_s/\beta)\big)/p_{s+1} \ge \beta.\end{aligned}$$

Hence, by induction, we have chosen from the independent family $X_i^{(n)} : 1 \le i \le m(n), n \ge 1\}$ of random vectors in $\boldsymbol{X}$, with zero-means, and q-moments bounded by 1, a sequence (X_n) such that $\limsup_n n^{-1}(\mathbf{E}\|X_1 + \cdots + X_n\|^p)^{1/p} \ge \beta$. A contradiction.

Hence $\lim_{n\to\infty} b_n = 0$. To complete the proof for an arbitrary bound M it is sufficient to use the homegeneity of the q-norm. QED

8.5 Laws of large numbers for weighted sums and not necessarily independent summands

We begin by proving a Law of Large Numbers for nonstandard weighted averages.

Theorem 8.5.1.[24] *Let $1 \le p < 2$, $p < q$, and let $\boldsymbol{X}$ be a (k, ϵ)-convex Banach space with $k(1-\epsilon) < k^{1/p}$. If (X_n) is a sequence of independent random vectors in $\boldsymbol{X}$, bounded in $\boldsymbol{L}_q(\boldsymbol{X})$, then*

$$\lim_{n\to\infty} \frac{1}{n^{1/p}} \Big(\mathbf{E}\Big\|\sum_{k=1}^{n} X_k\Big\|^q\Big)^{1/q} = 0, \tag{8.5.1}$$

and

$$\lim_{n\to\infty} \frac{1}{n^{1/p}} \sum_{i=1}^{n} X_i = 0, \qquad \text{a.s} \tag{8.5.2}$$

Proof. By our assumption, for each $\boldsymbol{x}_1, \ldots, , \boldsymbol{x}_k \in B_X$,

$$\inf\Big\{\Big\|\sum_{i=1}^{k} \varepsilon_i \boldsymbol{x}_i\Big\| : \varepsilon = \pm 1\Big\} \le k(1-\epsilon) < k^{1/p},$$

so that the embedding $\boldsymbol{l}_1 \mapsto \boldsymbol{l}_p$ is not finitely representable in $\boldsymbol{X}$. Therefore $\boldsymbol{X}$ is of stable-type p by Theorem 6.5.1, so that $\boldsymbol{L}_q(\boldsymbol{X})$

[24]Due to G. Pisier and B. Maurey (1976).

is of stable-type p by Theorem 6.5.3. Without loss of generality we can assume that $\boldsymbol{X}_i$'s are symmetric. Then, by Theorem 6.9.3,

$$\lim_{n\to\infty} \frac{1}{n^{1/p}} \Big(\mathbf{E}\Big\|\sum_{k=1}^{n} X_i\Big\|^q\Big)^{1/q} = \lim_{n\to\infty} \frac{1}{n^{1/p}} \Big(\mathbf{E}\Big\|\sum_{k=1}^{n} r_i X_i\Big\|^q\Big)^{1/q} = 0,$$

(r_i)-almost surely. This proves (8.5.1). Statement (8.5.2) can be proven in a similar fashion. QED

Below we discuss a Strong Law of Large Numbers for general weighted sums of independent random vectors in B-convex spaces. However, its validity is restricted by rather stringent condition on weights.

Theorem 8.5.2.[25] *Let* $\boldsymbol{X}$ *be a B-convex Banach space and let* $(a_{nk}) \subset \mathbf{R}$, $n, k \in \mathbf{N}$, *be an array such that*

$$\lim_{n\to\infty} a_{nk} = 0, \qquad \forall k \in \mathbf{N}, \tag{8.5.3}$$

$$\sum_{k=1}^{\infty} |a_{nk}| \le 1, \qquad \forall n \in \mathbf{N}, \tag{8.5.4}$$

$$a_{nk} \ge 0, \quad \text{and} \quad \lim_{n\to\infty} \Big[\sum_{k=1}^{n} a_{nk} - n \max_{1\le k\le n} a_{nk}\Big] = 0, \tag{8.5.5}$$

$$\max_{1\le k\le n} a_{nk} = O(n^{-\alpha}), \tag{8.5.6}$$

where $0 < 1/\alpha < p-1$, *for some* $p > 1$. *Then, if* (X_n) *is a sequence of zero-mean, independent random vectors in* $\boldsymbol{X}$, *with* $\sup_n \mathbf{E}\|X_n\|^p = M < \infty$, *then, almost surely,*

$$\lim_{n\to\infty} \sum_{k=1}^{n} a_{nk} X_k = 0.$$

Proof. Step 1. Suppose additionally that X_n's are symmetric and bounded by 1. Denote, $\beta_n = \min_{1\le k\le n} a_{nk}$. Then, we have

$$0 \le \operatorname{ess\,sup} \overline{\lim}_n \Big\|\sum_{k=1}^{n} a_{nk} X_k\Big\|$$

[25] Due to W.J. Padgett and R.L. Taylor (1975).

$$\leq \text{ess sup}\,\overline{\lim}_n \Big(\Big\|\sum_{k=1}^{n}(a_{nk}-\beta_n)X_k\Big\| + \Big\|\sum_{k=1}^{n}\beta_n X_k\Big\|\Big)$$

$$\leq \text{ess sup}\,\overline{\lim}_n \sum_{k=1}^{n}(a_{nk}-\beta_n)\|X_k\| + \text{ess sup}\,\overline{\lim}_n \Big\|\frac{1}{n}\sum_{k=1}^{n}X_k\Big\|,$$

because the inequality, $n\beta_n \leq a_{n1}+\cdots+a_{nn} \leq 1$, implies that $\beta_n \leq 1/n$. On the other hand, the Strong Law of Large Numbers (Theorem 8.4.1) implies that the last term is zero. Since $\|X_k\| \leq 1$, by (8.5.5) we also see that the next to the last term vanishes as well. Therefore,

$$\text{ess sup}\,\overline{\lim}_n \Big\|\sum_{k=1}^{n}a_{nk}X_k\Big\| = 0.$$

Step 2. Now, we drop the assumption of uniform boundedness of (X_n) while preserving the symmetry assumption. Suppose that $\mathbf{E}\|X_n\|^p \leq M = 1$. For a positive integer q, define

$$X_n' = X_n, \quad \text{and} \quad X_n'' = 0, \quad \text{if} \quad \|X_n\| \leq q,$$

$$X_n' = 0, \quad \text{and} \quad X_n'' = X_n, \quad \text{if} \quad \|X_n\| > q.$$

The random sequences (X_n'), and (X_n''), are independent, and

$$\frac{1}{q}\Big\|\sum_{k=1}^{n}a_{nk}X_n'\Big\| = \Big\|\sum_{k=1}^{n}a_{nk}\frac{X_n'}{q}\Big\| \longrightarrow 0$$

as $n\to\infty$, in view of Step 1. Also, for each $n\in\mathbf{N}$,

$$\mathbf{E}\|X_n''\| \leq q^{1-p}\mathbf{E}\|X_n\|^p \leq q^{1-p}. \tag{8.5.7}$$

Define a probability density function of a random variable ξ as follows: $f_\xi(s) = 0$, for $s \leq 1$, and $= p/s^{1+p}$, for $s > 1$. Then, for each $\alpha > 0$, $\mathbf{P}(|\xi| \geq s) = 1$, if $s \leq 1$, and $= 1/\alpha^p$, if $s > 1$. Also

$$\mathbf{E}|\xi|^{1+1/\alpha} = \int_1^\infty t^{1+1/\alpha}t^{p-1}p\,dt < \infty, \tag{8.5.8}$$

since $1/\alpha < p-1$. For $s \geq 1$, in view of the above tail estimate,

$$\mathbf{P}\big(\big|\|X_k''\| - \mathbf{E}\|X_k''\|\big| \geq s\big) \leq \mathbf{P}\big(\|X_k''\| \geq s\big)$$

$$\leq \mathbf{E}\|X_k''\|/s^p \leq \mathbf{E}\|X_k\|/s^p \leq s^{-p} = \mathbf{P}(|\xi| \geq s), \qquad (8.5.9)$$

since the inequality $(\mathbf{E}\|X_k''\| - \|X_k''\|) \geq s)$ is impossible if $a \geq 1$.

Now the symmetry assumption can be removed by the standard symmetrization procedure.[26] QED

To a certain extent the assumption of independence can also be removed from the Strong Law of Large Numbers in B-convex Banach spaces.

Definition 8.5.1. A family of random vectors $(X_\alpha), \alpha \in A$, is said to be *mutually symmetric* if, for every $\alpha \in A$, and every set $S \in \sigma(X_\beta, \beta \in A, \beta \neq \alpha)$, $I_S X_\alpha$ is symmetric. Equivalently, if for every $(n+1)$-tuple $X_0, \ldots, X_n \subset (X_\alpha))$, and every $B_0, \ldots, B_n \in \mathcal{B}(\boldsymbol{X})$

$$\mathbf{P}\Big(X_0^{-1}(B_0) \cap X_1^{-1}(B_1) \cap \cdots \cap X_n^{-1}(B_n)\Big)$$

$$= \mathbf{P}\Big(X_0^{-1}(-B_0) \cap X_1^{-1}(B_1) \cap \cdots \cap X_n^{-1}(B_n)\Big),$$

then the family (X_α) is mutually symmetric. Evidently, an independent family of symmetric random vectors is mutually symmetric.

Theorem 8.5.3. [27] *If $\boldsymbol{X}$ is a B-convex Banach space, and (X_i) is a sequence of mutually symmetric random vectors in $\boldsymbol{X}$ such that $\|X_i(\omega)\| \leq 1$, for each $i \in \mathbf{N}$, and $\omega \in \Omega$, then, almost surely,*

$$\lim_{n \to \infty} \frac{X_1 + \cdots + X_n}{n} = 0.$$

Proof. Consider first the case when each X_i takes on only countably many values. Let $\epsilon > 0$. We shall show that the set of points for which the lim sup of Cesaro averages is greater than ϵ has probability 0.

Let (p_n) be the sequence defined in Theorem 8.4.3. For each $0 \leq n \leq m$ consider every sequence $(\boldsymbol{y}_1, \ldots, \boldsymbol{y}_m)$ of possible values of $X_1, \ldots, X_m$, respectively. We define $F(\boldsymbol{y}_1, \ldots, \boldsymbol{y}_m)$ as the

[26]See, e.g., A. Beck (1963).

[27]Due to A. Beck (1976).

set of all sample points where $X_i = \pm \boldsymbol{y}_i$, for each $i = 1, \dots, m$. Note that there are at most countably many sets $F(\boldsymbol{y}_1, \dots, \boldsymbol{y}_m)$. In each such set there are exactly 2^m different m-tuples which X_i can take as values, corresponding to 2^m possible ways of assigning $+$, and $-$ signs to the $\boldsymbol{y}_i$'s. By the condition of mutual symmetry, each of these 2^m sets has the same probability, $2^{-m}\mathbf{P}(F(\boldsymbol{y}_1, \dots, \boldsymbol{y}_m))$. Thus, if we restrict our attention to the set $F(\boldsymbol{y}_1, \dots, \boldsymbol{y}_m)$, and the conditional probability on that set, the random vectors $X_1, \dots, X_n$, defined there satisfy the hypotheses of Theorem 8.4.3. Therefore,

$$\mathbf{P}\Big(\sup_{n<k\le m}\Big\|\frac{1}{k}\sum_{i=1}^{k} X_i\Big\| \le \epsilon \;\Big|\; F(\boldsymbol{y}_1, \dots, \boldsymbol{y}_m)\Big) \ge p_n.$$

Summing over all sets $F(\boldsymbol{y}_1, \dots, \boldsymbol{y}_m)$ we obtain the inequality,

$$\mathbf{P}\Big(\sup_{n<k\le m}\Big\|\frac{1}{k}\sum_{i=1}^{k} X_i\Big\| \le \epsilon\Big) \ge p_n.$$

The event on the left-hand side of this inequality increases monotonically with increasing m, which implies that

$$\mathbf{P}\Big(\sup_{n<k}\Big\|\frac{1}{k}\sum_{i=1}^{k} X_i\Big\| \le \epsilon\Big) \ge p_n.$$

Since $p_n \to 1$, as $n \to \infty$, we obtain the equality,

$$\mathbf{P}\Big(\overline{\lim}_{n\to\infty}\Big\|\frac{1}{n}\sum_{i=1}^{n} X_i\Big\| \le \epsilon\Big) = 1.$$

Since we have demonstrated the validity of the above equality for each $\epsilon > 0$, we also have the equality,

$$\mathbf{P}\Big(\overline{\lim}_{n\to\infty}\Big\|\frac{1}{n}\sum_{i=1}^{n} X_i\Big\| = 0\Big) = 1.$$

Thus $(X_1 + \cdots + X_n)/n$ converges to zero almost surely for the restricted class of sequences (X_n).

To show the Theorem in the general case choose $\eta > 0$. Let $(\boldsymbol{y}_i) \subset B_X$ bedense in B_X, with $\boldsymbol{y}_1 = 0$. For each $i \in \mathbf{N}$, let B_i be the open ball of radius η, centered at $\boldsymbol{y}_i$, and define

$$A_i = \big(B_i \cup (-B_i)\big) - \bigcup_{j<i} A_j.$$

Then, the A_i's cover B_X, A_i's are Borel, and $B_i \cup (-B_i) \subset B_1$.

Define a sequence (Y_i) of mutually symmetric random variables in $\boldsymbol{X}$ as follows:

$$Y_i(\omega) = \begin{cases} \boldsymbol{y}_j & \text{if } X_i(\omega) \in Aj \cap B_j,\ j \in \mathbf{N} \\ -\boldsymbol{y}_j & \text{if } X_i(\omega) \in Aj \cap (-B_j), \end{cases}$$

and let $Z_i(\omega) = X_i(\omega) - Y_i(\omega)$. Then Y_i's satisfy the assumptions of the first case considered above, and $\|Z_i(\omega)\| < \eta$, for each $i \in \mathbf{N}$, and each $\omega \in \Omega$. Now

$$\overline{\lim}_n \Big\| \frac{1}{n} \sum_{i=1}^{n} X_i(\omega) \Big\| \leq \overline{\lim}_n \Big\| \frac{1}{n} \sum_{i=1}^{n} Y_i(\omega) \Big\| + \overline{\lim}_n \Big\| \frac{1}{n} \sum_{i=1}^{n} Z_i(\omega) \Big\| \leq 0 + \eta,$$

for almost all $\omega \in \Omega$. Since η was arbitrary, this proves the Theorem. QED

8.6 Ergodic properties of B-convex spaces

Definition 8.6.1. (*i*) A sequence $(\boldsymbol{x}_n)$ in a Banach space $\boldsymbol{X}$ is said to be *stable* (with the limit $\boldsymbol{x}_\infty$) if there exists $\boldsymbol{x}_\infty \in \boldsymbol{X}$ such that

$$\big\| n^{-1}(\boldsymbol{x}_{k_1} + \boldsymbol{x}_{k_2} + \cdots + \boldsymbol{x}_{k_n}) - \boldsymbol{x}_\infty \big\| \to 0, \qquad \text{as} \qquad n \to \infty, \tag{8.6.1}$$

uniformly in the set $\mathcal{K}$ of all strictly increasing sequences $(k_n) \subset \mathbf{N}$. A Banach space is said to be *stable* if every bounded sequence in it contains a stable subsequence.

(*ii*) A Banach space $\boldsymbol{X}$ is said to have the *Banach-Saks property* if, for every bounded sequence $(\boldsymbol{x}_n) \subset \boldsymbol{X}$, there exists a subsequence $(\boldsymbol{x}_{k_n}$, and a $\boldsymbol{x}_\infty \in \boldsymbol{X}$, such that (8.6.1) is satisfied.

(*iii*) A Banach space $\boldsymbol{X}$ is said to be *ergodic* if, for each linear isometry $T : \boldsymbol{X} \mapsto \boldsymbol{X}$, and each $\boldsymbol{x} \in \boldsymbol{X}$, the following limit exists:

$$\lim_{n\to\infty} \frac{T^0\boldsymbol{x} + T^1\boldsymbol{x} + \cdots + T^{n-1}\boldsymbol{x}}{n}.$$

Remark 8.6.1. (*a*) In standard ergodic theory a measure preserving transformation τ is stable if for every $f \in \boldsymbol{L}_2$ the sequence $(f\tau^n)$ is stable in $\boldsymbol{L}_2$. If the measure space is finite the transformation τ is mixing if, and only if, it is stable and ergodic.

(*b*) In general, the stability is equivalent with the Banach-Saks property,[28] the Banach-Saks property implies the reflexivity.[29] It is a classical result that the reflexivity implies the ergodicity.[30]

Theorem 8.6.1.[31] *If a Banach space is* $(2, \epsilon)$*-convex then it is stable (and also superstable).*

Proof. The main idea of the proof is to construct for each non-stable space $\boldsymbol{X}$ a Banach space $\boldsymbol{G}$ which is finitely representable in $\boldsymbol{X}$, thus also $(2, \epsilon)$-convex, and which contains a copu of $\boldsymbol{c}_0$. This will lead to a contradiction in view of Example 8.1.1 ((c).

Recall that by Brunel-Sucheston Theorem (see Chapter 1), for every bounded sequence $(\boldsymbol{x}_n)$ there exists a subsequence $(\boldsymbol{e}_n) \subset (\boldsymbol{x}_n)$, and a norm $|.|$ on the (non-closed) span of $(\boldsymbol{e}_n)$, which is invariant under translations and spreading, and such that the completion $\boldsymbol{F}$ of the span of $(\boldsymbol{e}_n)$ under the norm $|.|$ is finitely representable in $\boldsymbol{X}$.

To complete the proof of the above theorem we shall have need for the following four Lemmas.

Lemma 8.6.1. *If the shift* $T : \boldsymbol{e}_n \mapsto \boldsymbol{e}_{n+1}$ *is ergodic in* $\boldsymbol{F}$, *then there exists a subsequence of* $(\boldsymbol{e}_n)$ *which is stable in* $\boldsymbol{X}$.

Proof. For each $\epsilon > 0$, there exists an integer, $N \in \mathbf{N}$, such

[28] See P. Erdös and M. Magidor (1976).

[29] See T. Nishiura and D. Waterman (1963).

[30] See, e.g., N. Dunford and J.T. Schwartz (1958).

[31] Due to A. Brunel (1973/74), and A. Brunel and L.Sucheston (1974), and (1975).

that, for all $p, q \geq N$,

$$\Big\|\frac{1}{p}\sum_{i=1}^{p} \boldsymbol{e}_i - \frac{1}{q}\sum_{i=p+1}^{p+q} \boldsymbol{e}_i\Big\| < \epsilon.$$

Also, for each $\epsilon > 0$, and all $p, q \geq N$, the exists an integer $M \in \mathbf{N}$ such that, for all sequences $(n_i) \subset \mathbf{N}, M \leq n_1 < n_2 < \cdots < n_p < n_{p+1} < \cdots < n_{p+q}$, we have

$$\Big\|\frac{1}{p}\sum_{i=1}^{p} \boldsymbol{e}_{n_i} - \frac{1}{q}\sum_{i=p+1}^{p+q} \boldsymbol{e}_{n_i}\Big\| < \epsilon. \tag{8.6.2}$$

Now, let $\epsilon = 2^{-n}$. Given N, choose a sequence $(P_n) \subset \mathbf{N}$ satisfying the conditions, $P_n \geq N, P_n > nP_{n-1}$. Finally, take $p = P_n, q = P_{n+1}$, and take the corresponding M, which we will call ν_n, and define

$$v_n = \sum_{j=1}^{n}(\nu_j + P_j).$$

Consider a sequence

$$\boldsymbol{a}_n = \frac{1}{P_n}\sum_{j\leq P_n} \boldsymbol{e}_{v_N+j} \in \boldsymbol{X}.$$

The inequality (8.6.2) implies that $\|\boldsymbol{a}_n - \boldsymbol{a}_{n+1}\| < 2^{-n}$, so that the sequence $(\boldsymbol{a}_n)$ converges to an $\bar{\boldsymbol{a}} \in \boldsymbol{X}$. Consider the terms, $\boldsymbol{e}_i - \boldsymbol{a}$, with the indices appearing in the sequence

$$v_1{+}1, v_1{+}2, \ldots, v_1{+}P_1, v_1{+}1, \ldots, v_2{+}P_2, \ldots, , v_n{+}1, \ldots, v_n{+}P_n, \ldots$$

and call them $\boldsymbol{e}_i - \boldsymbol{a}, \boldsymbol{y}_1, \boldsymbol{y}_2, \ldots$, taking them in that order. The inequality (8.6.2) shows that

$$\Big\|\frac{1}{p_n}\sum_{i=1}^{P_n} \boldsymbol{e}_{m_i} - \bar{\boldsymbol{a}}\Big\| \leq 2^{-n+3}, \tag{8.6.3}$$

whenever $v_n < m_1 < m_2 < \cdots < m_{P_n}$. Now, let $(i_n) \subset \mathbf{N}$ be strictly increasing, and consider the sequence $\boldsymbol{y}_{i_1}, \boldsymbol{y}_{i_2}, \ldots, \boldsymbol{y}_{i_n}$. Put $\boldsymbol{z}_l := \boldsymbol{y}_{i_l}$, for $l - 1, 2, \ldots, n$. Define $k \in \mathbf{N}$ by the inequality

$$P_1 + \cdots + P_k \leq n < P_1 + \cdots + P_{k+1},$$

and $m = n - (P_1 + \cdots + P_k)$. The Euclidean sieve,

$$m = d_k P_k + Q_k, \qquad Q_k < P_k,$$

$$Q_k = d'_k P_{k-1} + Q'_k, \qquad Q_k < P_{k-1},$$

defines the integers, $d_k, d'_k, Q_k.Q'_k$. Now, it follows from (8.6.3) that

$$\sum_{k=1}^{n} \boldsymbol{z}_k = (\boldsymbol{z}_1 + \cdots + z_{P_1}) + (z_{P_1+1} + \cdots + \boldsymbol{z}_{P_1+P_2})$$

$$+ \cdots + (z_{P_1+\cdots+P_{k-1}} + \cdots + z_{P_1+\cdots+P_k}) + \cdots + (\cdots + \boldsymbol{z}_n),$$

is bounded in the norm $\|.\|$ by

$$(P_1 2^{3-1} + \ldots P_k 2^{3-k}) + d_k P_k 2^{3-k} + d'_k P_{k-1} 2^{3-(k-1)} + Q'_k.$$

Hence, one obtains the inequality

$$\Big\| \frac{1}{n} \sum_{j=1}^{n} \boldsymbol{y}_{i_j} \Big\| \leq 8 \Big(\sum_{j=1}^{k} P_j 2^{-j} \Big) \;/\; \Big(\sum_{j=1}^{k} P_j \Big) + 2^{3-k} + 2^{3-(k-1)} + P_{k-1}/P_k,$$

which shows the stability of $(\boldsymbol{y}_n)$ in $\boldsymbol{X}$. QED

Before formulating the next three lemmas we need to introduce the new seminorm $|||.|||$.

Definition 8.6.2. Let $\boldsymbol{a} = (a_1, a_2, \dots)$ and, for any $n_1 m n_2, \in \mathbf{N}$, put

$$M(n_1, n_2; \boldsymbol{a}) = \Big| \frac{a_1}{n_1} \sum_{i=1}^{n_1} \boldsymbol{e}_i + \frac{a_2}{n_2} \sum_{i=n_1+1}^{n_1+n_2} \boldsymbol{e}_i \Big|.$$

By the convexity argument (see Section 1.2) for the definition of $L(\boldsymbol{a})$),

$$L(\boldsymbol{a}) \geq M(n_1, n_2, ; \boldsymbol{a}) \geq M(n_1 n'_1, n_2 n'_2; \boldsymbol{a}), \qquad n_1, n'_1, n_2, N'_2 \in \mathbf{N}.$$

In an analogous manner one defines $M(n_1, n_2, \dots, n_k; \boldsymbol{a})$ (for an arbitrary sequence $\boldsymbol{a} \in \boldsymbol{s}$ with finitely many non-zero terms) which has similar properties to $M(n_1, n_2; \boldsymbol{a})$, so that

$$|||\Phi(\boldsymbol{a})||| := \lim_{n_1 < n_2 < \cdots < n_k \to \infty} M(n_1, n_2, \dots, n_k; \boldsymbol{a})$$

exists and defines a seminorm on $\Phi(\boldsymbol{s}) \subset \boldsymbol{X}$. Recall that $\boldsymbol{s} \subset \boldsymbol{a} \mapsto \Phi(\boldsymbol{a}) = \sum_i a_i \boldsymbol{e}_i \in \boldsymbol{X}$. The seminorm $|||.|||$ is evidently invariant under spreading because the norm $|.|$ was also invariant under spreading, and one can show that the space $\boldsymbol{G}$ (the completion of span$[\boldsymbol{e}_i]$ under $|||.|||$) is also finitely representable in $\boldsymbol{X}$, which can be demonstrated in the same way it has been proved for the space $\boldsymbol{F}$.

Lemma 8.6.2. *The seminorm* $|||.|||$ *is equal-signs-additive on the* span$[\boldsymbol{e}_i]$, *i.e., if* $a_k a_{k+1} \geq 0, \boldsymbol{a} \in \boldsymbol{s}$, *then*

$$|||\Phi(\boldsymbol{a})||| = |||\Phi(\boldsymbol{a}) + a_{k+1}(\boldsymbol{e}_k - \boldsymbol{e}_{k+1})|||. \tag{8.6.4}$$

Proof. It is easy to see that by an approximation procedure one can restrict our attention to the case when a_{k+1}/a_k is rational (and $a_k a_{k+1} > 0$). Then one can write $a_k = p\alpha, a_{k+1} = q\alpha$, with $p, q \in \mathbf{N}$. Therefore, for some $n_1, n_2, \ldots, n_k$, one can write $M(n_1, n_2, \ldots, n_k; \boldsymbol{a})$ in the form

$$\left|\cdots + \frac{p\alpha}{pn}(\boldsymbol{e}_{j+1} + \cdots + \boldsymbol{e}_{j+pn}) + \frac{q\alpha}{pn}(\boldsymbol{e}_{j+pn+1} + \cdots + \boldsymbol{e}_{j+(p+q)n}) + \ldots\right|$$

$$= \left|\cdots + \frac{(p+q)}{(p+q)n}(\boldsymbol{e}_{j+1} + \cdots + \boldsymbol{e}_{j+(p+q)n}) + \ldots\right|,$$

and the term in the bottom expression is exactly a term that appears in $M(n_1, n_2, \ldots, n_k; \boldsymbol{a}')$ for $\Phi(\boldsymbol{a}') = \Phi(\boldsymbol{a}) + \boldsymbol{a}_{k+1}(\boldsymbol{e}_k - \boldsymbol{e}_{k+1})$. This proves the Lemma. QED

Lemma 8.6.3. *The seminorm* $|||.|||$ *satisfies the inequality,*

$$\left|\left|\left|\sum_{i=1}^{k-1} a_i \boldsymbol{e}_i + \Big(\sum_{i=k}^{l} a_i\Big)\boldsymbol{e}_k + \sum_{i \geq l+1} a_i \boldsymbol{e}_i\right|\right|\right| \leq \left|\left|\left|\sum_i a_i \boldsymbol{e}_i\right|\right|\right|. \tag{8.6.5}$$

Proof. It is sufficient to consider two consecutive terms, $a_k + \boldsymbol{e}_k + a_{k+1}\boldsymbol{e}_{k+1}$. Let us begin with the special case when $a_k = \alpha, a_{k+1} = -\alpha$, and denote

$$\boldsymbol{u} = \sum_{i<k} a_i \boldsymbol{e}_i, \qquad \boldsymbol{v} = \sum_{i>l} a_i \boldsymbol{e}_i.$$

Let $n, m \in \mathbf{N}$ be sufficiently large so that, in view of invariance under spreading,

$$|||\Phi(\boldsymbol{a})||| = |||\boldsymbol{u} + \alpha(\boldsymbol{e}_i - \boldsymbol{e}_{k+1}) + T^m\boldsymbol{v}|||$$

$$= |||\boldsymbol{u}+\alpha(\boldsymbol{e}_{k+1}-\boldsymbol{e}_{k+2})+T^m\boldsymbol{v}|||+|||\boldsymbol{u}+\alpha(\boldsymbol{e}_{k+n-1}-\boldsymbol{e}_{k+n})+T^m\boldsymbol{v}|||,$$

from which it follows that

$$|||\Phi(\boldsymbol{a})||| \geq |||\boldsymbol{u} + \frac{\alpha}{n}(\boldsymbol{e}_k - \boldsymbol{e}_{k+n}) + T^m\boldsymbol{v}|||,$$

and

$$|||\Phi(\boldsymbol{a})||| \geq |||\boldsymbol{u} + \boldsymbol{v}||| - 2|\alpha|/n.$$

Letting $n \to \infty$, we obtain (4). If $a_k + a_{k+1} = \beta \neq 0$, then either a_k, or a_{k+1}, have the same sign as $-\beta$. Suppose, it is the case for a_k. Then, by Lemma 8.6.2,

$$|||\Phi(\boldsymbol{a})||| = |||\boldsymbol{u} + a_k\boldsymbol{e}_k + (-a_k + \beta)\boldsymbol{e}_{k+1} + \boldsymbol{v}|||$$

$$= |||\boldsymbol{u} + a_k\boldsymbol{e}_k - a_k\boldsymbol{e}_k + \beta\boldsymbol{e}_{k+2} + \ T\boldsymbol{v}|||$$

$$\geq |||\boldsymbol{u} + \beta\boldsymbol{e}_{k+2} + T\boldsymbol{v}|| = |||\boldsymbol{u} + \beta\boldsymbol{e}_k + \boldsymbol{v}|||,$$

so that (8.6.5) is proven.QED

Lemma 8.6.4. *If T is not ergodic on the space $\boldsymbol{F}$, then $|||.|||$ is a norm.*

Proof. If the shift T does not satisfy the ergodic theorem then $|||\boldsymbol{e}_1-\boldsymbol{e}_2||| \neq 0$. Let $\boldsymbol{a} \in \boldsymbol{s}$, and suppose that $a_1 \neq 0$. If $|||\Phi(\boldsymbol{a})||| = 0$, then also $|||a_1\boldsymbol{e}_1+b\boldsymbol{e}_2||| = 0$, with $b = \sum_{i>l} a_i$, by Lemma 8.6.3. Also, $b \neq -a - i$, because $|||\boldsymbol{e}_1 - \boldsymbol{e}_2||| \neq 0$. On the other hand, $|||a - i\boldsymbol{e}_i + b\boldsymbol{e}_2||| \geq |a_1 + b||||\boldsymbol{e}_i|||$, and $|||\boldsymbol{e}_1||| = |||\boldsymbol{e}_2||| \neq 0$, imply that $a_1 + b = 0$. A contradiction. QED

Proof of Theorem 8.6.1, continued. Suppose that space $\boldsymbol{X}$ is not stable. Then, by Lemma 6.8.1, the shift T is not ergodic in $\boldsymbol{F}$ so that, by Lemma 8.6.4, the space $(\boldsymbol{G}, |||.|||)$ is a Banach space finitely representable in $\boldsymbol{X}$. The proof of the theorem will be complete if we show that, if $\boldsymbol{G}$ does not contain a subspace isomorphic to $\boldsymbol{c}_0$ then it is not $(2,\epsilon)$-convex for any $\epsilon > 0$. However, the former assumption implies that

$$\lim_{n\to\infty}\Big|\Big|\Big|\sum_{k=1}^{n}(-1)^k\boldsymbol{e}_k\Big|\Big|\Big| = +\infty. \tag{8.6.6}$$

Indeed, put $\boldsymbol{u}_1 = \boldsymbol{e}_1 - \boldsymbol{e}_2, \boldsymbol{u}_2 = \boldsymbol{e}_3 - \boldsymbol{e}_4, \ldots$, and consider the subspace $\boldsymbol{U} = \overline{\text{span}}[\boldsymbol{u}_i] \subset \boldsymbol{G}$. Let $\boldsymbol{a}, \boldsymbol{a}$" $\in \boldsymbol{s}$, and $\text{supp}(\boldsymbol{a}') \subset \text{supp}(\boldsymbol{a})$. It follows from Lemma 8.6.3 that

$$\Big|\Big|\Big|\sum_i a_i\boldsymbol{u}_i\Big|\Big|\Big| \geq \Big|\Big|\Big|\sum_i a_i'\boldsymbol{u}_i\Big|\Big|\Big|,$$

so that

$$\Big|\Big|\Big|\sum_i \varepsilon_i a_i\boldsymbol{u}_i\Big|\Big|\Big| \leq 2\Big|\Big|\Big|\sum_i a_i\boldsymbol{u}_i\Big|\Big|\Big|,$$

for all choices of $\varepsilon_i = \pm 1$, from which we deduce that were $|||\sum_{i=1}^n \boldsymbol{u}_i|||$ bounded, then $\boldsymbol{U}$ would be isomorphic to $\boldsymbol{c}_0$. Thus we obtain (8.6.6).

Now, let $n \in \mathbf{N}$, and

$$\boldsymbol{u} = \alpha_n(\boldsymbol{e}_1 - \boldsymbol{e}_3 + \boldsymbol{e}_5 + \cdots + \boldsymbol{e}_{4n-3} - \boldsymbol{e}_{4n-1}) \in \Phi(\boldsymbol{s}),$$

$$\boldsymbol{v} = \alpha_n(\boldsymbol{e}_2 - \boldsymbol{e}_4 + \boldsymbol{e}_6 + \cdots + \boldsymbol{e}_{4n-2} - \boldsymbol{e}_{4n}) \in \Phi(\boldsymbol{s}),$$

where α_n are chosen so that $|||\boldsymbol{u}||| = |||\boldsymbol{v}||| = 1$. Furthermore, for each $\epsilon \in (0,1)$, choosing n sufficiently large, one can find $\boldsymbol{u}, \boldsymbol{v} \in S_X$ such that

$$\inf_{\varepsilon=\pm 1}\|\boldsymbol{u} + \varepsilon\boldsymbol{v}\| > 2(1-\epsilon),$$

so that $\boldsymbol{G}$ is not $(2,\epsilon)$-convex for any $\epsilon > 0$. QED

Definition 8.6.3. A Banach space $\boldsymbol{X}$ is said to have the *alternate signs Banach-Saks property* if from every bounded sequence $(\boldsymbol{x}_n) \subset \boldsymbol{X}$ one can choose a subsequence $(\boldsymbol{y}_n)$ such that

$$\lim_{n\to\infty}\frac{1}{n}\big(\boldsymbol{y}_1 - \boldsymbol{y}_2 + \boldsymbol{y}_3 - \cdots + (-1)^{n+1}\boldsymbol{y}_n\big) = 0. \tag{8.6.7}$$

Theorem 8.6.2.[32] *If $\boldsymbol{X}$ is a B-convex Banach space then it has the alternate signs Banach-Saks property.*

Proof. We may assume that $(\boldsymbol{x}_n)$ is not stable since, otherwise, $(\boldsymbol{y}_n)$ satisfying (8.6.7) may be obtained as a union of two stable subsequences of $(\boldsymbol{x}_n)$. Let $\boldsymbol{F}_1$ be a subspace of the space $\boldsymbol{F}$ (defined earlier in this section) generated by $\boldsymbol{u}_1 = \boldsymbol{e}_1 - \boldsymbol{e}_2, \boldsymbol{u}_2 = \boldsymbol{e}_3 - \boldsymbol{e}_4, \ldots$. If $\boldsymbol{X}$ is B-convex then $\boldsymbol{F}_1$ is also B-convex. Therefore, by the Strong Law of Large Numbers proven earlier in this chapter there exist $\varepsilon_i = \pm 1$ such that

$$\lim_{n\to\infty} \left| \frac{1}{n} \sum_{i=1}^{n} \varepsilon_i \boldsymbol{u}_i \right| = 0. \tag{8.6.8}$$

Since the norm $|.|$ is invariant under spreading (see Chapter 1),

$$\lim_{n\to\infty} \left| \frac{1}{n} \sum_{i=1}^{n} \boldsymbol{u}_i \right| \le 2 \lim_{n\to\infty} \left| \frac{1}{n} \sum_{i=1}^{n} \varepsilon_i \boldsymbol{u}_i \right| = 0.$$

Repeating the proof of Lemma 8.6.1, with $(\boldsymbol{e}_n)$ replaced by $(\boldsymbol{u}_n)$, one obtains a stable subsequence of $(\boldsymbol{u}_n)$ wich proves (8.6.7). QED

Remark 8.6.2. There exist Banach spaces which are not B-convex and still have the alternate signs Banach-Saks property. One such example is the space $\boldsymbol{c}_0$. Indeed, if $\boldsymbol{x}_n = (x_n^{(i)}), i = 1, 2, \ldots, n = 1, 2, \ldots$, with $\|\boldsymbol{x}_n\| \le 1$, then, for each $\epsilon > 0$, there exists a subsequence $(\boldsymbol{y}_n) \subset (\boldsymbol{x}_n)$ such that, for each $n \in \mathbf{N}$,

$$\left\| \sum_{j=1}^{m} (-1)^{j+1} \boldsymbol{y}_j \right\| = \sup_i \left| \sum_{j=1}^{m} (-1)^{j+1} \boldsymbol{y}_j^{(i)} \right| \le 2 + \epsilon. \tag{8.6.9}$$

To see this (up to taking subsequences, and applying the diagonal procedure) put $a_i = \lim_{n\to\infty} x_n^{(i)}$, where also $|x_n^{(i)} - a_i| < 2^{-n}\epsilon$, if $|x_k^{(i)} > 2^{-k}\epsilon$, for some $k < n$. Then, for a subsequence $(\boldsymbol{y}_n)$, (8.6.9) is satisfied since, for each i we can replace each $x_n^{(i)}$ by a_i,

[32] Due to A. Brunel (1973/74), and A. Brunel and L.Sucheston (1974), and (1975).

and there exists a $k < n$ such that $|x_k^{(i)}| > 2^{-k}\epsilon$, thus implying the inequality,

$$\Big|\sum_{j=1}^{m}(-1)^{j+1}\boldsymbol{y}_j^{(i)}\Big| < \epsilon\sum_{n=1}^{\infty}2^{-n} + |x_k^{(i)}| + |a_i| \leq 2+\epsilon.$$

8.7 Trees in B-convex spaces

Definition 8.7.1. Let $\boldsymbol{X}$ be a Banach space. We say that $\boldsymbol{x}_1, \boldsymbol{x}_2 \in \boldsymbol{X}$ form a $(1,\epsilon)$-*symmetric branch* if

$$\|\boldsymbol{x}_1 - \boldsymbol{x}_2\| \geq \epsilon, \qquad \text{and} \qquad \|\boldsymbol{x}_1 + \boldsymbol{x}_2\| \geq \epsilon.$$

Proceed by induction and suppose that we have defined an $(n-1,\epsilon)$-symmetric branch. Then we say that the 2^n-tuple $\boldsymbol{x}_1, \ldots, \boldsymbol{x}_{2^n} \in \boldsymbol{X}$ forms an (n,ϵ)- *symmetric branch* if, for any choice of $\varepsilon = \pm 1, i = 1, \ldots, 2^{n-1}$, we have the inequality $\|\boldsymbol{x}_{2i-1} + \varepsilon\boldsymbol{x}_{2i}\| \geq \epsilon$,, and if the 2^{n-1}-tuple, $(\boldsymbol{x}_{2i-1} + \varepsilon_i\boldsymbol{x}_{2i})/2, i = 1, \ldots, 2^{n-1}$, forms an $(n-1,\epsilon)$-symmetric branch.

We say that a Banach space has the *finite symmetric tree property* if there exists an $\epsilon > 0$ such that, for each $n \in \mathbf{N}$ one can find an (n,ϵ)-symmetric branch in its unit ball.

Theorem 8.7.1.[33] *A Banach space $\boldsymbol{X}$ is B-convex if, and only if, it does not have the finite symmetric tree property.*

Proof. If $\boldsymbol{X}$ is not B-convex then, by Theorem 8.2.1, $\boldsymbol{l}_1$ is finitely representable in $\boldsymbol{X}$, and, evidently, it has the finite symmetric tree property because the canonical basis in $\boldsymbol{l}_1^{(2^n)}$ forms an (n,ϵ)-symmetric branch, for each $n \in \mathbf{N}$.

Conversely, if $\boldsymbol{X}$ has the finite symmetric tree property then we shall show that $\boldsymbol{X}$ is of Rademacher-type p, for no $p > 1$, that is, by Theorem 8.1.4, it is not B-convex. Indeed, for every $\epsilon > 0$, and $n \in \mathbf{N}$, and any (n,ϵ)-symmetric branch $\boldsymbol{x}_1, \ldots, \boldsymbol{x}_{2^n} \in \boldsymbol{X}$, one has the inequality

$$\mathbf{E}\Big\|\sum_{i=1}^{2^n} r_i\boldsymbol{x}_i\Big\| \geq 2^{n-1}\epsilon. \tag{8.7.1}$$

[33]Due to B. Beauzamy (1973/74).

We will prove (8.7.1) by induction. For $n = 1$, in view of the definition of a symmetric branch,

$$\mathbf{E}\|r_1\boldsymbol{x}_1 + r_2\boldsymbol{x}_2\| = 2^{-1}\|\boldsymbol{x}_1 + \boldsymbol{x}_2\| + 2^{-1}\|\boldsymbol{x}_1 - \boldsymbol{x}_2\| \geq \epsilon.$$

Assume that (8.7.1) is true for $n-1$. Putting

$$r_{ik} = r_i(t), \qquad \text{for} \qquad t \in (k/2^n, (k+1)/2^n),$$

one has

$$\mathbf{E}\Big\|\sum_{i=1}^{2^n} r_i\boldsymbol{x}_i\Big\| = 2^{-2^n}\sum_{k=1}^{2^{2^n}}\Big\|\sum_{i=1}^{2^n} r_{ik}\boldsymbol{x}_i\Big\| =: 2^{-2^n} I.$$

Decompose I as follows,

$$I = \sum_{\varepsilon_{2i}=\pm 1}\sum_{\varepsilon_i'=\pm 1}\Big\|\sum_{i=1}^{2^{n-1}} \varepsilon_i'(\boldsymbol{x}_{2i-1} + \varepsilon_{2i}\boldsymbol{x}_{2i})\Big\|.$$

By the inductive hypothesis, fixing ε_{2i}, one obtains the inequality,

$$\sum_{\varepsilon_i'=\pm 1}\Big\|\sum_{i=1}^{2^{n-1}} \varepsilon_i'(\boldsymbol{x}_{2i-1} + \varepsilon_{2i}\boldsymbol{x}_{2i})\Big\| \geq 2^{n-1}2^{2^{n-1}}\epsilon,$$

so that

$$\mathbf{E}\Big\|\sum_{i=1}^{2^n} r_i\boldsymbol{x}_i\Big\| = 2^{-2^n} I \geq 2^{n-1}\epsilon.$$

Now, in view of (8.7.1), the proof is immediate. Suppose that $\boldsymbol{X}$ has the finite symmetric tree property, and $\boldsymbol{X}$ is of Rademacher-type p, for some $p > 1$. Then

$$\mathbf{E}\Big\|\sum_{i=1}^{2^n} r_i\boldsymbol{x}_i\Big\| \leq C\Big(\sum_{i=1}^{2^n}\|\boldsymbol{x}_i\|^p\Big)^{1/p}, \qquad \in\in \mathbf{N},$$

and since $\|\boldsymbol{x}_i\| \leq 1$, by (8.7.1), we have the inequality, $\epsilon \cdot 2^{n-1} \leq C \cdot 2^{n/p}$, for all $n \in \mathbf{N}$. A contradiction, QED

Chapter 9

Marcinkiewicz-Zygmund Theorem in Banach spaces

9.1 Preliminaries

In this chapter we study[1] the asymptotics of almost sure and tail behavior of sums, $(S_n/n^{1/p}) = (X_1 + \cdots + X_n)/n^{1/p})$, $1 \leq p < 2$, for independent, centered random vectors $X_n, n = 1, 2, \ldots$, and of martingales, (M_n), taking values in a Banach space $\boldsymbol{X}$. The obtained results are in the spirit of classical theorems of Marcinkiewicz-Zugmund, Hsu-Robbins-Erdös-Spitzer, and Brunk, for real-valued random variables, and show the essential role played by the geometry of $\boldsymbol{X}$ in the infinite-dimensional case.

In particular, we will show that for independent (X_i) with uniformly bounded tail probabilities the implication

$$\mathbf{E}\|X_i\|^p < \infty, \quad \mathbf{E}X_i = 0 \implies S_n/n^{1/p} = 0,$$

depends in an essential way on $\boldsymbol{l}_p$ not being finitely representable in $\boldsymbol{X}$. We also prove that a Banach space analogue of the Brunk's strong law of large numbers depends on the Rademacher-type of $\boldsymbol{X}$. Recall that the Brunk's law is particularly useful in cases where one has information about existence of moments of X_i's

[1]The results of this chapter are based on the papers by W.A. Woyczyński (1980, 1982).

of order higher than 2. Such information may not be utilized in the framework of the more classical Kolmogorov-Chung's law. Extensions of these types of results to the case of Banach space valued martingales are also provided.

As far as the rates of convergence are concerned a number of simple observations is in order. Directly from definitions and from Chebyshev;s Inequality one obtains the following "trivial" rate:

Proposition 9.1.1. *Let $1 \leq p \leq 2$, and let $\boldsymbol{X}$ be of Rademacher-type p. If (X_i) are independent and identically distributed random vectors in $\boldsymbol{X}$ with $\mathbf{E}\|X_i\|^p < \infty$, and $\mathbf{E}X_1 = 0$, then, for every $\epsilon > 0$,*

$$\mathbf{P}(\|S_n/n\| \geq \epsilon) = O(n^{1-p}).$$

Also, some exponential rates can be immediately obtained without any restrictions on the geometric structure of the space $\boldsymbol{X}$.

Proposition 9.1.2. *If (X_i) are independent and identically distributed random vectors in $\boldsymbol{X}$ with $\mathbf{E}X_1 = 0$, and such that, for every $\epsilon > 0$, there exist constants C_ϵ, and β_ϵ, such that, for every $\beta \leq \beta_\epsilon$,*

$$\mathbf{E}\exp[\beta\|X_1\|] \leq C_\epsilon \exp[\beta\epsilon],$$

then, for every $\epsilon > 0$, there exists an $\alpha < 1$ such that

$$\mathbf{P}(\|S_n/n\| > \epsilon) = O(\alpha^n).$$

Proof. By Chebyshev's Inequality, and for $\delta < \epsilon$, we get

$$\mathbf{P}(\|S_n/n\| > \epsilon) \leq \exp[-\beta_\delta n\epsilon]\,\mathbf{E}\exp[\beta_\delta\|S_n\|]$$

$$\leq \exp[-\beta_\delta n\epsilon](\mathbf{E}\exp[\beta_\delta\|X_1\|])^n \leq C_\delta(\exp[(\delta-\epsilon)\beta_\delta])^n. \quad \text{QED}$$

It is also interesting to notice that a sufficiently rapid rate of convergence to zero of the tail probabilities, $\mathbf{P}(\|S_n/n\| > \epsilon)$, implies similar rates of convergence in the Strong Law.

Proposition 9.1.3. *Let $\mathbf{X}$ be a Banach space, and let (X_i) be a sequence of independent, symmetric random vectors in $\mathbf{X}$. Let $(a_i), (b_i), (c_i) \subset \mathbf{R}$ be such that*

$$0 < a_i \uparrow \infty, \quad b_i, c_i \downarrow 0, \quad \text{and} \quad \sum_{i=1}^{j} 2^i b_{2^i} = O(2^j c_{2_j}),$$

and let

$$\sum_{n=1}^{\infty} c_n \mathbf{P}(\|S_n/a_n\| > \epsilon) < \infty,$$

for every $\epsilon > 0$. Then, for every $\epsilon > 0$,

$$\sum_{n=1}^{\infty} b_n \mathbf{P}(\sup_{k \geq n} \|S_k/a_k\| > \epsilon) < \infty.$$

Proof. Grouping the terms in exponential blocks, $(n : 2^j < n \leq 2^{j+1})$ we get

$$A \equiv \sum_{n=1}^{\infty} b_n \mathbf{P}(\sup_{k \geq n} \|S_k/a_k\| > \epsilon) \leq \sum_{i=1}^{\infty} b_{2^i} \cdot 2^i \mathbf{P}(\sup_{k \geq 2^i} \|S_k/a_k\| > \epsilon)$$

$$\leq \sum_{i=1}^{\infty} \sum_{j=i}^{\infty} b_{2^i} \cdot 2^i \mathbf{P}(\max_{2^i < k \leq 2^{j+1}} \|S_k/a_k\| > \epsilon),$$

and, by Lévy's Inequality,

$$A \leq 2 \sum_{i=1}^{\infty} \sum_{j=i}^{\infty} b_{2^i} \cdot 2^i \mathbf{P}\big(\|S_{2^{j+1}}/a_{2^{j+1}}\| > \epsilon\big)$$

$$= 2 \sum_{j=1}^{\infty} \Big(\sum_{i=j}^{\infty} b_{2^i} \cdot 2^i \Big) \mathbf{P}\big(\|S_{2^{j+1}}/a_{2^{j+1}}\| > \epsilon\big)$$

$$\leq 2C \sum_{j=1}^{\infty} c_{2^j} 2^j \mathbf{P}\big(\|S_{2^{j+1}}/a_{2^{j+1}}\| > \epsilon\big).$$

Now, by the symmetry assumptions, grouping the terms again as follows,

$$S_n = S_{2^{j+1}} - X_{2^{j+1}} - X_{2^{j+1}-1} - \cdots - X_{n+1}, \qquad 2^{j-1} \le n < 2^j,$$

we obtain the inequality

$$A \le 8C \sum_{n=1}^{\infty} c_n \mathbf{P}(\|S_n/a_n\| > 2\epsilon). \qquad \text{QED}$$

Two special cases of the above Proposition will be of interest later on.

Corollary 9.1.1. *Let* $\boldsymbol{X}$ *be a Banach space, and let* (X_i) *be a sequence of independent symmetric random vectors in* $\boldsymbol{X}$. *Then:*

(i) For every $q > 1$, *there exists a constant* $C > 0$ *such that*

$$\sum_{n=1}^{\infty} n^{-q} \mathbf{P}(\sup_{k \ge n} \|S_k/a_k\| > \epsilon) \le C \sum_{n=1}^{\infty} n^{-q} \mathbf{P}(\|S_k/a_k\| > \epsilon);$$

(ii) There exists a constant $C > 0$ *such that*

$$\sum_{n=1}^{\infty} n^{-1} \mathbf{P}(\sup_{k \ge n} \|S_k/a_k\| > \epsilon) \le C \sum_{n=1}^{\infty} n^{-1} (\log n) \mathbf{P}(\|S_k/a_k\| > \epsilon).$$

9.2 Brunk-Prokhorov's type strong law and related rates of convergence

In Proposition 9.1.1 we could have used only moments of order $p, 1 \le p \le 2$, and in Proposition 9.1.2 exponential moments were needed. The following analogue of the classical Marcinkiewicz-Zygmund inequality permits us to use the information about moments of arbitrary order.

Proposition 9.2.1. *Let* $1 \le p \le 2$, *and* $q \ge 1$. *The following properties of a Banach space* $\boldsymbol{X}$ *are equivalent:*

(i) The space $\boldsymbol{X}$ *is of Rademacher-type* p;

(ii) There exists a constant $C > 0$ *such that, for every* $n \in \mathbf{N}$, *and for any sequence* (X_i) *of independent random vectors in* $\boldsymbol{X}$ *with* $\mathbf{E}X_i = 0$,

$$\mathbf{E}\Big\|\sum_{i=1}^{n} X_i\Big\|^q \le C\mathbf{E}\Big(\sum_{i=1}^{n} \|X_i\|^p\Big)^{q/p}.$$

Proof. $(i) \implies (ii)$ Let $(\tilde{X}_i) = (X_i - X_i')$ be a symmetrization of the sequence (X_i), and let the Rademacher sequence (r_i) be independent of (X_i), and (X_i'). Then

$$\mathbf{E}\Big\|\sum_{i=1}^{n} X_i\Big\|^q \le \mathbf{E}\Big\|\sum_{i=1}^{n} \tilde{X}_i\Big\|^q = \mathbf{E}\Big\|\sum_{i=1}^{n} r_i\tilde{X}_i\Big\|^q$$

$$\le C\mathbf{E}\Big(\Big\|\sum_{i=1}^{n} \tilde{X}_i\Big\|^p\Big)^{q/p} \le C \cdot 2^q\mathbf{E}\Big(\sum_{i=1}^{n} \|X_i\|^p\Big)^{q/p},$$

where the first inequality follows from the condition $\mathbf{E}X_i = 0$, and because (X_i') are independent of (X_i), the equality is a consequence of the symmetry of $(\tilde{X}_i)$, the second inequality is a consequence of the space $\boldsymbol{X}$ being of Rademacher-type p and the Fubini's Theorem, and the third simply, by the triangle inequality.

$(ii) \implies (i)$ This implications is obvious. QED

Corollary 9.2.1. *Let* $\boldsymbol{X}$ *be a Banach space of Rademacher-type* p, *and* $q \ge p$. *If* (X_n) *are independent, identically distributed random vectors in* $\boldsymbol{X}$ *with* $\mathbf{E}\|X_1\|^q < \infty$, *and* $\mathbf{E}X_1 = 0$, *then* $\mathbf{E}\|S_n\|^q = O(n^{q/p})$.

Proof. If $p = q$ then the estimate follows directly from the definition of Rademacher-type p. If $q > p$, then by Hölder's Inequality with exponents q/p, and $q/(q-p)$, and by proposition 9.2.1, we obtain the estimate,

$$\mathbf{E}\Big\|\sum_{i=1}^{n} X_i\Big\|^q \le C\mathbf{E}\Big(\sum_{i=1}^{n} \|X_i\|^p\Big)^{q/p}$$

$$\leq C\mathbf{E}\Big(\sum_{i=1}^{n}\|X_i\|^q\Big)n^{(q-p)/p} = Cn^{q/p}\mathbf{E}\|X_1\|^q. \qquad \text{QED}$$

Hence, by Chebyshev's Inequality we immediately obtain the following result.

Corollary 9.2.2. *Let $\boldsymbol{X}$ be a Banach space of Rademacher-type p, and $q \geq p$. If (X_n) are independent, identically distributed random vectors in $\boldsymbol{X}$ with $\mathbf{E}\|X_1\|^q < \infty$, and $\mathbf{E}X_1 = 0$, then, for every $\epsilon > 0$,*

$$\mathbf{P}\big(\|S_n/n\| > \epsilon\big) = O(n^{q(1/p-1)}).$$

The following result provides an extension to Banach spaces of the Brunk-Prokhorov's Law of Large Numbers[2] that is an extension of the classical Kolmogorov-Chung's Strong Law.

Theorem 9.2.1. *Let $1 \leq p \leq 2$, $\boldsymbol{X}$ be of Rademacher-type p, and $q \geq 1$. If (X_n) is a sequence of independent, zero-mean random vectors in $\boldsymbol{X}$ such that*

$$\sum_{n=1}^{\infty}\frac{\mathbf{E}\|X_n\|^{pq}}{n^{pq+1-q}} < \infty, \tag{9.2.1}$$

then, almost surely,

$$\lim_{n\to\infty}\left\|\frac{S_n}{n}\right\| = 0.$$

Proof. For $q = 1$ the theorem boils down to the classical Kolmogorov-Chung's Strong Law mentioned above. So, assume that $q > 1$. Then $\|S_n\|^{pq}$ is a real submartingale and, by the well known Hajek-Renyi-Chow type inequality, we get that for every $\epsilon > 0$,

$$\begin{aligned}\epsilon^{pq}\mathbf{P}\big(\sup_{j\geq n}\|S_j/j\| > \epsilon\big) &= \epsilon^{pq}\lim_{m\to\infty}\mathbf{P}\big(\sup_{m\geq j\geq n}\|S_j/j\|^{pq} > \epsilon^{pq}\big) \\ &\leq n^{-pq}\mathbf{E}\|S_n\|^{pq} + \sum_{j=n+1}^{\infty} j^{-pq}\mathbf{E}\big(\|S_j\|^{pq} - \|S_{j-1}\|^{pq}\big).\end{aligned} \tag{9.2.2}$$

[2]See H.D. Brunk (1948), and Yu.V. Prokhorov (1950).

By Proposition 9.2.1, and Hölder's Inequality,

$$\mathbf{E}\|S_j\|^{pq} \leq C\mathbf{E}\Big(\sum_{i=1}^{j}\|X_i\|^p\Big)^q \leq Cj^{q-1}\sum_{i=1}^{j}\mathbf{E}\|X_i\|^{pq},$$

so that, by (9.2.1), and Kronecker's Lemma, we obtain that

$$\lim_{j\to\infty} j^{-pq}\mathbf{E}\|S_j\|^{pq} = 0.$$

Also, the series on the right-hand side of (9.2.2) converges because of Proposition 9.2.1. Hence, summing by parts, we obtain the estimate,

$$\sum_{j=1}^{n}((j-1)^{-pq}+j^{-pq})\mathbf{E}\|S_j\|^{pq} \leq \sum_{j=1}^{n}((j-1)^{-pq}+j^{-pq})j^{q-1}\sum_{i=1}^{j}\mathbf{E}\|X_i\|^{pq}$$

$$\leq C\sum_{j=1}^{n}\mathbf{E}\|X_j\|^{pq}/j^{pq+1-q} + \sum_{i=1}^{n}\mathbf{E}\|X_i\|^{pq}/n^{pq+1-q}.$$

Therefore, for every $\epsilon > 0$,

$$\lim_{n\to\infty}\mathbf{P}\big(\sup_{j\geq n}\|S_j/j\| > \epsilon\big) = 0. \qquad \text{QED}$$

9.3 Marcinkiewicz-Zygmund type strong law and related rates of convergence

In this section we will prove a result that extends to some Banach spaces the classical theorem of Marcinkiewicz and Zygmund which was developed for real random variables.[3]

Theorem 9.3.1. *Let $1 < p < 2$. Then the following properties of a Banach space $\boldsymbol{X}$ are equivalent:*

(i) The space $\boldsymbol{l}_p$ is not finitely representable in $\boldsymbol{X}$;

---

[3]See, J. Marcinkiewicz and A. Zygmund (1937).

(ii) For any sequence (X_n) of zero-mean, independent random vectors in $\boldsymbol{X}$, with tail probabilities uniformly bounded by tail probabilities of a random vector $X_0 \in \boldsymbol{L}_p$, the series

$$\sum_{n=1}^{\infty} \frac{X_n}{n^{1/p}}$$

converges almost surely in norm.

(iii) For any sequence (X_n) satisfying the assumptions of statement (ii), $S_n/n^{1/p} \to 0$, almost surely.

The proof of the above Theorem will be based on the following Lemma.

Lemma 9.3.1. *Assume that $1 \le p < 2$, $\boldsymbol{l}_p$ is not finitely representable in $\boldsymbol{X}$, and the assumptions of Theorem 9.3.1 (iii) are satisfied. Then the series*

$$\sum_{n=1}^{\infty} (X_n - \mathbf{E}Y_n)/n^{1/p}, \qquad \text{where} \qquad Y_n = X_n I(\|X_n\| \le n^{1/p}),$$

converges almost surely.

Proof. Since

$$\sum_{n=1}^{\infty} \mathbf{P}(X_n \neq Y_n) = \sum_{n-1}^{\infty} \mathbf{P}(\|X_n\| > n^{1/p})$$

$$\le C \sum_{n=1}^{\infty} \mathbf{P}(|X_0| > n^{1/p}) \le C_1 \mathbf{E}|X_0|^p < \infty,$$

in view of the Borel-Cantelli Lemma, it suffices to show that the series $\sum_{n=1}^{\infty} (Y_n - \mathbf{E}Y_n)/n^{1/p}$ converges almost surely.

Let $r > p$. Then

$$\sum_{n=1}^{\infty} \mathbf{E}\|Y_n - \mathbf{E}Y_n\|^r/n^{r/p} \le 2^{r+1} \sum_{n=1}^{\infty} \mathbf{E}\|Y_n\|^r/n^{r/p}$$

$$= 2^{r+1} \sum_{n=1}^{\infty} n^{-r/p} \int_{\|X_n\| \le n^{1/p}} \|X_n\|^r d\mathbf{P}$$

$$= 2^{r+1}\sum_{n=1}^{\infty} n^{-r/p}\int_0^{n^{1/p}} t^r\|X_n\|^r d\mathbf{P}(\|X_n\|\le t)$$

$$= 2^{r+1}\sum_{n=1}^{\infty} n^{-r/p}\Big(n^{r/p}\mathbf{P}(\|X_n\|\le n^{1/p}) - r\int_0^{n^{1/p}} \mathbf{P}(\|X_n\|<t)dt\Big)$$

$$\le C_1\sum_{n=1}^{\infty}\Big(1 - rn^{-r/p}\int_0^{n^{1/p}} t^{r-1}\big(1-\mathbf{P}(|X_0|>t)\big)dt\Big)$$

$$= C_1\sum_{n=1}^{\infty} rn^{-r/p}\int_0^{n^{1/p}} t^{r-1}\mathbf{P}(|X_0|>t)dt$$

$$= C_1\sum_{n=1}^{\infty} rn^{-r/p}\int_0^{1}\mathbf{P}(|X_0 s^{-1/r}|>n^{1/p})ds$$

$$\le C_2\mathbf{E}|X_0|^p\int_0^1 s^{-p/r}ds = C_2\frac{r}{r-p}\mathbf{E}|X_)|^p<\infty.$$

By Theorem 6.5.1 and Corollary 6.5.2, and in view of our assumptions, there exists an $r>p$ such that $\boldsymbol{X}$ is of Rademacher-type r. Therefore, the above estimate, and Theorem 6.7.4, give the desired convergence of the series $\sum_{n=1}^{\infty}(Y_n-\mathbf{E}Y_n)/n^{1/p}$. QED

Proof of Theorem 9.3.1. (*i*) $\Longrightarrow$ (*ii*) In view of the above Lemma it is sufficient to prove the absolute convergence of the series $\sum_n \mathbf{E}Y_n n^{-1/p}$. Since $\mathbf{E}X_n = 0$, and $p>1$, we have the estimate,

$$\sum_{n=1}^{\infty}\|\mathbf{E}Y_n\|n^{-1/p}\le\sum_{n=1}^{\infty}n^{-1/p}\int_{n^{1/p}}^{\infty} t\,d\mathbf{P}(\|X_n\|\le t)$$

$$=\sum_{n=1}^{\infty}\mathbf{P}(|X_0|>n^{1/p})+\int_1^{\infty}\mathbf{P}(|X_0/s|>n^{1/p})ds\le C\,\mathbf{E}|X_0|^p.$$

which give (*i*) $\Longrightarrow$ (*ii*).

(*ii*) $\Longrightarrow$ (*iii*) This implication follows by a straightforward application of Kronecker's Lemma.

(iii) $\implies$ *(i)*[4] In view of Kronecker's Lemma it suffices to construct, in any Banach space $\boldsymbol{X}$ such that $\boldsymbol{l}_p$ is finitely representable in $\boldsymbol{X}$, a sequence $(\boldsymbol{x}_n) \subset \boldsymbol{X}$, $\|\boldsymbol{x} - n\| \leq 1$, such that, for a sequence $(n_k) \subset \mathbf{N}, N_k \to \infty$, for all choices $\varepsilon_n = \pm 1$, and for all $k \in \mathbf{N}$,

$$N_k^{-1/p} \Big\| \sum_{i=1}^{N_k} \varepsilon_i \boldsymbol{x}_i \Big\| > \frac{1}{2}. \tag{9.3.1}$$

Put $N_1 = 1$, and choose an arbitrary $\boldsymbol{x}_1 \in \boldsymbol{X}$ with $\|\boldsymbol{x}_1\| = 1$. Suppose that $N_1, \ldots, N_k$, and $\boldsymbol{x}_1, \ldots, \boldsymbol{x}_{N_k}$, have been chosen so that $\|\boldsymbol{x}_i\| \leq 1, i = 1, \ldots, N_k$, and for all choices of $\varepsilon_i = \pm 1$ the inequality (9.3.1) is satisfied. Next, choose $N_{k+1} \in \mathbf{N}$ large enough so that

$$N_{k+1}^{-1/p} \Big[\frac{2}{3} (N_{k+1} - N_k)^{1/p} - N_k \Big] > \frac{1}{2}.$$

Since $\boldsymbol{l}_p$ is finitely representable in $\boldsymbol{X}$, we can find $\boldsymbol{x}_{N_k+1}, \ldots, \boldsymbol{x}_{N_{k+1}}$ such that, for all $(\alpha_k) \subset \mathbf{R}$,

$$\frac{2}{3} \Big(\sum_{i=N_k+1}^{N_{k+1}} |\alpha_i|^p \Big)^{1/p} \leq \Big\| \sum_{i=N_k+1}^{N_{k+1}} \alpha_i \boldsymbol{x}_i \Big\| \leq \Big(\sum_{i=N_k+1}^{N_{k+1}} |\alpha_i|^p \Big)^{1/p}.$$

Therefore, for all $\varepsilon = \pm 1$,

$$N_{k+1}^{-1/p} \Big\| \sum_{i=1}^{N_{k+1}} \varepsilon_i \boldsymbol{x}_i \Big\| \geq N_{k+1}^{-1/p} \Big(\Big\| \sum_{i=N_k+1}^{N_{k+1}} \varepsilon_i \boldsymbol{x}_i \Big\| - \Big\| \sum_{i=1}^{N_k} \varepsilon_i \boldsymbol{x}_i \Big\| \Big)$$

$$> N_{k+1}^{-1/p} \Big[\frac{2}{3} (N_{k+1} - N_k)^{1/p} - N_k \Big] > \frac{1}{2}. \qquad \text{QED}$$

For Banach spaces $\boldsymbol{X}$ in which $\boldsymbol{l}_1$ is not finitely representable, that is for B-convex Banach spaces, Lemma 9.3.1 permits us to prove the following result:

Theorem 9.3.2. *The following properties of a Banach space $\boldsymbol{X}$ are equivalent:*

(i) $\boldsymbol{l}_1$ is not finitely representable in $\boldsymbol{X}$;

[4] This implication is essentially due to B. Maurey and G. Pisier (1976).

(ii) For any sequence (X_n) of zero-mean, independent random vectors in $\boldsymbol{X}$ with tail probabilities uniformly bounded by tail probabilities of an $X_0 \in \boldsymbol{L}\log^+\boldsymbol{L}$, the series $\sum_{n=1}^{\infty} X_n/n$ converges almost surely;

(iii) For any sequence (X_i) satisfying assumptions of statement (ii) $S_n/n \to 0$ almost surely, as $n \to \infty$.

Proof. (*i*) $\implies$ (*ii*) In view of Lemma 9.3.1 it suffices to prove that $\sum_n \|EY_n\|/n$ converges, whenever $X_0 \in \boldsymbol{L}\log^+\boldsymbol{L}$. Since, $\mathbf{E}X_n = 0$, integrating by parts we obtain the following estimates,

$$\sum_{n=1}^{\infty} \|\mathbf{E}Y_n\|/n \le \sum_{n=1}^{\infty} n^{-1}\int_N^{\infty} t\,\mathbf{P}(\|X_n\| \le t)dt$$

$$= \sum_{n=1}^{\infty}\Big[\mathbf{P}(\|X_n\| > n) + n^{-1}\int_n^{\infty} \mathbf{P}(\|X_n\| \le t)dt\Big]$$

$$\le C_1\Big[\mathbf{E}|X_0| + \sum_{n=1}^{\infty} n^{-1}\sum_{k=n}^{\infty}\mathbf{P}(|X_0| > k)\Big]$$

$$= C_1\Big[\mathbf{E}|X_0| + \sum_{k=1}^{\infty}\sum_{n=1}^{k} n^{-1}\mathbf{P}(|X_0| > k)\Big]$$

$$= C_1\Big[\mathbf{E}|X_0| + \sum_{k=1}^{\infty}(\log k)\mathbf{P}(|X_0| > k)\Big]$$

$$\le C_1\Big[\mathbf{E}|X_0| + \mathbf{E}|X_0|\log^+|X_0|\Big] < \infty.$$

The implication (*ii*) $\implies$ (*iii*) follows directly from the Kronecker's Lemma, and (*iii*) $\implies$ (*i*) can be proven exactly the same way the same implication in Theorem 9.3.1 was demonstrated. QED

Theorem 9.3.3. *(i) Let $\boldsymbol{X}$ be a Banach space, $1 < p < 2$, and let $\alpha \ge 1/p$. Then $\boldsymbol{l}_p$ is not finitely representable in $\boldsymbol{X}$ if, and only if, for each sequence (X_i) of zero-mean, independent random vectors in $\boldsymbol{X}$ with tail probabilities uniformly bounded by tail probabilities of an $X_0 \in \boldsymbol{L}^p$, we have, for every $\epsilon > 0$,*

$$\sum_{n=1}^{\infty} n^{\alpha p-2}\mathbf{P}\big(\max_{1\le i\le n}\|S_i\| > n^{\alpha}\epsilon\big) < \infty.$$

(ii) Let $\boldsymbol{X}$ be a Banach space, $1 \leq p < 2$. Then $\boldsymbol{l}_p$ is not finitely representable in $\boldsymbol{X}$ if, and only if, for each sequence (X_i) of zero-mean, independent random vectors in $\boldsymbol{X}$, with tail probabilities uniformly bounded by tail probabilities of an $X_0 \in \boldsymbol{L}^p \log^+ \boldsymbol{L}$, we have, for every ϵ,

$$\sum_{n=1}^{\infty} n^{-1}(\log n)\mathbf{P}\big(\|S_i\| > n^{1/p}\epsilon\big) < \infty.$$

Proof. (i) First, we shall prove the sufficiency of the condition that $\boldsymbol{l}_p$ is not being finitely representable in $\boldsymbol{X}$. By Theorem 9.3.1, $S_n/n^{1p} \to 0$, almost surely and, as is easy to see, also

$$M_n/n^{1/p} \to 0 \quad \text{a.s., where} \qquad M_u := \max_{1 \leq i \leq [u]} \|S_i\|, \; u \in \mathbf{R},$$

with $[u]$ denoting the integer part (entier) of u. Hence, if we introduce the Chow's delayed sums

$$S_{u,v} = \sum_{1 \leq j \leq v} X_{[u]+j}, \qquad u, v \in \mathbf{R},$$

we get

$$M_{n,n}n^{-1/p} \leq (M_n + M_{2n})n^{-1/p} \to 0, \qquad \text{a.s., as} \qquad n \to \infty.$$

Now, in the case $\alpha = 1/p$, since $M_{2^n,2^n}, n = 1, 2, \ldots,$ are independent, the Borel-Cantelli Lemma implies that, for every $\epsilon > 0$,

$$\infty > \sum_{n=1}^{\infty} \mathbf{P}(M_{2^n,2^n} > 2^{n/p}\epsilon) = \sum_{n=1}^{\infty} \mathbf{P}(M_{2^n} > 2^{n/p}\epsilon)$$

$$\geq \int_!^{\infty} \mathbf{P}(M_{2^t} > 2^{(t+1)/p}\epsilon)\, dt$$

$$> (\log 2)^{-1} \int_1^{\infty} u^{-1}\mathbf{P}(M_u > 2^{1/p}\epsilon u^{1/p})\, du.$$

So, $\sum_n n^{-1}\mathbf{P}(m_n > n^{1/p}\epsilon) < \infty$, for every $\epsilon > 0$.

In the case $\alpha > 1/p$, for $m \geq 1$, we have

$$(m+1)^{\alpha p/(\alpha p-1)} \geq m^{\alpha p/(\alpha p-1)} + \frac{\alpha p}{\alpha p - 1} m^{1/(\alpha p-1)}$$

$$\geq m^{\alpha p/(\alpha p-1)} + m^{1/(\alpha p-1)},$$

so that the random variables $M_{m^{\alpha p/(\alpha p-1}, m^{1/(\alpha p-1)}}$, $m = 1, 2, \ldots$, are independent. Moreover, by Theorem 9.3.1,

$$m^{-\alpha/(\alpha p-1)} M_{m^{\alpha p/(\alpha p-1)}, m^{1/(\alpha p-1)}}$$

$$\leq m^{-\alpha/(\alpha p-1)} M_{m^{\alpha p/(\alpha p-1)}, m^{\alpha p/(\alpha p-1)}} \to 0,$$

almost surely as $m \to \infty$. Therefore, again using the Borel-Cantelli lemma, we obtain that

$$\infty > \sum_{m=1}^{\infty} \mathbf{P}(M_{m^{\alpha p/(\alpha p-1)}, m^{1/(\alpha p-1)}} \geq m^{\alpha/(\alpha p-1)}\epsilon)$$

$$= \sum_{m=1}^{\infty} \mathbf{P}(M_{m^{1/(\alpha p-1)}} \geq m^{\alpha/(\alpha p-1)}\epsilon)$$

$$\geq \int_1^{\infty} \mathbf{P}(M_{t^{1/(\alpha p-1)}} \geq (t+1)^{\alpha/(\alpha p-1)}\epsilon)\, dt$$

$$\geq (\alpha p - 1) \int_1^{\infty} u^{\alpha p-1} \mathbf{P}(M_u \geq 2^{\alpha/(\alpha p-1)} u^{\alpha}\epsilon)\, du,$$

which gives the desired rate of convergence. The necessity of the condition of $\boldsymbol{l}_p$ not being representable in $\boldsymbol{X}$ follows directly from the example developed in the proof of the implication $(iii) \implies (i)$ in Theorem 9.3.1.

(ii) First, let us prove the sufficiency of the condition that $\boldsymbol{l}_p$ is not finitely representable in $\boldsymbol{X}$. Without loss of generality we can assume that X_n's are symmetric. The case of zero expectations can be handled by adapting, in a standard way, the method presented below.

Put $Y_{kn} = X_k I(\|X_k\| < n^{1/p})$. Then

$$\sum_{n=1}^{\infty} \frac{\log n}{n} \mathbf{P}(\|S_n\| > n^{1/p}\epsilon)$$

$$\leq \sum_{n=1}^{\infty} \frac{\log n}{n} \mathbf{P}\Big(\bigcup_{k=1}^{n} (\|X_k\| > n^{1/p}\epsilon)\Big) + \sum_{n=1}^{\infty} \frac{\log n}{n} \mathbf{P}\Big(\Big\|\sum_{k=1}^{n} Y_{kn}\Big\| > n^{1/p}\epsilon\Big).$$

The series on the right-hand side can be estimated from above by the quantity,

$$C\sum_{n=1}^{\infty} (\log n)\mathbf{P}(|X_0| > n^{1/p}\epsilon) \leq C_1 \mathbf{E}|X_0|^p \log^+ |X_0| < \infty,$$

and the convergence of the second series can be verified as follows:

Since $\boldsymbol{l}_p$ is not finitely representable in $\boldsymbol{X}$, by the Maurey-Pisier Theorem mentioned earlier, there exists a $\delta > 0$ such that $\boldsymbol{X}$ is of Rademacher-type $(p+\delta)$. Hence, making use of Chebyshev's Inequality, and integrating by parts, we obtain the estimates,

$$\sum_{n=1}^{\infty} \frac{\log n}{n} \mathbf{P}\Big(\Big\|\sum_{k=1}^{n} Y_{kn}\Big\| > n^{1/p}\epsilon\Big)$$

$$\leq C_1 \sum_{n-1}^{\infty} n^{-1-(p+\delta)/p} (\log n) \sum_{k=1}^{n} \mathbf{E}\|Y_{kn}\|^{p+\delta}$$

$$\leq C_2 \sum_{n=1}^{\infty} \frac{\log n}{n^{1+(p+\delta)/p}} \sum_{k=1}^{n} \int_0^{n^{1/p}} t^{p+\delta} d\mathbf{P}(\|X_k\| \leq t)$$

$$\leq C_2 \sum_{n=1}^{\infty} \frac{\log n}{n^{(p+\delta)/p}} \int_0^{n^{1/p}} t^{p+\delta-1} \mathbf{P}(|X_0| > t) dt$$

$$= C_2 \int_0^1 s^{\delta/p} \sum_{n=1}^{\infty} (\log n) \mathbf{P}(|X_0 s^{-1/p} > n^{1/p}) ds$$

$$\leq C_2 \int_0^1 s^{\delta/p} \mathbf{E}|X_0 s^{-1/p}|^p \log^+ |X_0 s^{-1/p}| ds$$

$$\leq C_3 \mathbf{E}(|X_0|^p \log^+ |X_0|) \int_0^1 s^{-1+\delta/p} ds < \infty.$$

This completes the proof of the sufficiency. The necessity can be obtained exactly as in part (i). QED

Corollary 9.3.1. *If $\boldsymbol{l}_p$ is not finitely representable in $\boldsymbol{X}, 1 < p < 2$, and (X_n) are zero-mean, independent and identically distributed random vectors in $\boldsymbol{X}$ with $\mathbf{E}\|X_1\|^p < \infty$, then, for every $\epsilon > 0$,*

$$\mathbf{P}(\|S_n/n\| > \epsilon) = o(n^{1-p}).$$

Corollary 9.3.2. *Let $\boldsymbol{X}$ be of Rademacher-type $p, 1 < p \leq 2$, and let (X_n) be a sequence of independent, zero-mean vectors in $\boldsymbol{X}$ such that, uniformly in k,*

$$\mathbf{P}(\|X_k\| > n) = o(n^{-p}), \tag{9.3.2}$$

then, for every $\delta, \epsilon > 0$,

$$\mathbf{P}(\|S_n/n\| > \epsilon) = o(n^{1-p+\delta}).$$

Proof. Since $\boldsymbol{X}$ is of Rademacher-type p, for every $\delta > 0$, the space $\boldsymbol{l}_{p-\delta}$ is not finitely representable in $\boldsymbol{X}$. Now, (9.3.2) implies that X_k's have tail probabilities uniformly bounded by tail probabilities of an $X_0 \in \boldsymbol{L}^{p-\delta}$. Therefore, by Theorem 9.3.3,

$$\sum_{n=1}^{\infty} n^{p-\delta-2}\mathbf{P}(\|S_n/n\| > \epsilon) < \infty,$$

so that

$$n^{p-\delta-2}\mathbf{P}(\|S_n/n\| > \epsilon) = o(n^{-1}). \qquad \text{QED}$$

From Corollary 9.1.1, and Theorem 9.3.3, we immediately obtain the final result:

Corollary 9.3.3. *If $1 \leq p < 2$, and $\boldsymbol{l}_p$ is not finitely representable in $\boldsymbol{X}$, then, for any sequence (X_i) of zero-mean, independent random vectors in $\boldsymbol{X}$, with tail probabilities uniformly bounded by tail probabilities of an $X_0 \in \boldsymbol{L}^p$, if $1 < p < 2$, and of an $X_0 \in \boldsymbol{L}\log^+\boldsymbol{L}$, in the case $p = 1$,*

$$\sum_{n=1}^{\infty} n^{p-2}\mathbf{P}(\sup_{k\geq n}\|S_k/k\| > \epsilon) < \infty, \qquad \forall \epsilon > 0.$$

9.4 Brunk and Marcinkiewicz-Zygmund type strong laws for martingales

For an $\boldsymbol{X}$-valued martingale $(M_n, \mathcal{F}_n)$, on a probability space $(\Omega, \mathcal{F}, \mathbf{P})$, consider their difference sequence $D_n = M_n - M_{n-1}, n = 1, 2, \ldots,$. In what follows we shall assume that $M_0 = 0$, a.s.

Proposition 9.4.1[5] *The Banach space $\boldsymbol{X}$ is p-smoothable if, and only if, for any $q \geq 1$, there exists a constant $C > 0$ such that, for each $|XX$-valued martingale (M_n)*

$$\mathbf{E}\|M_n\|^q \leq C\mathbf{E}\Big(\sum_{i=1}^n \|D_i\|^p\Big)^{q/p}. \tag{9.4.1}$$

Remark 9.4.1. A Banach space $\boldsymbol{X}$ is said to be ζ-convex if there exists a symmetric, biconvex function ζ on $\boldsymbol{X} \times \boldsymbol{X}$ satisfying the following two conditions: $\zeta(0,0) > 0$, and $\zeta(\boldsymbol{x}, \boldsymbol{y}) \leq \|\boldsymbol{x} + \boldsymbol{y}\|$, if $\|\boldsymbol{x}\| \leq 1 \leq \|\boldsymbol{y}\|$. It turns out[6] that $\boldsymbol{X}$ is ζ-convex if, and only if, for any $p, 1 < p < \infty$, there exists a constant C_p such that

$$\mathbf{E}\| \pm D_1 \pm \cdots \pm D_n\|^p \leq C_p^p\, \mathbf{E}\|M_n\|^p, \qquad n = 1, 2, \ldots, \tag{9.4.2}$$

for all $\boldsymbol{X}$-valued martingales (M_n), and all sequences ± 1.

In connection with ζ-convexity, and the Marcinkiewicz-Zygmund type inequality (9.4.1) it is worthwhile to observe that if $\boldsymbol{X}$ is of Rademacher-type p, and is also ζ-convex, then the inequality (9.4.1) is satisfied. Indeed, let (r_i) be a Rademacher sequence independent of (M_n). Then, by ζ-convexity and Fubini's Theorem,

$$\mathbf{E}\|M_n\|^q \leq C\mathbf{E}_r\mathbf{E}\Big\|\sum_{i=1}^n r_i D_i\Big\|^q \leq C\mathbf{E}\Big(\mathbf{E}_r\Big\|\sum_{i=1}^n r_i D_i\Big\|^q\Big)^{1/q\cdot q}$$

$$\leq C\mathbf{E}\Big(\mathbf{E}_r\Big\|\sum_{i=1}^n r_i D_i\Big\|^p\Big)^{1/p\cdot q} \leq C\mathbf{E}\Big(\sum_{i=1}^n \|D_i\|^p\Big)^{q/p},$$

[5]Due to P. Assuad (1975).

[6]This result is due to D.L. Burkholder (1981), who also introduced the concept of ζ-convexity.

where the last two inequalities follow, respectively, from Kahane's Theorem (see Chapter 1), and the definition of Rademacher-type p.

The above fact immediately implies that ζ-convex Banach spaces of Rademacher-type p are p smoothable. On the other hand, although p-smooth spaces are necessarily of Rademacher-type p, they need not be ζ-convex.

The next result is an extension of the Brunk's type Strong Law proven for independent summands in the previous section. The theorem, as well as the following later extension of the Marcinkiewicz-Zygmund Law, permit use of moments of order greater than 2 in establishing the asymptotics of Banach space valued martingales.

Theorem 9.4.1.[7] *Let $1 \le p \le 2$, $q \ge 1$, and assume that the Banach space $\mathbf{X}$ is p-smoothable. Then:*

(i) If (M_n) is an $\mathbf{X}$-valued martingale such that

$$\sum_{n=1}^{\infty} \frac{\mathbf{E}||D_n||^{pq}}{n^{pq+1-q}} < \infty, \tag{9.4.3}$$

then $||M_n|| = o(n)$, almost surely.

(ii) For every $\epsilon > 0$ there exists a positive constant C such that, for any $\mathbf{X}$-valued martingale (M_n)

$$\sum_{n=1}^{\infty} n^{-1}\mathbf{P}(||M_n/n|| > \epsilon) \le C \sum_{n=1}^{\infty} \frac{\mathbf{E}||D_n||^{pq}}{n^{pq+1-q}}. \tag{9.4.4}$$

Proof. (*i*) The case $q = 1$ is covered in Section 3.2 of Chapter 3. So, assume that $q > 1$. Then $||M_n||^{pq}$ is a real-valued submartingale, and by the well-known Hajek-Renyi-Chow's type inequality we get that, for every $\epsilon > 0$

$$\epsilon^{pq}\mathbf{P}\big(\sup_{j\ge n} ||M_j/j|| > \epsilon\big) = \epsilon^{pq} \lim_{m\to\infty} \mathbf{P}\Big(\sup_{n\le j\le m} ||M_j/j||^{pq} > \epsilon^{pq}\Big) \tag{9.4.5}$$

[7]Due to W.A. Woyczyński (1982).

$$\leq n^{-pq}\mathbf{E}\|M_n\|^{pq} + \sum_{j=n+1}^{\infty} j^{-pq}\mathbf{E}\Big(\|M_j\|^{pq} - \|M_{j-1}\|^{pq}\Big).$$

In view of Proposition 9.4.1, and Hölder's Inequality,

$$\mathbf{E}\|M_j\|^{pq} \leq C\mathbf{E}\Big(\sum_{i=1}^{n}\|D_i\|^p\Big)^q \leq Cj^{q-1}\sum_{i=1}^{j}\mathbf{E}\|D_i\|^{pq},$$

so that, by Kronecker's Lemma, $j^{pq}\mathbf{E}\|M_j\|^{pq} \to 0$, as $j \to \infty$. Also, the series on the right-hand side of (9.4.5) converges because of Proposition 9.4.1. Hence, summing by parts, we obtain the inequalities,

$$\sum_{j=1}^{n}\big((j-1)^{-pq} + j^{-pq}\big)\mathbf{E}\|M_j\|^{pq}$$

$$\leq \sum_{j=1}^{n}\big((j-1)^{-pq} + j^{-pq}\big)j^{q-1}\sum_{i=1}^{j}\mathbf{E}\|D_i\|^{pq}$$

$$\leq \text{const}\Big(\sum_{j=1}^{n}\frac{\mathbf{E}\|D_j\|^{pq}}{j^{pq+1-q}} + \frac{1}{n^{pq+1-q}}\sum_{i=1}^{n}\mathbf{E}\|D_j\|^{pq}\Big),$$

so that, for every $\epsilon > 0$, $\mathbf{P}(\sup_{j\geq n}\|M_j/j\| > \epsilon) \to 0$, as $n \to \infty$.

(*ii*) By Chebyshev's and Hölder's Inequalities, in view of Proposition 9.4.1,

$$\sum_{n=1}^{\infty} n^{-1}\mathbf{P}(\|M_n\| > \epsilon n) \leq \sum_{n=1}^{\infty} n^{-1}n^{-pq}\epsilon^{-pq}\mathbf{E}\|M_n\|^{pq}$$

$$\leq \epsilon^{-pq}C\sum_{n=1}^{\infty} n^{-1+(q-1)-pq}\sum_{k=1}^{n}\mathbf{E}\|D_k\|^{pq}$$

$$\leq C\epsilon^{-pq}\sum_{k=1}^{\infty}\mathbf{E}\|D_k\|^{pq}\sum_{n=k}^{\infty} n^{-pq+q-2}$$

$$\leq C\epsilon^{-pq}\sum_{k=1}^{\infty}\mathbf{E}\big(\|D_k\|^{pq}/k^{pq+1-q}\big).$$

QED

In the next step we will prove the analogue of the Marcinkiewicz-Zygmund Strong Law of Large Numbers for vector-valued martingales with uniformly bounded tail probabilities. Recall that a sequence (X_i) of random vectors is said to have uniformly bounded tail probabilities by tail probabilities of a positive real random variable X_0, if there exists a positive constant C such that, for all $t > 0$, and all $i = 1, 2, \dots$, $\mathbf{P}(\|X_i\| > t) \leq C\mathbf{P}(X_0 > t)$.

Theorem 9.4.2[8] *Let (M_n) be a martingale with values in a Banach space $\boldsymbol{X}$, with the difference sequence (D_n).*

(i) If the difference sequence (D_n) has uniformly bounded tail probabilities by an $X_0 \in \boldsymbol{L}\log\boldsymbol{L}$, and $\boldsymbol{X}$ is superreflexive, then

$$M_n = o(n), \quad a.s. \tag{9.4.6}$$

(ii) If the difference sequence (D_n) has uniformly bounded tail probabilities by an $X_0 \in \boldsymbol{L}^p, 1 < p < 2$, and $\boldsymbol{X}$ is r-smoothable for an $r > p$, then

$$M_n = o(n^{1/p}), \quad a.s. \tag{9.4.7}$$

The proof of the above theorem depends on the following two Lemmas:

Lemma 9.4.1. *Let $1 \leq p < 2$, and let (X_n) be a sequence of real-valued random variables with tail probabilities uniformly bounded by tail probabilities of $X_0 \in \boldsymbol{L}^p$. Then, if $X'_n := X_n I[|X_n| \leq n^{1/p}]$, and $r > p$, then*

$$\sum_{n=1}^{\infty} \frac{\mathbf{E}|X'_n|^r}{n^{r/p}} < \infty.$$

Proof. The proof involves a straightforward calculation:

$$\sum_{n=1}^{\infty} \frac{\mathbf{E}|X'_n|^r}{n^{r/p}} = \sum_{n=1}^{\infty} n^{-r/p} \int_0^{n^{1/p}} t^r d\mathbf{P}(|X_n| \leq t)$$

[8]Due to W.A. Woyczyński (1982).

$$= \sum_{n=1}^{\infty} n^{-r/p}\Big[n^{r/p}\mathbf{P}(|X_n| \le n^{1/p}) - r\int_0^{n^{1/p}} t^{r-1}\mathbf{P}(|X_n| \le t)\,dt\Big]$$

$$\le C\sum_{n=1}^{\infty}\Big[1 - rn^{-r/p}\int_0^{n^{1/p}} t^{r-1}\big(1-\mathbf{P}(|X_n| > t)\big)\,dt\Big]$$

$$= C\sum_{n=1}^{\infty} rn^{-r/p}\int_0^{n} t^{r-1}\mathbf{P}(|X_n| > t)\,dt$$

$$= C\sum_{n=1}^{\infty}\int_0^1 \mathbf{P}\big(X_0 s^{-1/r} > n^{1/p}\big)\,ds$$

$$\le C\mathbf{E}X_0\int_0^1 s^{-p/r}ds = C\frac{r}{r-p}\mathbf{E}X_0^p < \infty. \qquad \text{QED}$$

Lemma 9.4.2. *Let (X_n) be a sequence of real-valued random variables with tail probabilities uniformly bounded by tail probabilities of X_0.*

(i) If $X_0 \in \boldsymbol{L}\log\boldsymbol{L}$, and $X_n'' := X_n I[|X_n| > n]$, then

$$\sum_{n=1}^{\infty}\frac{\mathbf{E}|X_n''|}{n} < \infty.$$

(ii) If $X_0 \in \boldsymbol{L}^p, p > q \ge 1$, and $X_n'' := X_n I[|X_n| > n^{1/p}]$, then

$$\sum_{n=1}^{\infty}\frac{\mathbf{E}|X_n''|^q}{n^{q/p}} < \infty.$$

Proof. Again, the proof depends on straightforward calculations which we are including for the sake of completeness.

(i)

$$\sum_{n=1}^{\infty}\frac{\mathbf{E}|X_n''|}{n} = \sum_{n=1}^{\infty} n^{-1}\int_n^{\infty} t d\mathbf{P}(|X_n| \le t)$$

$$= \sum_{n=1}^{\infty}\Big[\mathbf{P}(|X_n| > n) + n^{-1}\int_n^{\infty}\mathbf{P}(|X_n| > t)\,dt\Big]$$

$$\le C\Big[\mathbf{E}X_0 + \sum_{n=1}^{\infty} n^{-1}\sum_{k=n}^{\infty}\mathbf{P}(X_0 > k)\Big]$$

$$= C\Big[\mathbf{E}X_0 + \sum_{k=1}^{\infty}\sum_{n=1}^{k} n^{-1}\mathbf{P}(X_0 > k)\Big]$$

$$= C\Big[\mathbf{E}X_0 + \sum_{k=1}^{\infty}\log k\mathbf{P}(X_0 > k)\Big] \le C\big[\mathbf{E}X_0 + \mathbf{E}X_0\log^+ X_0\big] < \infty.$$

(ii) Similarly,

$$\sum_{n=1}^{\infty}\frac{\mathbf{E}|X_n''|^q}{n^{q/p}} = \sum_{n=1}^{\infty} n^{-q/p}\int_{n^{1/p}}^{\infty} t^q d\mathbf{P}(|X_n| \le t)$$

$$= \sum_{n=1}^{\infty}\Big[\mathbf{P}(X_0 > n^{1/p}) + q\int_1^{\infty} s^{q-1}\mathbf{P}(X_0/s > n^{1/p})\,ds\Big] \le C\mathbf{E}X_0^p.$$

QED

Proof of Theorem 9.4.2. For $n = 1, 2, \ldots$, denote

$$D_n' = D_n I[\|D_n\| \le n^{1/p}], \qquad D_n'' = D_n I[\|D_n\| > n^{1/p}],$$

$$\Delta_n' = D_n' - \mathbf{E}(D_n'|\mathcal{F}_{n-1}), \qquad \Delta_n'' = D_n'' - \mathbf{E}(D_n''|\mathcal{F}_{n-1}).$$

Then (Δ_n'), and $\Delta_n'')$, are martingale difference sequences and, since $\mathbf{E}(D_n' + D_n''|\mathcal{F}_{n-1}) = 0$,

$$D_n = \Delta_n' + \Delta_N''. \tag{9.4.8}$$

To prove (i) set $p = 1$ in the above notation, and observe that since $\boldsymbol{X}$ is superreflexive it is r-smoothable for an $r > 1$. Thus, by (9.4.8), the triable inequality, and the Proposition 9.4.1,

$$\mathbf{E}\Big\|\sum_{k=1}^{n}\frac{D_k}{k}\Big\| \le \Big(\mathbf{E}\Big\|\sum_{k=1}^{n}\frac{\Delta_k'}{k}\Big\|^r\Big)^{1/r} + \sum_{k=1}^{n}\frac{\mathbf{E}\|\Delta_k''\|}{k} \tag{9.4.9}$$

$$\le C\Big(\sum_{k=1}^{n}\frac{\mathbf{E}\|\Delta_k'\|^r}{k^r}\Big)^{1/r} + \sum_{k=1}^{n}\frac{\mathbf{E}\|\Delta_k''\|}{k}$$

$$\le C\Big(\sum_{k=1}^{n}\frac{\mathbf{E}\|D_k'\|^r}{k^r}\Big)^{1/r} + \sum_{k=1}^{n}\frac{\mathbf{E}\|D_k''\|}{k}.$$

The first series converges by Lemma 9.4.1, and the second, by Lemma 9.4.2(*i*). Thus, the martingale $N_n := \sum_{k=1}^n D_k/k, n = 1, 2, \ldots,$ is $\boldsymbol{L}^1$-bounded, and since the superreflexive space is dentable, N_n converges almost surely. Now, an application of Kronecker's lemma completes the proof of (*i*).

To show (*ii*) we proceed exactly as in (9.4.9), except that now $p > 1$. Thus we get that

$$\mathbf{E}\Big\|\sum_{k=1}^n \frac{D_k}{k^{1/p}}\Big\| \le C\Big(\sum_{k=1}^n \frac{\mathbf{E}\|\Delta_k'\|^r}{k^{r/p}}\Big)^{1/r} + 2\sum_{k=1}^n \frac{\mathbf{E}\|D_k''\|}{k^{1/p}},$$

and an application of Lemma 9.4.1, and Lemma 9.4.2(*ii*), together with the dentability of $\boldsymbol{X}$, give us the almost sure convergence of $\sum_{k=1}^n D_k/k^{1/p}$, thus assuring, again via Kronecker's Lemma, that $M_n = o(n^{1/p})$. QED

Finally, we shall prove two results about integrability of the maximal function

$$M^{(p)}(\omega) := \sup_n \frac{\|M_n(\omega)\|}{n^{1/p}},$$

for Banach space-valued martingales.

Theorem 9.4.3. *Let (M_n) be a martingale with values in a Banach space $\boldsymbol{X}$.*

(i) If the difference sequence (D_n) has uniformly bounded tail probabilities by an $X_0 \in \boldsymbol{L}\log\boldsymbol{L}$, and $\boldsymbol{X}$ is superreflexive, then $M^{(1)} \in \boldsymbol{L}^1$.

(ii) If the difference sequence (D_n) has uniformly bounded tail probabilities by an $X_0 \in \boldsymbol{L}^p, 1 \le q < p < 2$, and $\boldsymbol{X}$ is r-smoothable for an $r > p$, then $M^{(p)} \in \boldsymbol{L}^q$.

Proof. By Kronecker's lemma, and the Closed graph Theorem, there exists a constant $C > 0$ such that, for any $(\boldsymbol{x}_n) \subset \boldsymbol{X}$,

$$\sup_n \frac{1}{n^{1/p}}\Big\|\sum_{k=1}^n \boldsymbol{x}_k\Big\| \le C \sup_n\Big\|\sum_{k=1}^n \frac{\boldsymbol{x}_k}{k^{1/p}}\Big\|. \tag{9.4.10}$$

Therefore, using the notation from the proof of Theorem 9.4.1, we get that

$$M^{(p)} \le C\Big(\sup_n \Big\|\sum_{k=1}^n \frac{\Delta_k'}{k^{1/p}}\Big\| + \sup_n \Big\|\sum_{k=1}^n \frac{\Delta_k''}{k^{1/p}}\Big\|\Big).$$

(i) Here, the assumption is that $\boldsymbol{X}$ is superreflexive and hence r-smoothable for an $r > 1$. So, by the Davis type inequality[9],

$$\mathbf{E}\sup_n \Big\|\sum_{k=1}^n \frac{\Delta_k'}{k}\Big\| \le C\sum_{k=1}^\infty \frac{\mathbf{E}\|\Delta_k'\|^r}{k^r}.$$

The latter series is finite by Lemma 9.4.1. On the other hand,

$$\mathbf{E}\sup_n \Big\|\sum_{k=1}^n \frac{\Delta_k''}{k}\Big\| \le C\sum_{k=1}^\infty \frac{\mathbf{E}\|D_k''\|^r}{k^{r/p}},$$

and the last series is also finite by Lemma 9.4.2(i). Thus $\mathbf{E}M^{(1)} < \infty$.

(ii) In this case, with $1 \le q < p < r < 2$, and $\boldsymbol{X}$ being r-smoothable, using the cited above Davis type inequality, we get the inequality

$$\Big(\mathbf{E}\sup_n \Big\|\sum_{k=1}^n \frac{\Delta_k'}{k^{1/p}}\Big\|^q\Big)^{r/q} \le C\sum_{n=1}^\infty \frac{\mathbf{E}\|\Delta_k'\|^r}{k^{r/p}},$$

which is finite by Lemma 9.4.1. Then again, applying the above Davis type inequality, and the fact that r-smoothablility implies q-smoothability for $q < r$, we get that

$$\mathbf{E}\sup_n \Big\|\sum_{k=1}^n \frac{\Delta_k''}{k^{1/p}}\Big\|^q \le C\sum_{n=1}^\infty \frac{\mathbf{E}\|\Delta_n''\|^q}{n^{q/p}} \le C\sum_{n=1}^\infty \frac{\mathbf{E}\|D_n''\|^q}{n^{q/p}},$$

and the last series converges in view of Lemma 9.4.2(ii). Thus $\mathbf{E}(M^{(p)})^q < \infty$. QED

[9]See W.A. Woyczyński (1976), Theorem 6.

Bibliography

Araujo, A, and Gine, E.,

(1978) Type, cotype, and Lévy measures in Banach spaces, *Annals of Probability* 6, 637-643.

(1979) On tails and domains of attraction of stable measures in Banach spaces, *Transactions Amer. Math.Soc.* 248, 105-119.

Asplund, E.,

(1967) Averaged norms, *Israel J. Math* 5, 227-233

Asplund, E., and Namioka, I.,

(1967) A geometric proof of Ryll-Nardzewski's fixed point theorem, *Bull. Amer. Math.Soc.* 73, 443-445.

Assuad, P.,

(1974) Martingales et rearrangement dans les espaces uniformement lisses, preprint, Orsay, Julliet 1974.

(1974/75) Espaces p-lisses, rearrangements, *Seminaire Maurey-Schwartz*, Exp. XVI.

(1975) Espaces p-lisses at q-convexes. Inegalites de Burkholder. *Seminaire Maurey-Schwartz*, Exp. XV.

Azlarov, T.A., and Volodin, N.A.,

(1981) The law of large numbers for identically distributed Banach space valued random variables, *Theorya veroyatnostey i prim* 26, 584-590.

Badrikian, A.,

(1976) Prolegomenes au calcul des probbilites dans les Banach, Springer *Lecture Notes in Math.* 539,1-167.

Beauzamy, B.,

(1973/74) Espaces de Banach uniformement convexifiables, *Seminaire Maurey-Schwartz*, XIII, 1-18.

(1974) Operateurs uniformement convexifiables, Preprint, Centre de Mathematiques, Ecole Polytechnique, Paris, Octobre 1974.

(1974/75) Proprietes geometriques des espaces d'interpolation, *Seminaire Maurey-Schwartz*, Exp. XIV, 1-17.

Beck, A.,

(1962) A convexity condition in Banach spaces and the strong law of large numbers, *Proc. Amer. Math. Soc.* 13, 329-334.

(1963) On the strong law of large numbers, *Ergodic Theory, Proc. Intern. Symp.*, Academic Press, New York, 21-53.

(1976) Conditional independence, *Z. Warsch.verw. Geb.* 33, 253-268.

Bessaga, C., and Pelczynski, A.,

(1958) On basis and unconditional convergence of series in Banach spaces, *Studia Math.* 17, 151-164.

Brown, D.R.,

(1974) B-convexity and reflexivity in Banach spaces, *Transactions Amer. Math. Soc.* 187, 69-76.

(1974) P-convexity and B-convexity in Banach spaces, *Transactions Amer. Math. Soc.* 187, 77-81.

Brunel, A.,

(1973/74) Espaces associes à une suite borné dans un espace de Banach, *Seminaire Maurey-Schwartz*, Exp. XV. et XVIII.

Brunel, A., and Sucheston, L.,

(1974) On B-convex Banach spaces, *Math Systems Theory* 7, 294-299.

(1975) Sur quelques conditions equivalentes a la superreflexivité dans les espaces de Banach, *Comptes Rendus Acad. Sci., Paris* A275, 993-994.

(1975) On J-convexity and some ergodic super-properties of Banach spaces, *Transactions Amer. Math. Soc.* 204, 21-33.

(1976) On sequences invariant under spreading in Banach spaces, Springer *Lecture Notes in Math.* 204, 21-33.

Brunk, H.D.,

(1948) The strong law of large numbers, *Duke Math. J.* 15, 181-195.

Buldygin, V.V.,

(1973) On random series in Banach spaces, *Teorya Veroyatnostey i Prim.* 18, 491-504.

Burkholder, D.L.,

(1981) A geometrical characterization of Banach spaces in which martingale difference sequences are unconditional, *Annals of Probability* 9, 997-1011.

Chatterji, S.D.,

(1960) Martingales of Banach-valued random variables, *Bull. Amer. Math. Soc* 66, 395-398.

(1964) A note on the convergence of Banach-space valued martingales, *Math. Ann.* 153, 142-149.

(1969) Martingale convergence and the Radon-Nikodym theorem in Banach spaces, *Math. Scandinavica* 22, 21-41.

Chevet, S., Chobanyan, S.A., Linde, W. , and Tarieladze, V.I.,

(1977) Characterization of certain classes of Banach spaces by Gaussian measures, *Comptes Rendus Acad. Sci. Paris* A 285 .

Chobanyan, S.A., and Tarieladze, V.I.,

(1977) Gaussian characterization of certain Banach spaces, *J. Multivariate Anal.* 7, 183-203.

Clarkson, J.A.,

(1936) Uniformly convex spaces, *Trans. Amer. Math. Soc.* 40, 396-414.

Davis, W.J.,

(1973/74) The Radon-Nikodym property, *Seminaire Maurey-Schwartz*, 0.1-0.12.

Davis, W.J., and Phelps, R.R.,

(1974) The Radon-Nikodym property and rentable sets in Banach spaces, *Proc. Amer. Math. Soc.* 49, 119-122.

Davis, W.J., Johnson, W.B., and Lindenstrauss, J.,

(1976) The l_2^n problem and degrees of non-reflexivity, *Studia Math.* 55, 123-139.

Davis, W.J., , and Lindenstrauss, J.,

(1976) The l_2^n problem and degrees of non-reflexivity II, *Studia Math.* 58, 1179-196.

Day, M.M.,

(1944) Uniform convexity in factor and conjugate spaces, *Ann. Math.* 45, 375–385.

(1973) *Normed Linear Spaces*, Third Edition, Springer-Verlag, New York.

de Acosta, A.,

(1981) Inequalities of B-valued random vectors with applications to the strong law of large numbers, *Annals of Probability* 9, 157-161.

de Acosta, A., and Samur, J,

(1979) Infinitely divisible probability measures and the converse Kolmogorov inequality in Banach spaces. *Studia Math.* 66, 143-160.

Diestel, J.,

(1975) *Geometry of Banach Spaces*, Lecture Notes in Mathematics 485, Springer-Verlag, New York.

Diestel, J., and Uhl, Jt, J.J.,

(1976) The Radon-Nikodym Theorem for Banach space valued measures, *Rocky Mountains J.* 6, 1-46.

Dubinsky, E., Pelczynski, A., and Rosenthal, H.P.,

(1972) On Banach spaces X for which $\Pi_2(L_\infty, X) = B(L_\infty, X)$, *Studia Math.* 44, 617-648.

Dunford, N., and Schwartz, J.T.,

(1958) *Linear Operators, Part I: General Theory*, Interscience, New York.

Dvoretzky, A.,

(1961) Some results on convex bodies in Banach spacers, *Proc. International Symposium on Linear Spaces*, New York, 123-160.

Elton, J.,

(1981) A law of large numbers for identically distributed martingale differences, *Annals of Probability* 9, 405-412.

Enflo, P.,

(1972) Banach spaces which can be given an equivalent uniformly convex norm, *Israel J. Math.* 13, 281-288.

Enflo, P., Lindenstrauss, J., and Pisier, G.,

(1975) On the "three space problem", *Math. Scand.* 36, 199-210.

Erdös, P., and Magidor, M.,

(1976) A note on regular methods of summability and the Banach-Saks property, *Proc. Amer. Math. Soc.* 59, 232-234.

Farahat, J.,

(1974/75) Examples d'espaces B-convexes non-reflexifs, d'apres James et Lindenstrauss, *Sem. Maurey-Schwartz*, Exp. XX.

Fernique, X.,

(1970) Integrabilite des vecteurs Gaussiens, *Comptes Rendus Acad. Sci. Paris* 220, 1698-1699.

Figiel, T.,

(1974/75) A short proof of the Dvoretzky's theorem, *Sem. Maurey-Schwartz*, Exp. XXIII.

(1976) On the moduli of convexity and smoothness, *Studia Math.* 56, 121-155.

Garling, D.J.H.,

(1975) Lattice bounding, Radonifying and summing mappings, *Math. Proc. Cambridge Phil. Soc.* 77, 327-333.

(1976) Functional central limit theorem in Banach spaces, *Ann. Prob.* 4, 600-611.

Giesy, D.P.,

(1969) On a convexity condition in normed linear spaces, *Transactions Amer. Math. Soc.* 125(1966), 114-146. Additions and corrections, ibid. 140, 511-512.

(1972) Super-reflexivity, stability, and B-convexity, *Western Michigan Math Report* 29.

(1973) B-convexity and reflexivity, *Israel J. Math.,* 15, 430-436.

(1973) The completion of a B-convex normed Riesz space is reflexive, *J. Functional Anal.* 12, , 188-198.

Giesy, D.P., and James, R.C.,

(1973) Uniformly non -l_1 and B-convex Banach spaces, *Studia Math.* 48, 61-69.

Giladi, O., Prochno, J., Schütt, C., Tomczak-Jaegermann, N., and Werner, E.,

(2017) On the geometry of projective tensor products, *J. Functional Analysis* 273, 471-495.

Hoffmann-Jorgensen, J.,

(1972/73) On the modulus of smoothness and the G_α-conditions in B-spaces, *Aarhus Universitet, Matematisk Institut, Preprint Series.*

(1972/73) Sums of independent Banach space valued random variables, *Aarhus Universitet, Matematisk Institut, Preprint Series* 15 , 1-96.

(1974) Sums of independent Banach space valued random variables, *Studia Math.* 52, 159-186.

(1974) The strong law of large numbers and the central limit theorem in Banach spaces, , *Aarhus Universitet, Matematisk Institut, Preprint Series*, September 1974.

(1973/74) Kwapien's proof of "$c_0 \subset L^1(E) \Leftrightarrow c_0 \subset E$" , *Aarhus Universitet, Matematisk Institut, Preprint Series* 13.

Hoffmann-Jorgensen, J., and Pisier, G.,

(1976) The law of large numbers and the central limit theorem in Banach spaces, *Ann. Prob.* 4, 587-599.

Huff, R.E.,

(1974) Dentability and the Radon-Nikodym Property, *Duke Math. J.* 41, 111-114.

Huff, R.E., and Morris, P.D.,

(1975) Dual spaces with the Krein-Milman property have the Radon-Nikodym property, , *Proc. Amer. Math. Soc.* 49, 104-108.

(1976) Geometric characterization of the Radon-Nikodym Property in Banach spaces, *Studia Math.* 56, 157-164.

Ito, K., and Nisio, S.,

(1968) On the convergence of sums of independent Banach space valued random variables, *Osaka J. Math.* 5, 35-48.

Jain, N.C.,

(1975) Tail probabilities for sums of independent Banach space valued random variables, *Zeit. Wahr. verw. Geb.* 33, 155-166.

(1976) Central limit theorem in Banach spaces, *Springer Lecture Notes in Math.* 526, 113-131.

(1976) An example concerning CLT and LIL in Banach spaces, *Ann. Prob.* 4, 690-694.

(1976) Central limit theorem and related questions in Banach spaces, *Proc. AMS Probability Symposium*, Urbana.

Jain, N.C., Marcus, M.B.,

(1975) Integrability of infinite sums of independent vector-valued random variables, *Trans. Amer. Math. Soc.* 212, 1-36.

James, R.C.,

(1964) Uniformly non-square Banach spaces, *Ann. Math.* 80, 542-550.

(1972) Some self-dual properties of normed linear spaces, *Ann. Math. Studies* 69, 159-176.

(1972) Super-reflexive spaces with bases, *Pacific J. Math.* 4, 409-420.

(1974) A non-reflexive Banach space that is uniformly non-octahedral, *Israel J. Math.* 18, 145-155.

James, R.C., and Lindenstrauss, J.,

(1975) The octahedral problem, *Proc. Sem. "Random Series Convex Sets and Geometry of Banach Spaces"* Aarhus, 100-120.

Johnson, W.B.,

(1974) On finite dimensional subspaces of Banach spaces with local unconditional structure, *Studia Math.* 51, 225-240.

Kadec, M.I.,

(1954) On conditionally convergent series in spaces L^p, *Uspekhi Mat. Nauk* 9, 107-109.

Kahane , J.P.,

(1968) *Some Random Series of Functions*, Heath, Lexington, MA.

Kantorovich, L.V., Vulikh, B.Z., and Pinsker, A.G.,

(1950) *Functional Analysis in Semiordered Spaces*, Moscow.

Kelley, J., and Namioka, I.,

(1963) *Linear Topological Spaces*, Van Nostrand, New York.

Kottman, C.A.,

(1970) Packing and reflexivity in Banach spaces, *Trans. Amer. Math. Soc.* 150, 565-576.

Krivine, J.L.,

(1973/74) Théorèmes de factorization dans les espaces réticulés, *Sem. Maurey-Schwartz*, Exp. XXII et XXIII.

(1976) Sous-espaces de dimension finie des espaces de Banach réticulés, *Ann. Math.* 104, 1-29.

Kuelbs, J.,

(1976) A counterexample for Banach space valued random variables, *Ann. Prob.* 4, 690-694.

(1977) Kolmogorov's law of the iterated logarithm for Banach space valued random variables, *Illinois J. Math.*21, 784-800.

Kuelbs, J., abd Woyczyński, W.A.,

(1978) Lacunary series and exponential moments, *Proc. Amer. Math. Soc.* 68, 281-291.

Kvaracheliya, V.A., and Tien, N.Z.,

(1976) The central limit theorem and the strong law of large numbers in $l_p(X)$-spaces, $1 \leq p < \infty$, *Teorya Veroyat. i Prim.*, 21, 802-812.

Kwapień, S.,

(1972/73) Isomorphic characterization of Hilbert spaces by orthogonal series with vector valued coefficients, *Sem. Maurey-Schwartz*, Exp. VIII.

(1974) On Banach spaces containing c_0, *Studia Math.* 52, 187-188.

(1976) A theorem on the Rademacher series with vector valued coefficients, *Springer Lecture Notes in Math.* 526, 157-158.

Kwapień, S., and Woyczyński, W.A.,

(1992) *Random Series and Stochastic Integrals: Single and Multiple*, Birkhäuser. Boston.

Lai T.L.,

(1974) Convergence rates in the strong law of large numbers for random variables taking values in Banach spaces, *Bull. Inst. Math. Academia Sinica* 2, 67-85.

Landau, H.J., and Shepp, L.A.,

(1970) On the supremum of a Gaussian process, *Sankhya* 32, 369-378.

LeCam, L.,

(1970) Remarque sur le theoreme limite central dans les espaces localement convexes, *Les Probabilities sur les Structures Algebriques*, Colloq. CNRS Paris, 233-249.

Lewis, D.R.,

(1972) A vector measure with no derivative, *Proc. Amer. math. Soc.* 32, 535-536.

Linde, V., and Pietsch, A.,

(1974) Mappings of Gaussian cylindrical measures in Banach spaces, *Theorya Veroyat. i Prim.* 19, 472-487.

Lindenstrauss, J.,

(1963) On the modulus of smoothness and divergent series in Banach spaces, *Michigan Math. J.* 10, 241-252.

(1975/76) The dimension of almost spherical sections of convex bodies, *Seminaire Mayrey-Schwartz*, Exp. XIX.

(1976) Type and superreflexivity, *Bull. London Math. Soc.* 8, 15.

Lindenstrauss, J., and Tazfriri, L.,

(1973) *Classical Banach Spaces*, Lecture Notes in Math. 338, Springer.

Mandrekar, V.,

(1977) Characterization of Banach spaces through validity of Bochner theorem, *Proc. Conf. Dublin*, Lecture Notes in Math. Springer, Berlin.

Marcinkiewicz, J. and Zygmund, A.,

(1937) Sur les fonctions independantes, *Fundamenta. Math.* 29, 60-90.

Marcus, M.B., and Woyczyński, W.A.,

(1977) Domaines d'attraction normale dans léspace de tupe p-stable, *Comptes Rendus Aced. Sci. Paris*, 285, 915-917.

(1978) A necessary condition for the central limit theorem on spaces of stable type, *Proc. Conf. Dublin, Lecture Notes in Math.* 644, 327-339.

(1979) Stable measures and central limit theorems in spaces of stable type, *Transactions Amer. math.Soc* 251, 71-102.

Maurey, B.,

(1972/73) Espaces de cotype $p, 0 < p \leq 2$, *Sem. Maurey-Schwartz*, Exp. VII.

(1972/73) Theoremes de Nikishin: Theoremes de factorization pour les applications lineaires a valeus dans un espace L^0, *Sem. Maurey-Schwartz*, Exp. X, XI.

(1972/73) Une nouvelle demonstration d'un theoreme de Grothendiesk, *Sem. Maurey-Schwartz*, Exp. XXII.

(1974) Theoremes de factorization pour les operateurs lineaires à valeurs dans les espaces L^p, *Asterisque 11.*

(1973/74) Nuveaux theoremes de Nikishin, *Sem. Maurey-Schwartz*, Exp. IV, V.

(1973/74) Type et cotype dans les espaces munis de structures locales unconditionelles, *Sem. Maurey-Schwartz*, Exp. XXIV, XXV.

Maurey, B., and Pisier, G.,

(1973 Characterization d'une classe d'espaces de Banach par desproprietes de sésries aléatoires vectorielles, *Comptes Rendus Acad. Sci, Paris.* 277), 687-690.

(1974/75) Remarques sur léxpose d'Assuad, *Sem. Maurey-Schwartz*, Annexe 1.

(1976) Series de variables aleatoires vectorielles independantes et proprietes geometriques des espaces de Banach, *Studia Math.* 58, 45-90.

Maynard, H.B.,

(1973) A geometrical characterization of Banach spaces with the Radon-Nikodym Property, *Trans. Amer. Math. Soc.* 185, 493-500.

Metivier, M.,

(1963) Limites projectives de mesures; martingales; applications, *Ann. Math. Pura Appl.* 63, 225-352.

Milman, V.D.,

(1971) Geometric theory of Banach spaces, Part II; Geometry of the unit sphere, *Uspekhi Mat. Nauk* 26, 73-149.

Mouchtari, D.,

(1975/76) Sur l'existence d'une topolgie du type de Sazonov sur un espace de Banach, *Sem. Maurey-Schwartz*, Exp. XVII.

(1973) Some general questions of the theory of probability measures in linear spaces, *Teorya Veroyat. i Prim.* 18, 66-77.

Musial, K., Ryll-Nardzewski, C., and Woyczyński, W.A.,

(1974) Convergence presque sure des series aleatoires vectorielles a multiplicateur bornes, *Comptes Rendus Acad. Sci Paris* 279, 225-228.

Neveu, J.,

(1972) *Martingales a temps discret*, Masson et Cie, Paris.

Nielsen, N.J.,

(1973) On Banach ideals determined by Banach lattices and their applications, *Dissertationes Math.* 109, 1-66.

Nishiura, T. and Waterman, D.

(1963) Reflexivity and summability, *Studia Math.* 23, 53-57.

Nordlander, G.

(1961) On sign-independent and almost sign-independent convergence in normed linear spaces, *Arkiv Mat.* 21, 287-296.

Padgett, W.J., and Taylor, R.L.,

(1975) Stochastic convergence of weighted sums in normed linear spaces, *J. Multivariate Analysis*, 5, 434-450.

(1973) *Laws of Large Numbers for Normed Linear Spaces and Certain Frechet Spaces*, *Springer Lecture Notes in Math.* 360.

(1973) Weak laws of large numbers in Banach spaces and their extensions, *Springer Lecture Notes in Math.* 360, 66-83.

(1976) Almost sure convergence of weighted sums of random elements in Banach spaces, *Springer Lecture Notes in Math.* 526, 187-202.

Parthasarathy, K. R.,

(1967) *Probability Measures on Metric Spaces*, Academic Press, New York.

Paulauskas V.,

(1976) Infinitely divisible and stable probability measures on separable Banach spaces, *University of Göteborg Report*-15.

Pelczynski, A.,

(1975) On unconditional bases and Rademacher averages, *Springer Lecture Notes in Math.* 472, 119-130.

Petrov, V.V.

(1972) *Sums of independent random variables*, Moscow.

Pietsch, A.,

(1967) Absolut p-summierende Abbildungen in normierten Raumen, *Studia Math.* 28, 333-353.

Pisier, G.,

(1972/73) Bases, suites lacunaires dans les espaces L^p *Sem. Maurey-Schwartz*, Exp. XVIII, XIX.

(1972/73) Sur les espaces qui ne contiennent pas de ∞_n uniformement, *Sem. Maurey-Schwartz*, Annexe.

(1973/74) Sur les espaces qui ne contiennent pas de l_n^1 uniformement, *Sem. Maurey-Schwartz*, Exp. VII., and *Comptes Rendus Acad. Sci. Paris* 277(1973), 991-994.

(1973/74) Type des espaces normées, *Sem. Maurey-Schwartz*, Exp. III, and *Comptes Rendus Acad. Sci. Paris* 276(1973), 1673-1676.

(1973/74) Une propriete du type p-stable, *Sem. Maurey-Schwartz*, Exp. VIII.

(1975) B-convexity, superreflexivity and the 3-space problem, *Proc. Sem. "Random Series, Convex Sets and Geometry of Banach Spaces*, Aarhus, 151-175.

(1975) Martingales with values in uniformly convex spaces, *Israel J. Math.* 20, 326-350.

(1975/76) Le theoreme limite centrale et la loi du logarithme iteree dans les espaces de Banach, *Sem. Maurey-Schwartz* , Exp. III, IV.

(1976) Sur la loi du logarithme iteree dans les espaces de Banach, *Springer Lecture Notes in Math.* 526, 203-210.

(2016) *Martingales in Banach Spaces,* Cambridge University Press.

Pisier, G., and Zinn, J.,

(1978) On the limit theorems for random variables with values in the spaces $L_p (2 \leq p < \infty)$, *Z. Wahr. verw. Geb.* 41, 289-304.

Prokhorov, Yu,V.,

(1950) On a strong law of large numbers (in Russian), *Izv. Akad. Nauk SSSR, Ser. Mat.* 14, 523-536.

Retherford, J.R.,

(1975) Applications of Banach ideals of operators, *Bull. Amer. Math. Soc.* 81, 978-1012.

Revesz, P.,

(1968) *Laws of Large Numbers* , Academic Press, New York.

Phelps, R.R.,

(1974) Dentability and extreme points in Banach spaces, *J. Functional Anal.* 16, 78-90.

Rieffel, M.A.,

1967 Dentable subsets of Banach spaces with application to a Radon-Nikodym Theorem, *Functional Analysis*, Thompson Book Co.

(1968) The Radon-Nikodym Theorem for the Bochner integral, *Trans. Amer. Math. Soc.* 131, 466-487.

Rosenthal, H.P.,

(1970) On the subspaces of $L^p (p > 2)$ spanned by sequences of independent random variables, *Israel J. Math.* 9, 272-303.

(1973) On subspaces of L^p, *Annals of Math.* 97, 344-373.

(1976) Some applications of p-summing operators to Banach space theory, *Studia Math.* 58, 21-43.

Ryll-Nardzewski, C., and Woyczyński, W.A.,

(1975) Bounded muliplier convergence in measure of random vector series, *Proceedings of the American Mathematical Society* 53, 96-98.

Scalora, F.S.,

(1961) Abstract martingale convergence theorem, *Pacific J. math.* 11, 347-374.

Schaffer, J.J., and Sundaresan, K.,

(1970) Reflexivity and the girth of spheres, *Math. Ann.* 184, 163-168.

Schwartz, L.,

(1969/70) La theoreme de dualite pour les applications radonifiantes, *Sem. Schwartz*, Exp. XXIV.

(1969/70) Les applications o-radonifiantes dans les espaces de suites, *Sem. Schwartz* , Exp. XXVI.

(1974) *Radon Measures on Arbitrary Topological Spaces and Cylindrical Measures*, Tata Institute Monographs on Mathematics and Physics, Bombay.

(1974) Les espaces de type et cotype 2 et leurs applications, d'apres Bernard Maurey, *Ann. Inst. Fourier* 24, 179-188.

Sundaresan, K, and Woyczyński, W.A.,

(1980) Laws of large numbers and Beck convexity in metric linear spaces, *Journal of Multivariate Analysis* 10, 442-459.

Szankowski, A.,

(1974) On Dvoretzky's theorem on almost spherical sections of convex bodies, *Isreal J. mamth.* 17, 325-338.

Szarek, S.J.,

(1976) On the best constant in the Khinchin inequality, *Studia Math.* 58, 197-208.

Szulga, J.,

(1977) On the L_r-convergence, $r > 0$, for $n^{-1/r}S_n$ in banach spaces, *Bull. Polon. Acad. Sci.* 25, 1011-1013.

Szulga , J., and Woyczyński, W.A.,

(1976) Convergence of submartingales in Banach lattices, *Annals of Probability* 4, 464-469.

Tortrat, A.,

(1976) Sur les lois $e(\lambda)$ dans les espaces vectorielles, Applications aux lois stables, *Z. Wahr. verw. Geb.* 27, 175-182.

Tulcea, A.I., and Tulcea, C.I.,

(1962) Abstract ergodic theorems, *Proc. Nat. Acad. Sci* 48, 204-206.

Tzafriri, L.,

(1972) Reflexivity in Banach lattices and their subspaces, *J. Functional Anal.* 10, 1-18.

Uhl, Jr, J.J.,

(1972) A note on the Radon-Nikodym Property for Banach spaces, *Rev. Roum. Mat.* 17, 113-115.

Vakhania, N.N.,

(1971) *Probability Distributions in Linear Spaces*, Tibilisi.

Warren, P., and Howell, J.,

(1976) A strong law of large numbers for orthogonal Banach spaces valued random variables, *Springer Lecture Notes in Math.* 526, 253-262.

Woyczyński, W.A.,

(1973) Random series and laws of large numbers in some Banach spaces, *Teorya Veroyatnostej i Prim.* 18, 371-377.

(1974) Strong laws of large numbers in certain linear spaces, *Ann. Institute Fourier* 24, 205-223.

(1975) Geometry and martingales in Banach spaces, *Springer Lecture Notes in Math.* 472, 229-276.

(1975) A few remarks on the results of Rosinski and Suchanecki concerning unconditional convergence and C-spaces, *Séminaire Maurey-Schwartz (Paris)*, XXVII, 1-9.

(1975) Laws of large numbers for vector valued martingales, *Bulletin de l'Academie Polonaise des Sciences* 23, 1199-1201.

(1975) Asymptotic behavior of martingales in Banach spaces, in *Probability in Banach Spaces, Oberwolfach, Springer's Lecture Notes in Mathematics* 526(1976), 273-284.

(1975) A central limit theorem for martingales in Banach spaces, *Bulletin de l'Academie Polonaise des Sciences* 23, 917-920.

(1977) Weak convergence to a Gaussian measure of martingales in Banach spaces, *Symposia Math.* 21, 319-331.

(1978) Geometry and martingales in Banach spaces, Part II: Independent increments, *Probability on Banach Spaces*, J. Kuelbs, Editor, Marcel Dekker, New York, 267 -518.

(1980) Tail probabilities of sums of random vectors in Banach spaces and related mixed norms, *Conference on Measure Theory, Oberwolfach 1979, Springer's Lecture Notes in Mathematics* 794, 455-469.

(1980) On Marcinkiewicz-Zygmund laws of large numbers in Banach spaces and related rates of convergence, *Probability and Mathematical Statistics* 1, 117-131.

(1983) Survey of asymptotic behavior of independent random vectors and general martingales in Banach spaces, *Fourth Conference on Probability Theory in Banach Spaces, Springer's Lecture*

Notes in Mathematics 990, 215-220.

Zinn, J.,

(1977) A note on the central limit theorem in Banach spaces, *Ann. Prob.* 5, 283-286.

Index